Electricity Production from Renewable Energies

Revised and Updated 2nd Edition

Series Editor
Bernard Multon

Electricity Production from Renewable Energies

Benoît Robyns
Arnaud Davigny
Bruno François
Antoine Henneton
Jonathan Sprooten

iSTE

WILEY

First edition published 2012 in Great Britain and the United States by ISTE Ltd and John Wiley & Sons, Inc., © ISTE Press Ltd 2012.

This edition published 2021 in Great Britain and the United States by ISTE Ltd and John Wiley & Sons, Inc.

ISTE Ltd
27-37 St George's Road
London SW19 4EU
UK

www.iste.co.uk

John Wiley & Sons, Inc.
111 River Street
Hoboken, NJ 07030
USA

www.wiley.com

Library of Congress Control Number: 2021941643

British Library Cataloguing-in-Publication Data
A CIP record for this book is available from the British Library
ISBN 978-1-78630-661-6

Contents

Foreword

The first edition of *Electricity Production from Renewable Energies* was a great success. Undoubtedly, it was well received by an audience looking for concise knowledge of the means of generating electricity from renewable resources. This success is the result of the collaboration of five teacher-researchers – Benoît Robyns, Arnaud Davigny, Bruno François, Antoine Henneton and Jonathan Sprooten – whose experience and practice are recognized internationally. Within this team, Benoît Robyns took the role of lead author, as well as that of the animated conductor, whose coordination efforts greatly contributed to the overall success of the book.

Preparation for the first edition, published in 2012, was long (more than three years) and clearly revealed the difficulty and extent of the work. However, almost 10 years later, the extremely dynamic sector of renewable electricity generation has led to the need for an update. The most promising sectors of the global energy transition are photovoltaics and wind power, whose growth remains particularly strong thanks to their superior, exploitable technical potential, capable of meeting most of humanity's energy requirements in the medium term. This dynamic is thus accompanied by significant progress – enough to warrant an update – leading to the publication of this new edition, by the same team of authors.

In this book, a contextual introduction presents the immense potential of renewable resources, undoubtably the only resources able to offer all of humanity a truly sustainable future, notably through intensifying the electrification of many sectors. The authors treat the subject as concisely as possible in order to make this work accessible to those who have studied science to a post-high school level. They describe the different conversion principles and associated technologies in a very educational way, covering the following forms of energy across five chapters: photovoltaic, wind, onshore and offshore, hydraulic (onshore and marine solutions, including wave generators and tidal turbines) and thermal (from biomass, geothermal energy, solar radiation), including electricity and heat cogeneration. By way of conclusion, Chapter 6 deals with the integration of highly decentralized electricity generation into electrical grids. At the end of Chapters 2 to 4, exercises and their respective answers have been included, allowing for self-assessment.

I warmly thank the five authors of this book for their contribution to the dissemination of knowledge essential to the energy transition which, particularly in France, is struggling to take off. This book will thus be able to support both teachers wishing to refresh their teaching practice, and their post-graduate students.

Bernard MULTON
July 2021

Introduction

From the beginning of the 21st century energy and environmental challenges have led to increasing electricity production from renewable energies. The concept of sustainable development and the concern for future generations challenge us daily, leading to the emergence of new energy production technologies and new behavior usage for these energies. The share of renewable energies in the global electricity production mix now exceeds 25% [RAB 20]. The quick emergence of new technologies can make its understanding and perception difficult. The purpose of this book is to contribute to a better understanding of these new electricity production technologies by targeting a large audience. It presents the challenges, sources and their conversion process into electricity by following a general approach. It also develops basic scientific notions to comprehend their main technical characteristics with a global view.

The objectives of this book are:

– to present electricity production systems from renewable energy sources from small to mean powers (up to 100 to 200 MW);

– to introduce basic electrical notions that are necessary for the understanding of the operational characteristics of these energy converters;

– to discuss integration constraints and issues in the electrical networks of these production units;

– to set a few exercises for self-assessment.

Chapter 1 introduces the concept of decentralized electricity production from renewable energy resources. It presents the challenges that have led to the development of electricity production, not only just from the 20th century

Introduction written by Benoît ROBYNS.

centralized approach, but also dispersed throughout the territories. After all, the available resources that are managed by various actors in competition are also dispersed. This chapter also presents the challenges that led to the development of electricity production from renewable resources. It introduces the various exploitable energies and describes the basic principles of their conversion into electrical energy. Self-generation and self-consumption of energy, well suited to renewable energies, have also been introduced.

Chapter 2 presents the direct production (photovoltaics) of electricity from solar energy. It describes the characteristics of photovoltaic cells and panels. On average, this sector has experienced an annual growth rate of over 50% for the last 10 years [RAB 20]. It explains the operational principles of power electronic converters, which help to control the energy extracted from solar radiation and transform it into the form required by consumers. The chapter ends with some exercises.

Chapter 3 develops the conversion principles of converting wind kinetic energy into electrical energy. It describes the main wind turbine technologies. The annual growth rate of onshore wind has averaged over 20% over the past 10 years [RAB 20]. Is also explains electro-mechanical conversion from synchronous and induction generators, at fixed and variable speed. Examples of the characteristics of effective high and low power wind turbines are also provided. Wind power at sea or offshore, which offers good development prospects, has been introduced. Exercises concerning various types of wind turbines at fixed and variable speed, the characterization of a wind turbine and the estimate of the generated power are also proposed.

Chapter 4 introduces electrical energy production from the potential or kinetic energy of water, whether in a terrestrial or marine environment. At first, the principles of hydroelectricity (the first renewable source producing electricity, which has been implemented for more than a century) are developed, by focusing more specifically on the running of river hydraulics. Secondly, water power coming from marine sources are presented: wave, marine current and tidal power. The exploitation of these energies is still not very developed and most of the associated technologies are just emerging, except for tidal power production, which is quite mature but still marginal. A few examples of these technologies will be described in this chapter. Some exercises in the context of small hydroelectric power stations and a tidal power plant are also proposed.

Chapter 5 introduces the concept of thermal electricity production, in which heat is produced from renewable resources. This is the case for geothermal power, for concentrated thermodynamic solar power and for cogeneration, whose principles are described. The operational principles and characteristics of the synchronous alternators directly coupled to the electrical network are also presented.

Chapter 6 raises the question of the integration of renewable energy sources and more generally, of the decentralized production into electrical network. The latter are indeed confronted with a new paradigm, because of the random and unpredictable nature of some of these sources, due to their scattering on the territory, and because of the rules of a liberalized electricity market. The main connection constraints of these sources are also briefly described. Perspectives for a better integration into the networks of these sources are identified by considering actions on the levels of the sources, networks and consumers. Developments and incentives are initiated, so that the future electrical networks become smarter.

References

[RAB 20] RABAIN A., MULTON B., PAILLARD M., "Mais pourquoi diable chercher à convertir des énergies renouvelables en mer ?", *Connaissance des Energies*, available at: www.connaissancedesenergies.org, April 2020.

1

Electricity Production from Renewable Energy

1.1. Decentralized or centralized production?

1.1.1. *Decentralized production*

There is no clear official definition of decentralized production. Generally, decentralized production is defined as the opposite of centralized production [CRA 08, JEN 00]. To simplify, let us say at first that decentralized units:

– are not planned in a centralized way;

– are not controlled (or dispatched) in a centralized manner;

– have a power, which does not exceed 50 to 100 MW;

– are generally connected to the distribution network and not to the transportation network.

Another characteristic of decentralized production is that it is scattered over a territory, in contrast to conventional production, which is concentrated on a limited number of well-defined sites.

The development of decentralized production over the last few years has been especially favored by the opening of the electricity markets (which has spread in Europe from the beginning of the 2000s) and the development of renewable energies, especially wind energy, driven by a real commitment to the environment on a European scale. Decentralized production is thus developing in many countries

---

Chapter written by Benoît ROBYNS.

on the basis of cogeneration units, renewable energy systems or conventional productions, which have been installed by independent producers.

The development of this type of production can contribute to solving technical, economic and environmental problems [CRA 08, JEN 00], even if it is not the only answer to these multiple challenges.

Let us make a list of elements favoring decentralized production:

– the desire to reduce greenhouse gas emissions (mainly CO_2) encourages the development of renewable energies;

– the energy efficiency increase, which has been obtained thanks to cogeneration systems;

– the opening up of the electricity market enabling the emergence of independent producers;

– the desire to widen the range of energy supply, in order to limit the energy dependence of the European Union, which results from the use of fossil fuels;

– technological progresses contributing to the reliability and availability of 100 kW to 150 MW units;

– the greater facility to find sites able to accommodate a reduced power production;

– shorter construction periods and lower investments than for large conventional power plants;

– a production that can be carried out at the proximity of its use, thus reducing transportation costs.

Depending on the profile of the historical generation system of each country, the structure of their transport and distribution network and the organization of the electrical system, these various points can be more or less important, depending on the countries, especially within Europe.

1.1.2. *Centralized production*

The historical organization of an electricity grid is based on its centralized management at the level of the transmission grid to which the traditional production units of nuclear, thermal or hydraulic types are connected. In the case of the latter, it is a large hydraulic power plant equipped with a dam that retains a body of water in a mountain valley or along a river. These large power plants, often with a capacity of over a hundred MW, contribute to the stability of the electricity grid by, among other things, participating in ancillary services including frequency and voltage control, as explained in Chapter 6.

Certain power plants which exploit fluctuating renewable energies with huge capacity can be connected to the transmission grid. This is the case for offshore wind farms, with a capacity of several hundred MW, proliferating in the North and Baltic Seas. Due to the very intermittent upstream source, these wind turbines are barely able to contribute to ancillary services, aside from voltage control, unless they are combined with energy storage systems [ROB 15].

It should be noted that tidal power stations, similar to hydraulic power stations, can also be connected to the transmission grid, for example the La Rance Tidal Power Station in France, with a capacity of 240 MW (see Chapter 4). Certain high-power thermodynamic solar plants (see Chapter 5) could also be connected to the transmission grid, such as the 540 MW project at Ouarzazate in Morocco [CAS 20]. These plants provide some, albeit limited, energy storage capacity by storing water in a basin, or heat in a body of water or other materials with favorable properties [ROB 15].

1.2. The issue of renewable energies

1.2.1. *Observations*

The growing interest in the development of renewable energies is caused by several elements: climate change, increasing energy demand, limits of fossil fuel reserves, low global efficiency of the energy system and energy dependence, especially in the case of Western countries [CHA 04].

Climate change

The growing "greenhouse effect" leads to an increase in the global temperature at the surface of the planet. And yet, because of human activities, the concentration of greenhouse gases has soared since the pre-industrial era (1750-1800), with the concentration of carbon dioxide (CO_2) (the main greenhouse gas) increasing by more than 30%. The combined effects of all of the greenhouse gases (CO_2, methane, ozone, etc.) nowadays amounts to more than a 50% increase in CO_2 compared with this period.

Between 1860 and 2010, the mean temperature at the surface of the Earth rose by 0.6°C. Several prospective scenarios are predicting that by 2100, this temperature will increase further between 1.5 and 6°C, if energy systems and current consumption habits do not change. This significant increase would be accompanied by a sea level rise from 20 cm to 1 m. If the climate change seems non-reversible, this evolution can however be slowed down, by significantly reducing greenhouse gas emissions.

The natural CO_2 wells, such as land, trees and oceans, would only be able to absorb a little less than half of the CO_2 production resulting from human activities (produced in 2000). In order to stabilize the CO_2 concentration at its current level, we thus would have to immediately reduce the gas emissions from 50 to 70%. This drastic reduction is clearly impossible. However, it is urgent to start acting, because this is a cumulative issue. Indeed, the carbon dioxide lifespan in the atmosphere is of about one century and, therefore, the stabilization of the CO_2 concentrations to an acceptable level will take several generations.

CO_2 is produced by the combustion of all fossil fuels: oil, gas and coal. CO_2 emissions from coal are twice as high as the emissions from natural gas. Oil emissions are in-between.

At the beginning of the 21st century, the distribution by sectors of CO_2 emissions in the world was as follows: electricity production 39%, transport 23%, industry 22%, residential 10%, service sector 4% and agriculture 2%. However, this distribution varies from one country to another. For example, in France where only one tenth of the electricity is produced from fossil fuels, the transport sector is responsible for more than 40% of the CO_2 emissions into the atmosphere.

Increasing energy demand

At the beginning of the 21st century, the global energy consumption was about 10 Gtoe (toe = ton of oil equivalent; 1 toe corresponds to the energy produced by the combustion of one ton of oil). Fossil fuels represent about 8 Gtoe.

Many energy scenarios are developed each year by specialized organizations in the energy field. These scenarios plan an energy demand in 2050 of about 15 to 25 Gtoe. These prospective scenarios are based on various parameters, such as economic growth, increased by world population increase, the progressive access to electricity of the billion people still without any access to it at present, the growing needs of developing countries and the implementation of energy-saving policies in order to protect the environment. The uncertainties in relation to the evolution of these parameters explain the significant gap between extreme scenarios.

However, it seems quite reasonable to predict that by the middle of the century, the energy demand will have doubled.

Limits of the fossil fuel reserves

The R/P oil ratio (known reserves to the annual production) is about 50 years [WIK 20a]. This piece of data (which is equivalent to a period) should not be mixed up with the period during which we will still dispose of oil, nor to the one during which it will still be cheap enough. These two periods are completely unpredictable, because they depend on too many parameters. Let us note that since the 1980s, each year we consume more oil than we discover.

For natural gas, the R/P ratio is about 50 years as well. However, if we wanted to replace oil and coal with gas, in order to reduce greenhouse gas emissions, the R/P ratio would be reduced to less than about twenty years. When some countries give up nuclear energy for the benefits of gas, it could increase the consumption of resources.

Coal is the fossil fuel with the most significant reserves. Its R/P ratio is estimated at almost 200 years.

The R/P ratio of uranium is about 90 years (on the basis of "reasonably assured resources" added to "recoverable resources" for a cost between 80 and 260 $/kg of natural uranium and a conventional fission exploiting isotope 235) [WIK 20a].

The energy demand until 2050 (then predicted to be between 15 and 25 Gtoe, compared to nearly 14 Gtoe in 2018) could still be met mainly (at present) by non-renewable raw energy materials. This would have dramatic consequences for the climate in particular, and for the environment and would not really take into account the needs of future generations.

In order to limit the rise in temperature to a range from 1 to 3°C, the total emissions for centuries would have to be only a third of the current emissions, caused by the combustion of the accessible resources of gas, oil and coal. Humanity would then have to stop burning two-thirds of a relatively cheap and accessible energy source. It is thus not reasonable to bank on an early exhaustion of the resources, in order to naturally reduce greenhouse gas emissions. This is particularly true, because the relatively low price of fossil fuels (despite regular explosions) are disrupting the emergence of new technologies, which are inevitably more expensive until they become integrated into a mass production process.

Low global efficiency of the energy system

The global efficiency of our energy system is quite low: for example, in 2008, to satisfy the French requirements for the final energy (marketed) of 168 Mtoe, 262 primary Mtoe were needed to produce them, which corresponds to a 63% efficiency, all the while knowing that effectively useful energy is much lower. The energy transformation losses alone when making marketable energy are about 27%. 94 Mtoe have thus been lost in energy transformations (refining, electricity production, etc.). These losses of 94 Mtoe, associated with bad uses of the final energy (bad building insulation, low efficiency of the car heat engines, etc.) are the main item of expenditures and finally the most important cause of CO_2 emissions. For example, at the beginning of the 2000s global efficiency was about 34% and it has since evolved very slowly.

Energy dependence

About 50% of the energy consumed within the European Union comes from resources located in countries outside the EU. If nothing changes in the energy production field, and taking into account the expected consumption increase, this energy dependence will go up to 70% by 2030.

The global dependence on Middle Eastern countries (which possess 65% of the known oil reserves) should increase. The dependence is even higher for uranium (100% for France). Economic and political tensions could arise from the diminution of fossil resources, which are easily exploitable, and from their concentration in politically unstable zones. This would question the supply security of the countries in the European Union.

1.2.2. *The sustainable development context*

The concept of sustainable development was defined in 1986 as follows: "meeting the needs of the present without compromising the ability of future generations to meet their own needs".

This concept implies the exploitation of renewable energy sources, which are the only sustainability guarantees, and the minimization of environmental impacts, associated with their conversion and the manufacturing of their converters. Fossil and fissile fuels are appearing as a finite and economically limited resource, causing emissions that affect the environment and/or contribute to climate change in the case of fossil fuels. A sustainable energy system must include renewable energy sources and/or conversion chains exploiting low emission renewable fuels, which are available at an acceptable price. Despite the fact that the implementation of new energy facilities takes several decades, an increasing number of large companies are involved in the development and marketing of these new technologies.

Sustainable development requires good management of the balance between economic development, social equity and environmental protection in all regions of the world. This concept can only become effective with real political will from an increasing number of countries.

1.2.3. *Commitments and perspectives*

The concept of sustainable development is an answer to the observations above. To implement it, several decisions and associated objectives have been progressively made on the European and international level.

Kyoto protocol and other international conferences

In 1997, the Kyoto protocol set the objective to have reduced global greenhouse gas emissions by 5.2% around 2010, in comparison to the levels in 1990. The European Union promised an 8% reduction of its emissions by 2010, and each member was allocated their own reduction quota of emissions, by taking into account the specificities of each country. More than half of the countries had to reduce their emissions (Germany, Austria, Belgium, Denmark, Italy, Luxemburg, Netherlands), some others needed to stabilize their emissions (France, Finland), while other countries were authorized to increase their emissions (Greece, Ireland, Portugal, Spain, Sweden).

To stop the increase of the carbon dioxide concentration in the atmosphere by 2050, our current emissions will have to be halved worldwide, and thus reduced to between $\frac{1}{3}$ to $\frac{1}{5}$ in developed countries. The Bali conference in December 2007 re-launched the negotiations between the States, in order to increase commitments to countries reducing CO_2 emissions.

During the 2010s, global CO_2 emissions tended to continue increasing and reached 33 GT (gigatonnes) in 2019, mainly due to the production of electricity from fossil fuels, including coal [WIK 20b]. The Paris Climate Change Conference in 2015 (COP21) set targets for reducing CO_2 emissions, in order to limit the temperature rise to less than 2°C in 2100. However, these targets appear insufficient, given the levels of greenhouse gas emissions estimated for 2030, on the basis of planned contributions at the national level, which are not compatible with least-cost scenarios allowing for a temperature rise of 2°C, but result in an emissions level of 55 GT projected for 2030, and that far greater emission reduction efforts will be required in order to reduce emissions to a 40 GT target [WIK 20c]. The Madrid Climate Change Conference in 2019 (COP25) highlighted the fact that there were more disagreements than points of convergence between participating countries on the importance of carbon reduction and strategies to help limit CO_2 emissions, doing little to encourage optimism in the short term [WIK 20d].

European Union and sustainable energy development

At the beginning of the 21st century, the European Commission made the development of renewable energies a political priority, this is described in the White paper "Energy for the future: renewable sources of energy" and the Green paper "Towards a European strategy for the security of energy supply".

The objective set by the Commission was to double the proportion of renewable energies in the global energy consumption, from 6% in 1997 to 12% in 2010. This doubling objective fits within a strategy of supply security and sustainable development; a particularly significant effort has to be made in the electrical field. Within the European Union, the proportion of electricity produced from renewable energy sources should reach 22.1% in 2010, compared to 14.2% in 1999. This objective was defined for the EU-15, and was significantly lowered for the EU-27, in order to reach 21%.

In 2007, the European Council promised to reduce CO_2 emissions by 20% within the European Union by 2020. The objective was that 20% of the final energy consumption should be ensured by renewable energies, with a 10% biofuel use in transport, and a 20% energy efficiency improvement.

The European Union's target for 2030 is for at least 32% of gross final energy consumption to come from renewable sources [EUR 18].

Electricity market opening

Since the beginning of the 2000s, the electricity sector has been the scene of a deep restructuring, resulting from the European Directive CE 96-92. This directive imposes a management of the activities inherent in the transport of electricity, which is independent from the electric energy production activities. The backbone of the electrical supply remains the transport network, which is managed in each state by one or a reduced number of managers appointed by the government involved.

One of the consequences of the opening of the electricity market is the development of a decentralized production, on the basis of cogeneration units, renewable energy sources or traditional production, which is installed by independent producers.

The integration to the electrical networks of renewable energy sources, more specifically those subjected to climate vagaries, such as wind and solar power, and more generally of decentralized production, will require significant network upgrading, as well as the implementation of new equipment and new management methods. The challenge is then to maintain the reliability and quality of the power supply for private individuals and businesses, despite market liberalization and the growing use of random renewable energy sources.

Technological prospects

It is quite difficult to identify the technologies that will play a crucial role in the future for the fight against the greenhouse effect. A future energy system with low greenhouse gas emissions will probably rely on a combination of energies, of energy vectors and converters, which will take on various forms in the different regions of the world.

It is however possible to determine five trends of our energy future:

– The proportion of renewable energies is progressively increasing, but this progress will notably depend on the reduction of their costs and on the progress made in terms of massive energy storage, which could integrate important quantities of intermittent and scattered production into electrical networks [ROB 15, ROB 16, ROB 19]. However, according to the most optimistic predictions, the combination of different renewable energies could enable them to reach 30 to 50% of the market around the middle of the century (at the beginning of the 2000s, all renewable energies represented about 10% of the energy production, but nearly 20% in 2018 [WIK 20a]). However, some scenarios aim for 100% renewable energies as of 2050, due not only to a considerable increase in the efficiency of our energy system, but also thanks to a significant decrease in consumption, i.e. practicing energy sobriety [CAS 20];

– Fossil energies will still be used for several decades, all the while favoring energies with a low carbon content, such as gas. However, the capture and storage of carbon dioxide in economically bearable conditions seems to be the only technological option, which is likely to authorize the use of fossil resources, all the while limiting the CO_2 concentration in the atmosphere, while waiting for significant technological developments.

– Nuclear power does not generate CO_2, except for the CO_2 emissions during the plant construction, deconstruction, and during uranium enrichment. This type of energy will continue to be developed in some countries, such as France, by means of a well developed waste management process, the development of a new generation of safer reactors, knowing that fissile resources are also limited, and then in the long-term by the development of nuclear fusion. However, nuclear fusion is only considered on the horizon after 2050.

– The spreading of fuel cells could lead to the development of a "hydrogen economy". The production of hydrogen does not generate CO_2 if the hydrogen is produced from renewable, nuclear or fossil energies with CO_2 sequestration. The USA did not ratify the Kyoto protocol, because they consider it to be too restrictive for their economy. In 2003 they launched an ambitious research program aiming to reduce the hydrogen production cost, while controlling greenhouse gas emissions, mastering hydrogen storage and reducing the fuel cell cost. In recent years, Europe has been developing its efforts in regards to research and development on H_2, its production and its uses.

– Finally, controlling greenhouse gas emissions will not be possible without significant progress in energy efficiency in the construction, industrial and transport sectors. The challenge is to use less energy to satisfy the same needs, all the while knowing that the potential energy savings are huge.

1.3. Renewable energy sources

In our time scale, renewable energies are those continuously provided by nature. They come from solar radiation, the core of the Earth and from the gravitational forces of the Moon and the Sun with oceans. We can distinguish several types of renewable energies: wind power, solar power, biomass, hydraulics and geothermics [MUL 03, REV].

1.3.1. *Wind energy*

The available wind resources are evaluated on a global scale at 57,000 TWh/year. The contribution of offshore wind power (at sea) is estimated to be 25,000 to 30,000 TWh/year, if we limit ourselves to sites whose depth does not exceed 50 m. The global production of electricity in 2018 was about 26,700 TWh (which corresponds to a fossil or nuclear primary consumed energy of about 66,750 TWh, related to the low efficiency of the thermo-mechanical cycles, often ranging between 30 and 40%). In theory, wind energy could satisfy the global electricity demand. However, the main disadvantage of this energy source is its instability. There is often not much or no wind during very cold or very hot periods; and yet there is an increased energy demand during these periods. This is why we could envisage an important development of wind power, all the while associating it with other renewable energy sources, which would be less random or complementary, or have thermal sources or electrical energy storage devices. There are many ideas for storing large quantities of electricity (notably pumped storage power stations), their implementation is the subject of numerous research projects aimed at increasing performance and reducing costs [ROB 15].

Europe represents 9% of the wind energy potential available in the world. In 2017, the member countries of the European Union produced 362 TWh of electricity from wind energy. The wind energy technically available in Europe, not including offshore, would be 5,000 TWh/year.

1.3.2. *Solar energy*

The projected lifespan of the sun is still five billion years, which, in our time scale, makes it an inexhaustible and thus renewable energy. The total energy received at the surface of the Earth is 720 million TWh/year, i.e. 6,000 times the current primary consumption of all human activities. But the availability of this energy depends on the day-night cycle, the latitude of where this energy is captured, the seasons and the cloud cover.

Solar thermal energy consists of producing hot water usable in construction or enabling the operation of turbines, by exploiting concentration phenomena to increase temperatures in thermal power stations with thermodynamic cycles, in order to produce electricity. This electricity generation technique is the subject of experimental power stations, whose net efficiency is around 15%. Sea surfaces are naturally heated by the sun and there is thus a gigantic energy reservoir in the tropical zone. Projects for the extraction of this "ocean thermal energy" have been carried out by implementing thermodynamic machines, which operate on the small difference found between the surface (25 to 30°C) and the depth (5°C at 1,000 m). In order for this solution to be exploitable, the temperature difference has to be higher than 20°C. However, the efficiency obtained (around 2%) is very low. It should be noted that low yields mainly translate into larger energy conversion machines, but do not have the gravity associated with the consumption of non-renewable raw materials, which are irreparably consumed and therefore lost.

Photovoltaic solar energy consists of directly producing electricity via silicon cells. When the sun shines and weather conditions are favorable, the sun supplies a peak force of 1 kW/m^2 at ground level. Photovoltaic cells were developed in the 1960s to power satellites; at that time, the cost was in the order of $1,500 per peak watt. Industrial developments have since reduced this cost to $50 per peak watt in the 1970s, and around $0.5 per peak watt today [BAR 14]. Marketed photovoltaic panels help to directly convert 10 to 15% of this power into electricity. The productivity of a photovoltaic panel varies with the level of sunshine: about 100 kWh/m^2/year in Northern Europe and twice this amount in the Southern Mediterranean region. A photovoltaic roof of 5 x 4 meters has a power of 3 kW and produces from 2 to 6 MWh/year, depending on the sunshine. If the 10,000 km^2 of roof in France were used as photovoltaic generators, the production would be 1,000 TWh/year, i.e. a little less than double the yearly final electricity consumption of France in 2017 (562 TWh).

The main "brakes" to the massive use of photovoltaic solar (and thermal) energy are the intermittence of the supplied power (which requires electricity storage for an autonomous use [ROB 19] or the use of additional energy sources) on the one hand and the economic competitiveness on the other hand. Outside the zones not connected to the network, where it is already profitable, the parity between photovoltaic production costs and electricity sale prices starts to be found in countries where electricity is the most expensive and where sunshine levels are at the highest. It should spread to all territories before 2050. This development is encouraged by the self-consumption of locally produced electrical energy [ROB 19].

1.3.3. *Hydraulics*

Hydraulics is currently the first exploited renewable source of electricity. However, the global potential could be better exploited. Global production in 2018 was 4,239 TWh/year, with an installed capacity of 1 291.8 GW. It could go up to 8,100 TWh by 2050. 14,000 TWh would technically be exploitable and the theoretical potential would be 36,000 TWh.

Large hydraulics (with a power higher than 10 MW) are exploited at almost the maximum of their potential in industrialized countries. Dams store the energy and supply it in peak energy demand. In some cases, high and low storage pools enable actual energy storage by using groups of reversible turbo-generators, which are pumping in off-peak periods. This form of storage is frequently used around the world. In France, 4,200 MW are installed for this function.

Small hydraulics (of a power lower than 10 MW) are partly made from run-of-river power plants, which depend highly on the river flow rate. The small power plants are quite interesting for decentralized production. The global production is estimated at 85 TWh. In France, while large hydraulics has almost reached saturation, the development potential of small hydraulics remains, which is estimated at 4 TWh/year, ⅓ of which comes from the improvement of the existing facilities and the remainder from new facilities.

Tidal power can be used to produce electricity. In France, the Rance Tidal Power Station (240 MW) has proved the feasibility of this electricity production technique. Other significant projects are currently studied in Canada and England; however, whether or not these studies will be put into practice remains uncertain, because of the considerable changes that would occur in local ecosystems.

Hydro turbines were developed on the same principle as wind turbines, but are completely submerged at sea, so as to harness the energy of sea currents. Since water density is 800 times greater than that of air, the blades will be that much smaller than those of wind turbines. Since they are submerged, they are very discreet and have the added benefit of a more constant source of energy than wind power.

Wave motion is an important source of energy. The average annual power on the Atlantic coast ranges between 15 and 80 kW/m of coast. However the marine environment is very restrictive and wave energy recuperators are still being researched; they are not exploitable on a large scale. However, prototypes of wave-energy power plants are in the testing stages.

Finally, osmotic energy is obtained by bringing fresh water and salt water into contact with one another at river mouth estuaries, through a semi-permeable membrane. This results in a pressure difference which makes it possible to rotate a turbine. This technology is still at the research stage; a 4 kW demonstrator exists in Norway [BAR 14].

1.3.4. *Geothermal energy*

The temperature of our planet increases considerably as we get closer to the center. In some zones of our planet we can find, at depth, water at a high temperature. High temperature geothermal energy (150 to 300°C) consists of pumping this water towards the surface, producing vapor via exchanges, then turbining this vapor as in conventional thermal power stations, so as to produce electricity.

Low temperature geothermal resources (lower than 100°C) are upgraded with heat pumps for heating requirements.

However, the potential of natural geothermal energy is limited, because there are many sites where the temperature is high (higher than 200°C), but where there is no water. This thermal resource might be exploited with the help of the so-called "hot-dry rock" technology, which is currently under development. It consists of injecting, into a well, pressurized water in in-depth zones (deeper than 3,000 m) of fractured rocks. This reheated water returns to the surface up a second well and helps to produce electricity, as in conventional thermal power stations. However, the proportion of this potential, which would technically be accessible, has not yet been specified.

1.3.5. *Biomass*

Provided a sustainable exploitation of resources, biomass is a renewable energy that supplies biofuels, which are generally solid or liquid.

Wood covers more than 10% of the primary energy demand in many countries of Asia, Africa and Latin America and in some European countries (Sweden, Finland and Austria). The use of wood in developing countries has strongly increased over the last few decades, but this resource has not always been exploited sustainably and has thus led to deforestation. Emissions coming from wood combustion in a modern industrial boiler are advantageous in comparison to fossil fuels. If the forests where wood comes from are sustainably dealt with, the CO_2 emissions from the wood energy chain are only those corresponding to the gas and oil used during plantation, crop and marketing operations. This represents about 5% of the fuel sold. We can note that a renewable energy is not necessarily a completely non-polluting energy.

The consumption of biomass in France in primary energy was 10-11 Mtoe at the beginning of the 2000s; mainly wood. Without any specific energy crop, the biomass potential could be doubled by a systematic repercussion of all organic waste: non-recyclable household and industrial waste, methanization process of the sewage sludge and agriculture waste, which generates biogas. Global biomass electricity production in 2008 amounted to 669 TWh. The French energy potential is 60 TWh/year, i.e. 15% of the final electricity consumption in France.

Biomass is frequently used in cogeneration systems that produce electricity, such as conventional power stations, all the while upgrading the heat that is usually lost in various applications: heating of the facilities, industrial needs, agriculture, etc. This technology helps to increase the efficiency of energy conversion.

Liquid biofuels are more expensive to obtain and are industrially produced from energy crops (rape, sunflower, beet, wheat, barley, corn, etc.), and are better upgraded in transport applications. They are currently mainly used in engines and are mixed in small quantities in conventional fuels, in order to improve their characteristics. Nevertheless, this type of energy use competes with the potential of these crops for food use.

1.3.6. *Contribution of the various renewable energies*

Table 1.1 presents the contribution of each renewable energy as a percentage of global energy production in 2018 [WIK 20a]:

Hydro	16%
Wind	5%
Biomass	2.5%
Solar	2%
Other	0.5%

Table 1.1. *Contribution of each renewable energy to global energy production in 2018*

In comparison, of the non-renewable sources, coal contributed 38%, gas 23%, nuclear 10% and oil 3% of global electricity generation for 2018.

The contribution of each renewable energy source in the production of renewable electricity within the European Union in 2017 is shown in Table 1.2, for a total of 975.2 TWh.

Wind	37.2%
Hydro	30.8%
Biomass	19%
Solar	12.2%
Geothermal	0.7%
Marine energy	0.1%

Table 1.2. *Contribution of each renewable energy source in the production of renewable electricity within the European Union in 2017*

The growing rates of these sectors are really high, which contributes to an improvement in the penetration rate from one year to another.

1.4. Production of electricity from renewable energies

1.4.1. *Electricity supply chains*

To carry out energy conversions to produce electricity, several supply chains can be considered, depending on the use or not of electronic power converters.

The most frequently used electricity generation cycle requires a heat source to heat the water, in order to obtain vapor under pressure. By expanding in a turbine, this vapor drives an alternator, which generates electricity. After passing through the turbine, this vapor is condensated with the help of a cold source, which is generally water (river, sea) or air in cooling towers. Figure 1.1 represents the conventional cycle of electricity generation.

When the heat generated by the vapor condensation is recovered for heating needs, we can speak about cogeneration.

The heat source is generally obtained by the combustion of fossil fuels (oil, gas, coal), or by a nuclear fission reaction in reactors designed to control the extent of this reaction.

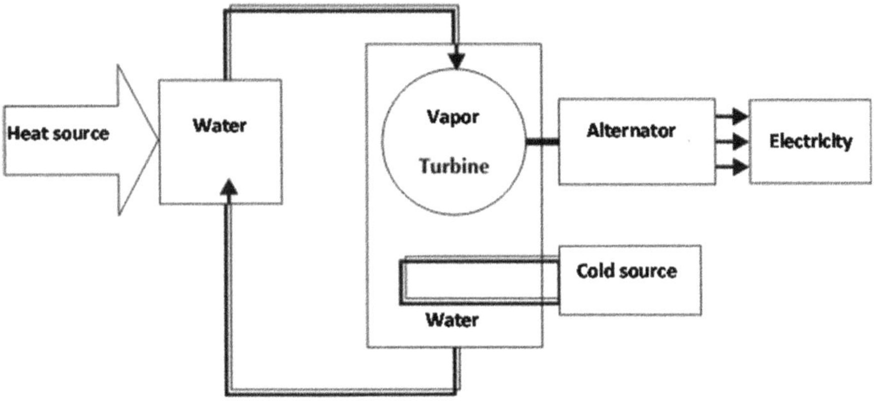

Figure 1.1. *Conventional cycle of electricity generation*

The fossil fuels or uranium used in conventional cycles can be replaced by some renewable energy sources. The heat source can then be obtained by:

– biomass combustion (wood, biogas, organic waste);

– the heat found in the depths of our planet, either by directly pumping hot water towards the surface or by exploiting the high temperature of the deep rocks by injecting them with water from the surface;

– the sun by concentrating its rays with the help of mirrors or by exploiting the water heated at the surface of the seas in tropical zones.

With some renewable energies, the electricity supply chain does not require a heat source. This is the case for wind power, hydraulics and photovoltaic solar energy.

In the case of wind power and hydraulics, the wind or water pressure drives the rotation of a turbine, which in turn drives an alternator, which produces electricity. Figure 1.2 represents this energy conversion chain.

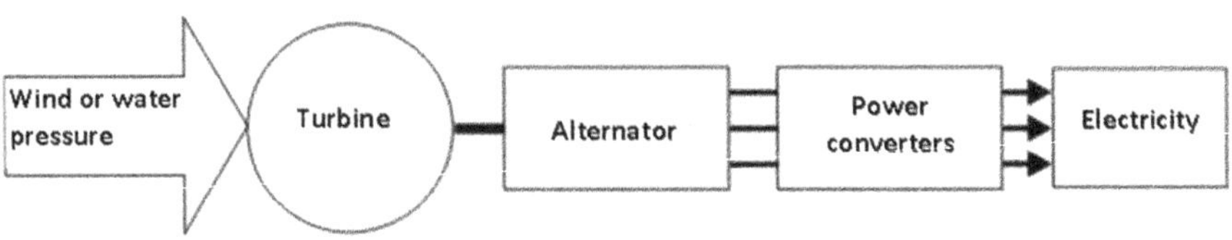

Figure 1.2. *Wind power or hydraulics electricity supply chain*

Wind pressure results from its kinetic energy. Water pressure results from its potential energy and its kinetic energy.

The electricity generated by the alternator can be directly sent along the electrical network without going through power converters, as indicated in Figure 1.2. However, in this case, in order to maintain the frequency of the voltages and the constant generated currents at 50 or 60 Hz, the alternator speed must be maintained as constant by acting on the direction of the turbine blades, or in the case of hydraulics, by winnowing upstream of the turbine.

The interesting aspect of power converters is that they enable alternators to operate at variable speed, thus increasing the energy conversion efficiency, all the while reducing the need for turbine mechanical control and for winnowing in the case of hydraulics. This variable speed operation is developing in the field of hydraulics (especially in small hydraulics), and tends to impose itself in wind powers, where this type of operation seems natural, because of the strong variations in the wind speed.

Electronic converters help us to convert power from one form to another. They can include rectifiers, inverters and choppers, or just a single inverter. The converter must be compatible with the frequency of the network and is equipped with filters, in order to satisfy power quality standards. Power electronics also ensure the protection functions of the production unit and the local network to which it is connected.

In the case of photovoltaic solar power, electricity is produced directly with the help of silicon cells using the energy from solar radiation. Power converters are generally used to ensure the optimization of energy conversion. Figure 1.3 illustrates this energy conversion chain.

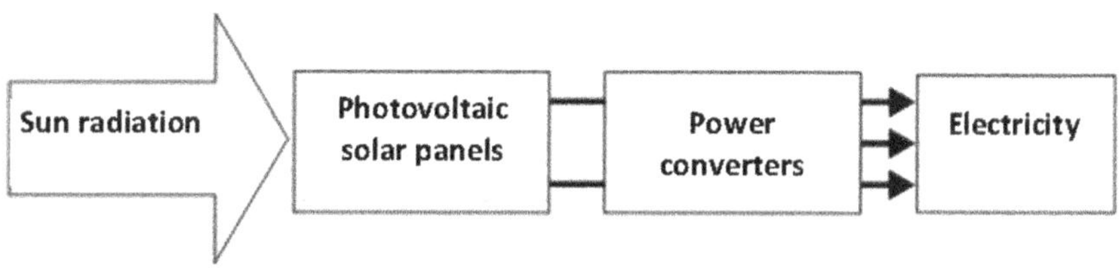

Figure 1.3. *Photovoltaic solar power chain of production generation*

Electricity can also be produced via a diesel engine or a gas turbine (derived from a jet engine) driving an alternator. The source of primary energy is generally made up of fossil fuels, but we can consider replacing them with biofuel or biogas.

1.4.2. *Efficiency factor*

The key factor for the competitiveness of energy production systems, based on renewable energies, is the cost of the kWh product. This cost is calculated from the investment cost of the generation system, its lifespan, the interest rate of the loan that may have been required and the operating costs related to maintenance and primary energy – which is free when it is the sun, wind... and not free in the case of fossil fuels, nuclear power...

In systems relying on a changeable nature (wind and solar power, run-of-river hydraulics, etc.), the system productivity fundamentally depends on natural conditions (number of hours of sunshine for example), whereas the investment cost mainly depends on the peak power. A 1 MW wind turbine will be able to supply a maximum power of 1 MW, but it will not permanently produce this power, because of the fluctuating nature of the wind speed, which is on the contrary to conventional power stations using fossil fuels or nuclear power. For this wind turbine, as well as for solar power and small hydraulics, it is the energy produced that is important.

Table 1.3 presents the efficiency factor, or energy return rate, of the electricity production chains from renewable energies, which are not based on the conventional water-vapor cycle. The efficiency factor is the ratio between the supplied energy and the production system throughout its entire lifecycle and the consumed energy, in order to build the production system.

Installation	**Efficiency factor**
Large hydraulics	100-200
Small hydraulics	80-100
Wind power	14-30
Photovoltaic solar energy	4-12

Table 1.3. *Efficiency factor, or energy return rate, of the systems producing electrical energy from renewable energies*

The efficiency factor strongly depends on the natural productivity of the accommodation site of the conversion system. For example, if the number of hours in the full sunshine equivalent goes from 1,000 to 2,000, the efficiency factor will double. Moreover, it is related to the lifespan of the facilities and it is often better for large facilities, thanks to generally favorable scale effects. This is notably the case for large hydraulics (lifespan of 30 to 50 years) in relation to small hydraulics (lifespan of 20 to 50 years).

The power of wind turbines has gone from a few hundred kW before 2000 to several MW after 2000, and prototypes are capable of reaching 14 MW. The lifespan of a wind turbine is 20 to 25 years.

Photovoltaic systems present the lowest efficiency factor, because the manufacturing of silicon cells requires a lot of energy. In four to five years a cell reimburses the energy spent during its manufacturing. As the lifespan of a photovoltaic system is 20 to 30 years, the efficiency factor could thus be, in the best case, slightly higher than six. Let us note however a continuous improvement of this criterion, notably with the arrival of thin film technologies, with which we reach times of return on energy investment shorter than one year, i.e. efficiency factors higher than 20.

1.5. Self-production and self-consumption of energy

The development of renewable energy sources encourages local production, close to the loads needed for supply, and hence, local consumption. This is particularly true for energies whose primary source varies independently of demand, without natural storage (for example in hydraulic or fuel form) and whose production forecasts are difficult, as is the case with wind and photovoltaics. In this context, self-production and self-consumption of renewable energy are defined, on the basis of Figure 1.4, as follows [ROB 19]:

– self-production is the share of total consumption provided by local renewable energies, over a day for example, or area-C/area-B (generally calculated over one year);

– self-consumption is the share of renewable production consumed in real time, over a day for example, or area-C/area-A (generally calculated over a year).

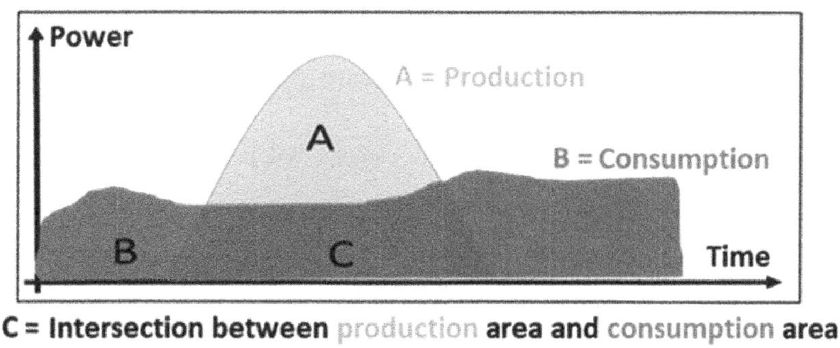

Figure 1.4. *Self-production and self-consumption over a twenty-four hour period [ROB 19]. For a color version of this figure, see www.iste.co.uk/robyns/electricity.zip*

A local energy network will only be able to operate autonomously (as an island outside of the main electricity network) if production A is, at any time, equal to or greater than consumption C, in the absence of an energy storage device.

1.6. References

[BAR 14] BARRE N., ROUBAUD M., *Les énergies renouvelables*, Editions 10/18, Usbek & Rica, 2014.

[CAS 20] CASSORET B., *Transition énergétique, Ces vérités qui dérangent*, De Boeck Supérieur, Louvain-la-Neuve, Paris, 2020.

[CHA 04] CHAMBOLLE T., MEAUX F., Rapport sur les nouvelles technologies de l'energie, Ministère délégué à la recherche et aux nouvelles technologies, Paris, 2004.

[CRA 08] CRAPPE M., *Electric Power Systems*, ISTE Ltd, London and John Wiley & Sons, New York, 2008.

[EUR 18] EUROBSERV'ER, Etat des énergies renouvelables en Europe, Edition 2018, EurObserv'ER.

[JEN 00] JENKINS N., ALLAN R., CROSSLEY P. *et al.*, *Embedded Generation*, The Institution of Electrical Engineers (IEE), London, 2000.

[MUL 03] MULTON B., "Production d'énergie électrique par sources renouvelables", *Techniques de l'Ingénieur*, Traité de Génie Electrique, D 4 005 and D 4 006, May 2003.

[REV] REVUE SYSTÈMES SOLAIRES, available at: www.energies-renouvelables.org.

[ROB 15] ROBYNS B., FRANÇOIS B., DELILLE G. *et al.*, *Energy Storage in Electric Power Grids*, ISTE Ltd, London and John Wiley & Sons, New York, 2015.

[ROB 16] ROBYNS B., SAUDEMONT C., HISSEL D. *et al.*, *Electrical Energy Storage in Transportation Systems*, ISTE Ltd, London and John Wiley & Sons, New York, 2016.

[ROB 19] ROBYNS B., DAVIGNY A., BARRY H. *et al.*, *Electrical Energy Storage for Buildings in Smart Grids*, ISTE Ltd, London and John Wiley & Sons, New York, 2019.

[WIK 20a] WIKIPEDIA, Ressources et consommation énergétiques mondiales, Wikipedia, accessed 26/03/2020.

[WIK 20b] WIKIPEDIA, Émission de dioxyde de carbone, Wikipedia, accessed 27/03/2020.

[WIK 20c] WIKIPEDIA, Accord de Paris sur le climat, Wikipedia, accessed 27/03/2020.

[WIK 20d] WIKIPEDIA, Conférence de Madrid de 2019 sur les changements climatiques, Wikipedia, accessed 27/03/2020.

[WIK 20e] WIKIPEDIA, Production d'électricité, Wikipedia, accessed 27/03/2020.

2

Solar Photovoltaic Power

2.1. Introduction

Solar photovoltaic power is the direct conversion of solar power into electric power. The primary energy thus comes from the Sun, which is located around 150 million km from Earth. This star is mostly made up of hydrogen. Thermonuclear reactions occur on the Sun, causing the temperature to reach several million degrees. At the same time as helium is produced, solar radiation in an electromagnetic form is emitted: visible radiation (from 380 nm to 780 nm), infrared radiation (higher than 780 nm) and ultraviolet (UV) radiation (from 100 nm to 400 nm). Part of the UV radiation reaches Earth [BER 04, GER 08]. Each year, the Earth receives about $1{,}600\times10^{15}$ kWh of energy from the sun, 70% of which goes through the upper atmosphere [MUL 11]. In comparison, humanity's primary energy consumption is about 160.10^{12} kWh per year (2018).

The photovoltaic effect was discovered in 1839. In 1930, cuprous oxide and then selenium cells appeared. It was only in 1954 that the potential of supplying electrical energy to systems through the development of the first silicon photovoltaic cells (with an efficiency of around 5% to 6%) became a possibility, thanks to research carried out by laboratories of the Bell Telephone Company. Their development and quick progress were encouraged by the conquest of space: they were very quickly used for the power electrical supply of spaceships (satellites in 1958) [PAT 99] (cell yield of around 9%). Throughout the 1990s, terrestrial photovoltaic technology has regularly progressed with the installation of photovoltaic roofs and several power stations. It even started to become familiar to consumers with its integration into many low-power products: watches, calculators, radios and weather beacons, solar-powered pumps and refrigerators. Events such as solar vehicle races have contributed to this progress by providing a futuristic and ecological high-technology

Chapter written by Arnaud DAVIGNY.

representation. Nowadays, the main applications of photovoltaic panels are as follows:

– isolated installations, where access to an electrical network is impossible or too expensive (developing countries, offshore platforms, mountain chalets, etc.);

– facilities from a few kW to several MW, on a support structure, or not, which are connected to an electrical network, in order to profit from the resale price of the kWh, which is advantageous in the framework of an environmental policy (Figure 2.1); or for self-consumption (section 1.5), often associated with a battery system in order to be less dependent on the electricity network and to limit the CO_2 footprint of a building [ZHA 12a, ROB 19];

– supply of systems requiring little energy: solar terminals, signs, parking ticket machines, toll booths, calculators, watches, etc.;

– supply for solar vehicles (Figure 2.2);

– space crafts: satellites, space station, etc.

Figure 2.1. *A photovoltaic roof (source: François Gionco, HEI). For a color version of this figure, see www.iste.co.uk/robyns/electricity.zip*

Figure 2.2. *Solar vehicle Hélios IV – Association HELIOS, Ecole HEI. For a color version of this figure, see www.iste.co.uk/robyns/electricity.zip*

2.2. Characteristics of the primary resource

The production of photovoltaic electricity depends on the following:

– The level of sunshine and the temperature of the place, and thus on its geographical location (especially the latitude) (Figure 2.3). By taking into account geographical and weather conditions, we can notice that the cumulative annual radiated energy, which is harnessed by a plane facing south and inclined to the latitude of the place (in the Southern Hemisphere this positioning is reversed), varies between less than 1220 kWh/m^2/year (average of 3 kWh/m^2/day with strong seasonal dispersions) in the north of France and more than 1760 kWh/m^2/year (average per day up to 5.2 kWh/m^2) in the south.

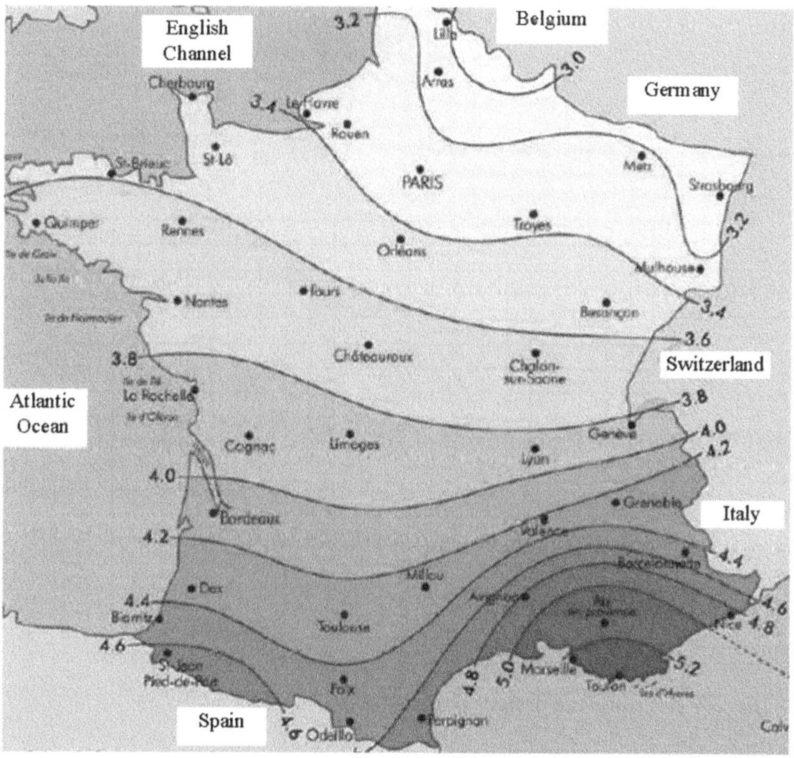

Figure 2.3. *Average irradiation in France in kWh/m^2/day on a tilted plan of the latitude value (fixed slope) and facing south (source: TECSOL). For a color version of this figure, see www.iste.co.uk/robyns/electricity.zip*

– The season and the hour of the day (Figure 2.4). The production is maximum at noon by a sundial (sun at its zenith) with a clear sky, because the thickness of the atmosphere crossed is the least significant.

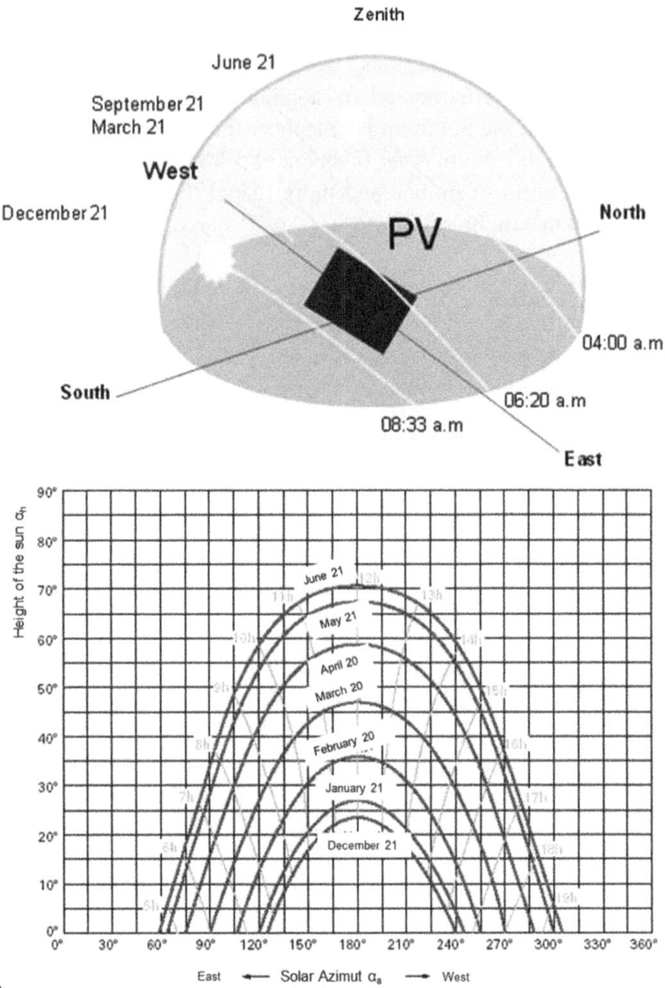

a)

b)

Figure 2.4. *Sun path over a year (Northern Hemisphere): (a) [GER 08]; (b) [AST 08a]. For a color version of this figure, see www.iste.co.uk/robyns/electricity.zip*

– The orientation and slope of photovoltaic sensors (Figures 2.5 and 2.6). The best solution would be to follow the path of the Sun from East to West and to change the slope, so that rays remain perpendicular to the plane of capture.

Figure 2.6 shows the optimal slope angle, depending on the day of the year and the condition that it remains constant throughout.

– The degree of pollution of the place.

– Weather conditions (cloud layers).

Let us note that in addition to the direct light of solar radiation, diffused light is also converted into electricity by photovoltaic generators.

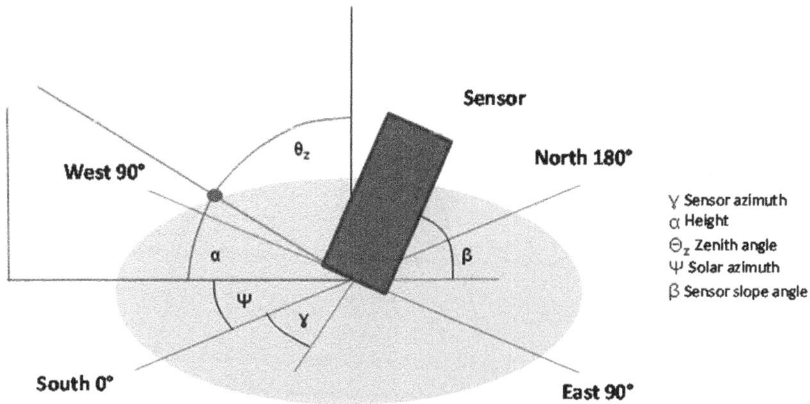

Figure 2.5. *The main angles defining the position of the sensor in comparison to the Sun (Northern Hemisphere) [GER 08]. For a color version of this figure, see www.iste.co.uk/robyns/electricity.zip*

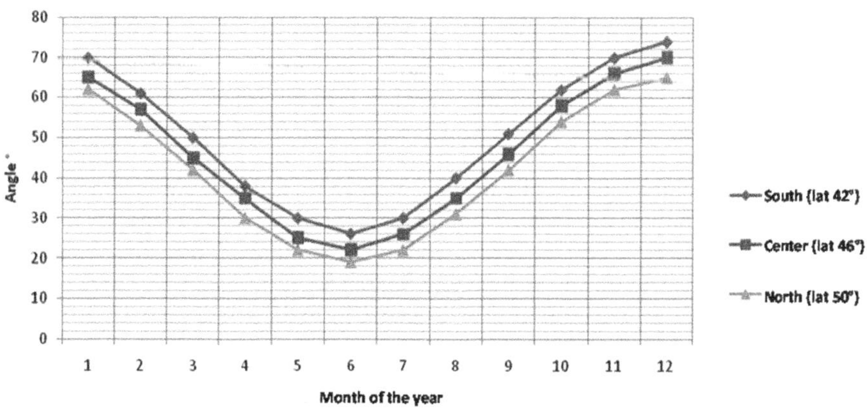

Figure 2.6. *Evolution of the optimum slope angle of the sensor according to the month of the year and the geographical situation in mainland France. For a color version of this figure, see www.iste.co.uk/robyns/electricity.zip*

The combination of all these parameters produces some variability in the space and time of the daily irradiation. Figures 2.7 and 2.8 show a record, with a 10 minute sampling (averaging), of the output power of one of the 31 inverters (associated with a group of panels) of Auchan's "COLIBRI" power station located in Villeneuve d'Ascq (north of France) (Figure 2.9), during a sunny day with and without clouds.

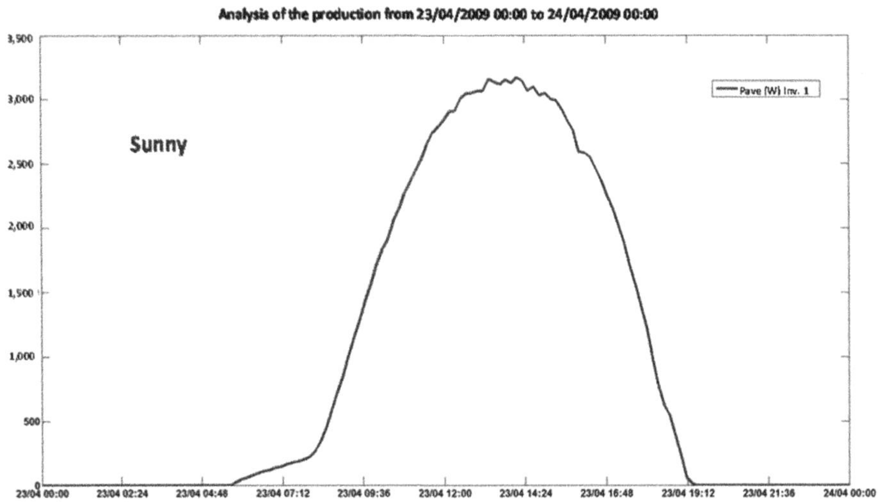

Figure 2.7. *Profile of a sunny day (source: Auchan). For a color version of this figure, see www.iste.co.uk/robyns/electricity.zip*

This primary resource was difficult to predict until recent years because, even if the sun were continuously transmitting a constant stream of solar radiation, knowledge of the next day's cloud layers relies on weather forecasts. When photovoltaics first emerged as part of electricity networks, knowing the level of production of electrical energy from photovoltaics was relatively unimportant. This was due to strong political financial incentives which allowed producers to make their installations profitable (minimal interest in having exact knowledge of the energy produced over the year) and the low impact that this electricity production could have on the economical operation of an electricity network due to the still few power installations; for example in France, less than 7 GWp at the end of 2016 against around 131 GW of installed power of all the energies combined [RTE 16].

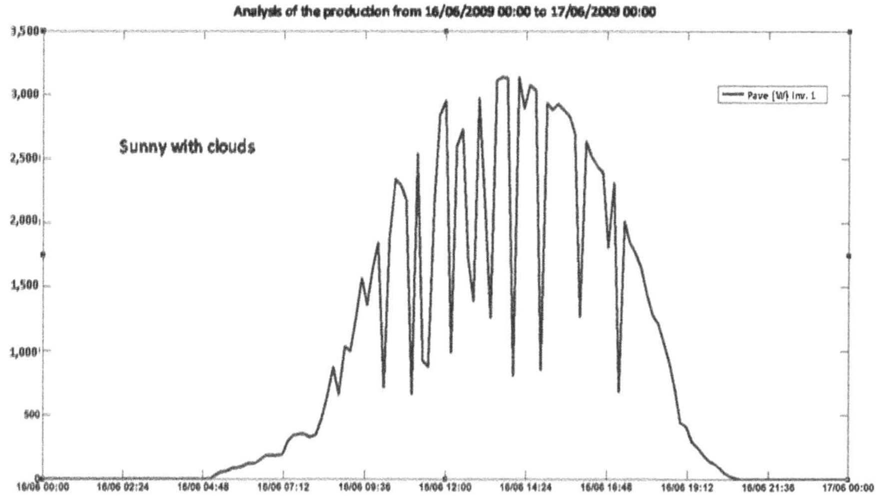

Figure 2.8. *Profile of a sunny day with clouds (source: Auchan). For a color version of this figure, see www.iste.co.uk/robyns/electricity.zip*

Figure 2.9. *Auchan photovoltaic power plant (source: Auchan). For a color version of this figure, see www.iste.co.uk/robyns/electricity.zip*

Following the decline in financial support in recent years (reduction in public funding for the implementation of installations and the encouraging feed-in tariff, in conjunction with discussions on the end of the injection buy-back obligation, etc.), competition from the recent integration of other alternative sources of electrical energy production to the electricity market, and the evolution of installed photovoltaic power on the metropolitan territory (9.4 GWp in France at the end of 2019 [RTE 19], approximately +2.7 GW in three years), it has become necessary to know the power produced, as is the case for wind power, where forecasts need to be projected at least for the day before (day D-1) the next day (day D) or on larger timescales as needed. This requirement is mandatory for island networks (island power grids) given that they are more sensitive to variations in power [LÉG 08, CRE 16, LÉG 15, SEI 16]. This forecasting of photovoltaic electricity production allows producers to optimize the operations of their facilities and, in particular, the cost of the energy produced, thereby making the facilities more profitable. For electricity network managers, forecasting makes it possible to ensure the security of the electricity network by planning a program for any shortfall through the production of electrical energy by conventional plants (often thermal), with the added benefit of minimizing the cost of production but also minimizing greenhouse gas emissions (CO_2). Finally, for a building, forecasting allows optimal management between the building's electricity production, the supply of energy from the electrical network, consumption devices and storage, thus improving upon the self-consumption of energy being generated locally and, at the same time, the environmental impact of the energy being consumed by the building [LIV 20, ZHA 12a, ROB 19]. This photovoltaic energy forecast should allow for better integration of photovoltaic installations into a mixed energy landscape, where competition between the means of production depends predominantly on the cost and characteristics of the primary resource (random or not random for example); it also depends on the technical characteristics of the production installation (possible shutdown of the plant, production forecast, possibility of helping the network, capacity to increase and decrease power, etc.).

With improved meteorological forecasting, as well as research and development in the power grid and the photovoltaic systems sector, it is possible, with varying degrees of confidence, to forecast the solar resource at ground level and thus the production of photovoltaic electric energy several days in advance (Figure 2.10) [RTE 20].

Forecasts for photovoltaic energy production are estimated using algorithms similar to those that estimate a country's electricity consumption for the day ahead, through the use of cameras and satellite imagery. Several methods have been developed to predict the production of renewable energy, in particular forecasts based on artificial intelligence or hybrid methods (a combination of two or more different forecasting methods) [ZHA 12b, YAN 14, GHA 17, MAT 17, COU 19]. Predicting photovoltaic energy is a complex problem to solve since it is affected by diverse meteorological factors such as solar irradiance, temperature, dew point, wind speed, cloud cover, etc. [INM 13]. In other words, the accuracy of photovoltaic power forecasting is dependent on the accuracy of weather observations and forecasts, which is still difficult to ever fully realize, especially for parameters such as cloud cover [ZHA 12b, COU 19].

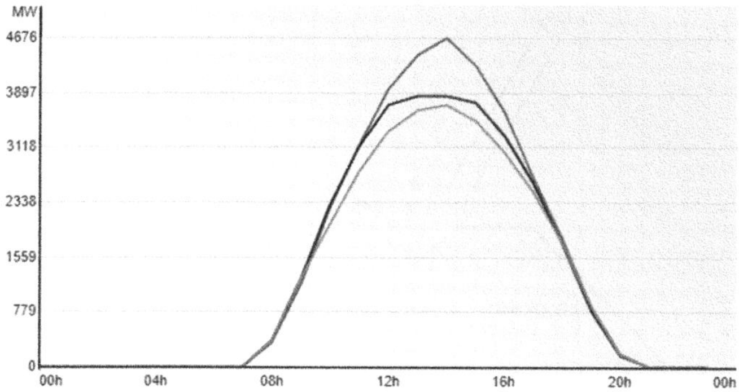

Figure 2.10. *Produced power curve planned for D-1 (blue), updated on D-day (purple), and the actual amounts that took place on D-day (orange) [RTE 20]. For a color version of this figure, see www.iste.co.uk/robyns/electricity.zip*

2.3. Photovoltaic conversion

2.3.1. *Introduction*

The transformation of solar radiation into electricity by the photovoltaic process is one of the means of exploiting insolation. The word "photocell" is sometimes used when referring to the photovoltaic (PV) cell. However, we have to note that despite this terminology, no energy is stored in a cell or in any other form, chemical or otherwise. This is not a cell but an instant converter which will be able to provide electric energy. A cell in complete darkness will behave as a passive component (diode).

Moreover, the solar cell cannot be compared to any other conventional generator of continuous electric power. This is explained by its current–voltage characteristic, which is highly non-linear. The cell is neither a source of constant voltage, nor of constant current.

2.3.2. *Photovoltaic effect*

When a photovoltaic cell is subjected to the incident light flux, it will interact so that one part of the flux is (Figure 2.11):

– reflected;

– diffused;

– absorbed;

– transmitted.

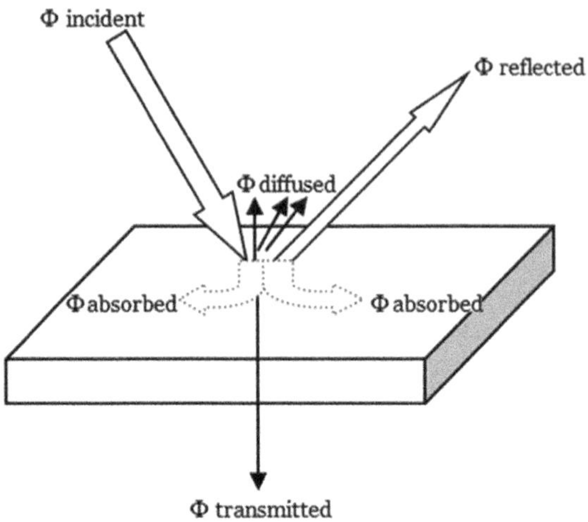

Figure 2.11. *Interaction of a photovoltaic cell with light flux*

This incident flux can come from direct radiation from the sun, from radiation reflected by the ground, from diffuse radiation coming from solar rays passing through clouds, or a combination of the three (Figure 2.12) [HOR 20].

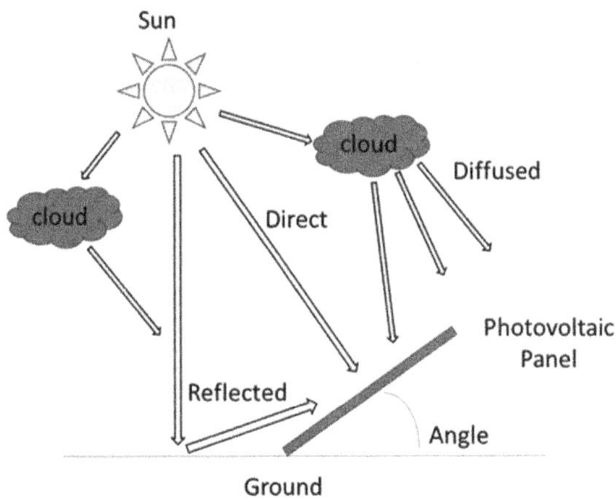

Figure 2.12. *Decomposition of the total radiation being received by a photovoltaic panel. For a color version of this figure, see www.iste.co.uk/robyns/electricity.zip*

The photovoltaic cell must absorb the incident flux at maximum. This is carried out by reducing the reflection and transmission factors. However, the cell will not absorb all the solar radiation. Depending on the technology, it is influenced by wavelengths of all or part of the spectrum.

A photovoltaic cell can be compared to a photosensitive diode. Its operation is based on the properties of the semiconductor materials [AST 08a, BER 00, LET 99]. It enables the direct conversion of luminous energy into electric energy, using operating principles that rely on the photovoltaic effect.

Indeed, a cell is made up of two thin semiconductor layers. These two layers are doped differently:

– for the N layer, contribution of outer electrons;

– for the P layer, holes or lack of electrons.

These two layers thus present a difference in their potential. The energy of luminous photons received by outer electrons (layer N) enables them to cross the potential barrier separating the N and P layers, to be attracted by the positively charged layer P and thus to generate a direct electric current. In order to accomplish this current, some electrodes are placed, via screen-printing, onto the two semiconductor layers (Figure 2.13). The upper electrode is a gate that allows for the passage of the luminous flux. An anti-reflection layer is then placed on this electrode, in order to increase the quantity of absorbed light.

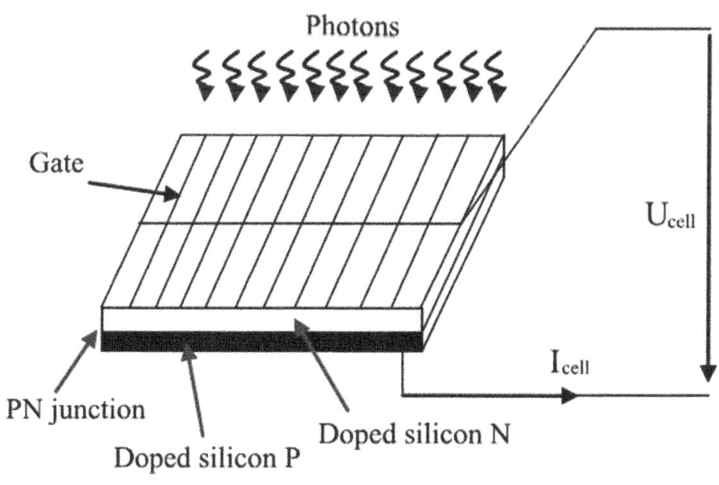

Figure 2.13. *Diagram of a basic cell*

2.3.3. *Photovoltaic cells*

2.3.3.1. *Solar cell technologies [GER 08, EQU 09]*

Today, there are four families of photovoltaic cell technologies:

– crystalline silicon-based;

– thin-films;

– organic;

– hybrids.

Table 2.1 gives the materials used in each of the technological families.

Technological families			
Crystalline silicon-based (1st Generation)	Thin-films (2nd Generation)	Organic (3rd Generation)	Hybrids (3rd Generation)
Monocrystalline silicon (mono-si) or (mono c-si)	Cadmium telluride-based CdTe	Dye solar cells	Multi-junctions
Polycrystalline/Multicrystalline Silicon (multi-si) or (poly c-si)	CIS based on copper, indium and selenium/ CIGS made of the same materials as those at CIS except that they use an alloy of indium with gallium	Polymer cells	Concentrated cells
Silicon tape (si-ribbon)	Amorphous silicon	Perovskite cells	
	Gallium arsenide (GaAs)		
	Materials III–V (single junction or multi-junction of these materials)		

Table 2.1. *The families of photovoltaic cell technologies*

The most frequently used material in photovoltaic cells is silicon (Figure 2.14), a type IV semiconductor. It is supposed to be tetravalent. This means that a silicon atom can bind with four other atoms of a similar nature. The transformation of this material in order to reach the finished product (the cell) requires a lot of energy. We estimate that a photovoltaic cell should operate from 2 (1 year for an ideal geographic location) to 4 years, depending on its technology, in order to produce the same amount of energy that was necessary for its manufacture. This is not long when taking into account an expected lifespan of up to 30 years for some technologies.

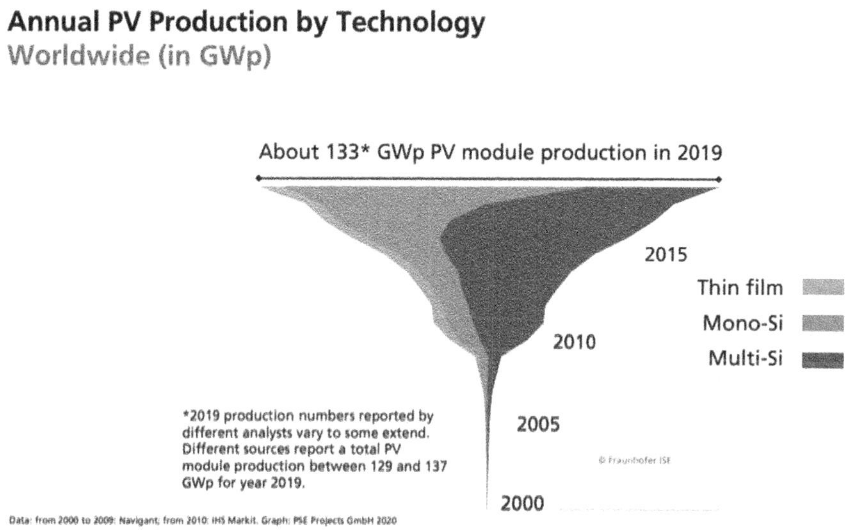

Figure 2.14. *Distribution of technologies on the world market in 2019 [FRA 20]. For a color version of this figure, see www.iste.co.uk/robyns/electricity.zip*

The objectives of research and development in recent years are to reduce the environmental impact, the cost and quantity of photovoltaic material per Wp, to increase the lifespan and efficiency of the cells, and to develop recycling.

In the following paragraphs several types of solar cells are described in detail.

Single crystal silicon

These are the first cells designed from a silicon block which is crystallized in a single crystal. As it cools, the molten silicon will solidify. It will then form a single large crystal. This crystal will then be cut into thin slices which will allow the cells to develop. These cells are generally of a uniform color (often black), which means they are more aesthetically pleasing when integrated into rooftops. Monocrystalline cells are installed in colder or temperate regions due to decreased performance, which is dependent on temperature, but which is greater than for polycrystalline cells.

They are available in the form of small round (direct cutting from a cylindrical ingot without any discards), square or almost square plates. The assembly of these cells results in a rigid end product in the form of a rectangular panel, which is heavier than other technologies.

The cells' standardized efficiency is 16 to 24%. They have a long lifespan of approximately 30 years. The data concerning lifespan is related to the fact that the conversion productivity of a photovoltaic cell decreases with age. Manufacturers generally provide a production decrease of 0.5% per year and a minimum performance of about 85% after 25 years. Usually, they can produce electrical energy for about 40 years. However, these cells have two disadvantages:

– their high price;

– high energy consumption during manufacturing with a pronounced CO_2 impact.

Polycrystalline silicon

Cells are designed from a block of silicon, crystalized into several crystals with different orientations. During silicon cooling, several crystals form. This type of cell is often blue with patterns of large tangled crystals, which can impact the overall aesthetic of the rooftop. Polycrystalline cells are installed in warmer regions, over monocrystalline cells, due to a decrease in efficiency as a function of lower temperatures. Their lifespan is also approximately 30 years (no significant difference from monocrystalline), with an efficiency of about 10 to 18%. However, their production cost is lower than single-crystal cells. The assembly of these cells also results in a rigid end product in the form of a rectangular panel which is also heavier than other technologies.

Silicon ribbon

This type of cell was developed with the objective of producing thin (light), high-efficiency cells at a reduced manufacturing cost. These thin ribbons are based on crystalline silicon and several manufacturing processes exist. One of the principles is the deposition of a ribbon of molten silicon on both sides of a flexible graphite support which is then removed (the process used by the company

Solarforce). The thickness of the silicon depends on the running speed of the support. This manufacturing process produces a very thin polycrystalline silicon ribbon which is available in the form of a roll. Efficiency is of the same degree as that of polycrystalline silicon cells. This sector, however, has encountered difficulties in making a place for itself in the photovoltaic market.

Amorphous silicon

These cells are made up of a glass or synthetic support on which a thin amorphous silicon layer is positioned (the organization of the atoms is no longer uniform as it is in a crystal). The assembly of these cells gives a flexible end product, 100 to 500 times thinner than the crystalline types, marketed in the form of rolls which allows for simple, fast installation as it can be cut to the desired shape and size, and with a very low weight compared to the other technologies. This low weight makes it possible to eliminate any constraints that will impact on the resistance forces of the roof frame. The efficiency of this technology is about 5 to 10%, which is lower than that of the crystal cells, but their price is low. Although they require more area, this price enables them to produce electricity quite cheaply.

Originally they were often used in small manufactured products in large series and low requirements (watches, calculators, etc.), but are less frequently used in the framework of solar installations. They are beginning to find their use in photovoltaic systems integrated into buildings, thanks to recent advancements in this type of cell.

They also have the advantage of having a better reaction to diffused light and fluorescent light. They are thus more efficient at low luminous intensities, such as those that are encountered in inside environments. They also have an efficiency that is less sensitive to high temperatures. They have a lifespan of about 10 years, which is highly conditioned by the exposure to high intensity radiation.

CIS (CuInSe2) and CIGS cells

CIS cells are based on copper, indium and selenium. This type of cell has the property of being stable under radiation. They have excellent absorption properties relative to other cells, particularly in the early and late parts of the day; they are also non-toxic unlike CdTE cells. Their efficiency is about 11 to 13% and they are marketed as thin films. Progress in manufacturing has made it possible to greatly reduce the cost of these cells (previously a barrier to commercialization). The energy used in their manufacture is less than that of polycrystalline silicon cells. The increasing use of indium, a rare element with criticality low reserves on Earth, used in certain industrial sectors, such as the manufacturing of LCD screens, for example, could slow their expansion.

The properties of CIS cells were improved by adding gallium atoms in place of some of the indium atoms. These cells are called CIGS. The enhanced properties include absorption and performance, which reached laboratory records of over 20% (12 to 17% for industrial cells).

This technology also performs well in panel shading where it continues to produce, whereas a single crystal cell would no longer produce energy. Like all thin-film technologies, its performance is also less sensitive to temperature.

CIGS cells are made up of the same materials as CIS cells, with the addition of a gallium indium alloy, enabling better properties to be obtained.

CdTe cells

CdTe thin-film solar cells are based on cadmium telluride, which is an interesting material because of its absorption. Telluride is a rare element on Earth. The presence of cadmium, a toxic material, requires us to prove the feasibility of recycling. Cadmium toxicity is present in the extraction, manufacturing and recycling of panels, but not considered in their usage (i.e. glass-sealed materials during a fire). The manufacture of these cells has a low carbon and energy footprint. The manufacturer, First Solar, has industrialized the product and developed a recycling process for cadmium and a recovery processes for its modules. The main advantage is their very low cost, (announced to be 2 to 3 times less critical than those used by the crystalline sector), coupled with quite reasonable performances, which makes it a strong contestant against silicon technologies, most notably for applications in large farms on the ground. However, limited cadmium resources will not allow a massive penetration rate in the long-term. These cells have a low mass. They also exhibit robustness. In the future, they may also be flexible. They have a high absorption capacity in low light such as dawn, dusk or on an overcast day. The yield of commercially available cells is in the order of 17%. They have a performance behavior with respect to high temperatures that is far better than that of crystalline technology.

Cells based on III–V materials

Cells made from III–V semiconductor materials, such as gallium arsenide (GaAs), under development, can achieve high efficiencies close to 30% for single junctions and over 30% for multi-functions. In 2020, due to the high cost of manufacturing these cells, their use is limited to unmanned aviation and space exploration.

Organic cells and hybrid cells

These two technological categories are under development.

Hybrid cells are made up of different layers that are sensitive to different wavelengths of the solar spectrum. The goal is to increase the yield significantly. The cost of manufacturing these cells makes them reserved for the space industry. In this family we find different types of cells:

– Tandem cells, which combine two layers of simple semiconductor (for example polycrystalline silicon and amorphous silicon or perovskite and silicon) with a yield close to 26% under laboratory conditions.

– Multijunction cells, which combine several layers of simple semiconductor. It is then possible to obtain returns of 30% to upwards of 40%.

– Concentrating cells for which lenses or mirrors will concentrate the sun's rays towards multi-junction cells. It is then possible to reduce the amount of material in III–V materials (reduction in the size of the cells by a factor of 100 to 1,000) and therefore to reduce the cost of the panels, compared to multi-junction panels not using concentration techniques.

Organic cells or OPV (organic photovoltaics), are made up of polymers which are semiconductor plastic materials (at the research stage) [DES 04]. They also have the property of absorbing photons and generating a current. In this family we find dye, polymer or perovskite cells. These have a reduced manufacturing cost and the energy impact during manufacturing is low. The same is true with a view to their ecological impact in terms of manufacture, recycling and toxicity. They have appreciable characteristics: low thickness, light, solid, flexible and replaceable. It is possible to integrate them into flexible structures (tents, clothing, etc.) or rigid structures (walls, roofs, etc.). Thanks to pigments, a panel of different colors is also available, unlike silicon technology which offers a few limited colors (black or blue). They are customizable (several design possibilities and different colors on the same panel). They have yields in the range of 10 to 13%. However, there is still room for improvement in terms of service life. They are already installed on demonstrators and could soon be found on the photovoltaic market in the years to come.

Table 2.2 shows the typical (panels on the production chain) and theoretical (efficiency of panels manufactured in laboratories) efficiencies that we can obtain from these various technologies.

Figure 2.15 gives the maximum yields achieved for different technologies. Figure 2.16 provides an overview of the race to improve the efficiency (laboratory value) of different cell technologies over recent years.

Technology	Efficiency	Lifespan	Main uses
Single crystal silicon	16 to 24%	30 years	Space, modules for roofs, façades, etc.
Polycrystalline silicon	10 to 18%	30 years	Modules for roofs, façades, generators, farms on the ground, etc.
Amorphous	5 to 10%	10 years	Electronic instruments (watches, calculators), integration in the building
CIS	11 to 13%	>20 years	Integration in the building
Multi-junctions	Up to 40%		Space
CdTe	6 to 10%	>20 years	Electronic instruments (watches, calculators), integration in the building, farms on the ground
CIGS	19.9% in laboratories	–	Space, integration in the building
Organic	5.9% in laboratories	–	Under development

Table 2.2. *Summary of the different technologies*

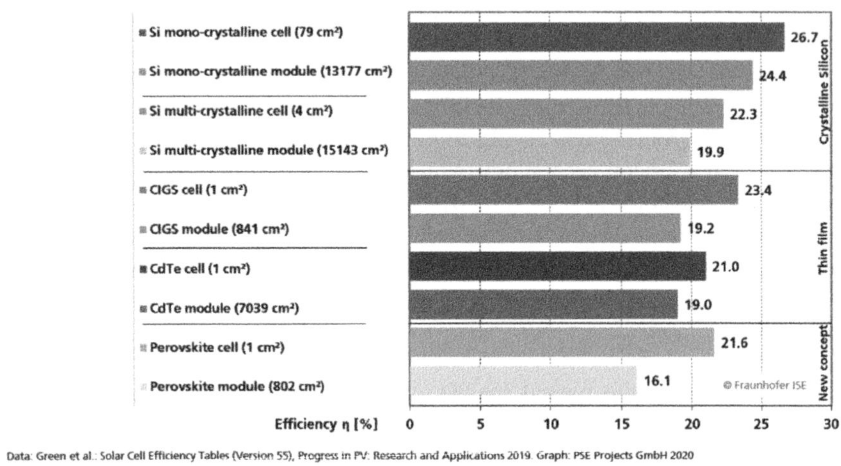

Figure 2.15. *Performance record for different technologies [FRA 20]. For a color version of this figure, see www.iste.co.uk/robyns/electricity.zip*

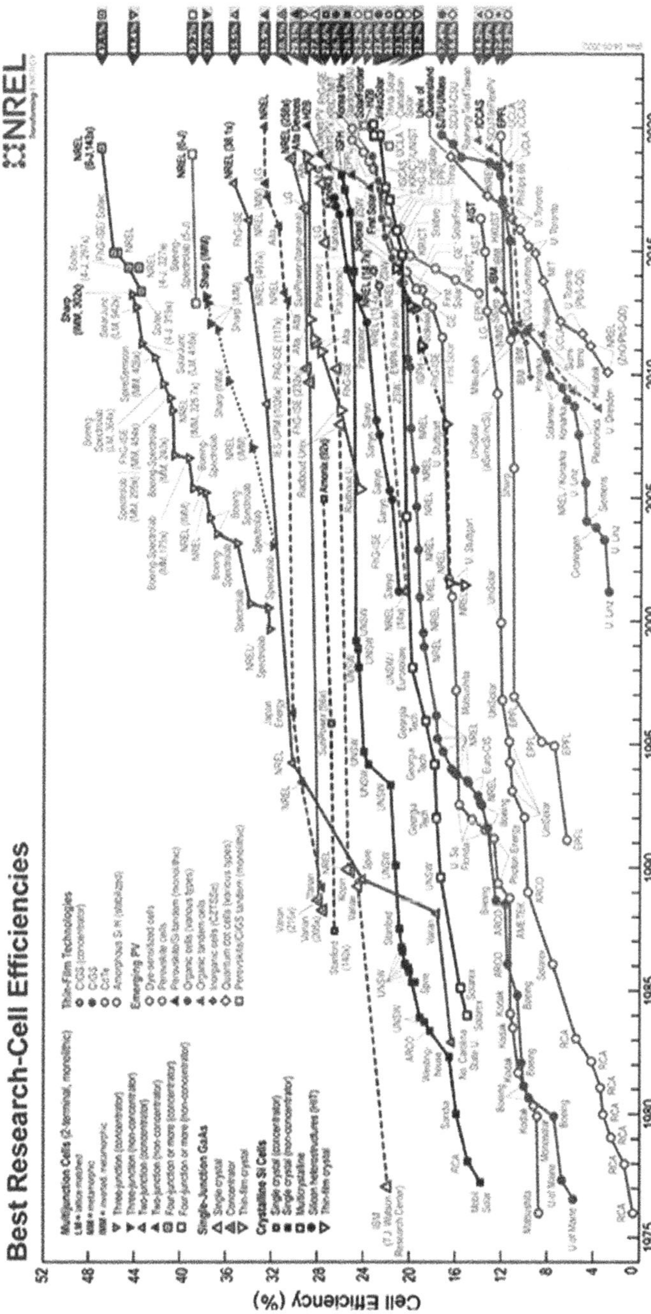

Figure 2.16. *Evolution of the yield for different families of cells [NRE 20]. For a color version of this figure, see www.iste.co.uk/robyns/electricity.zip*

Recycling

In 2020, it was possible to recycle between 85% and 95% of a photovoltaic panel. The European directive 2002/96/EC relating to waste electrical and electronic equipment and used electrical and electronic equipment (WEEE) of February 2003 regulates this recycling. It is taken into account in French law by Decree No. 2014-928 of August 19, 2014. The seller has the obligation to ensure the treatment and recycling of waste and to bear the cost. This obligation is satisfied by the payment of an eco-contribution by the buyer. The decree of May 9, 2017 clarifies and simplifies the process for scrapping photovoltaic panels for owner-producers.

Each panel technology offers recycling.

A photovoltaic panel consists of:

– an aluminum frame: infinitely recyclable and sold on the recycled raw materials market;

– glass (approximately 75%): 100% infinitely recyclable and sold on the raw materials market;

– plastics: polyester or transparent recyclable plastic film, transformed into granules to be remelted or burned to produce electrical energy;

– connections in copper, silver, aluminum or nickel. Directed towards the treatment of *Déchets d'Équipements Électriques et Électroniques* (Waste Electrical and Electronic Equipment) or DEEE;

– junction boxes (recyclable printed circuits). Directed towards the treatment of DEEE;

– crystalline-type silicon cells (a large majority of panels installed in the years 2000 to 2020 and still today): recyclable up to four times.

In France, the eco-non-profit organization, PV CYCLE France [PVC 20], a European organization with ministerial approval and financed with the help of the eco-tax levied on each panel when it is purchased, is a leading actor in the recycling of photovoltaic panels. It carries out three key roles:

– informs users in order to allow for the controlled management of used equipment;

– free-of-charge collection of all used photovoltaic solar panels regardless of the technology, brand or year it was introduced onto the market, dedicated to the recycling of photovoltaic solar panels;

– recycling of PV panels: in 2020, it recycled 94.7% of crystalline silicon-based photovoltaic panel with an aluminum frame.

Due to the growing recycling market, it joined forces with Véolia, following a call for tenders, in order to perpetuate its achievements, and in July 2018 was the first large-scale recycling plant for photovoltaic panels in the vicinity of Aix en Provence.

2.3.3.2. *Comparison with the diode [PRO 97]*

The photovoltaic cell is the basic element for photovoltaic conversion. In darkness, it behaves like a PN junction (diode). Under these conditions, we can again find the current–voltage characteristic of a PN junction for the cell (Figure 2.17).

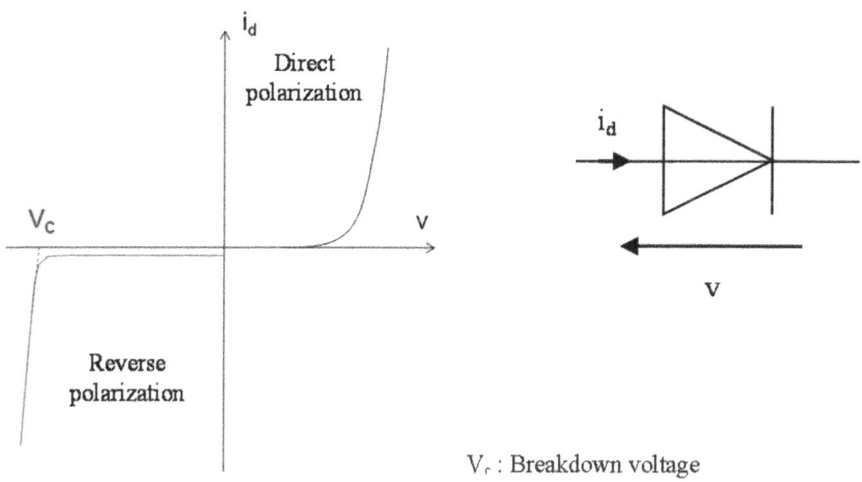

Figure 2.17. *Characteristic of a PN junction in receiver convention. For a color version of this figure, see www.iste.co.uk/robyns/electricity.zip*

A cell characteristic can be determined with the help of the characteristics of a PN junction. When the cell is illuminated, the produced current rises as the illumination becomes more intense. This current, which is proportional to the illumination, is called the photocurrent, I_{ph}. The resulting current i would then be equal to I_d-I_{ph}. By always considering the receiver convention, we would obtain, with constant illumination, the characteristic illustrated in Figure 2.18.

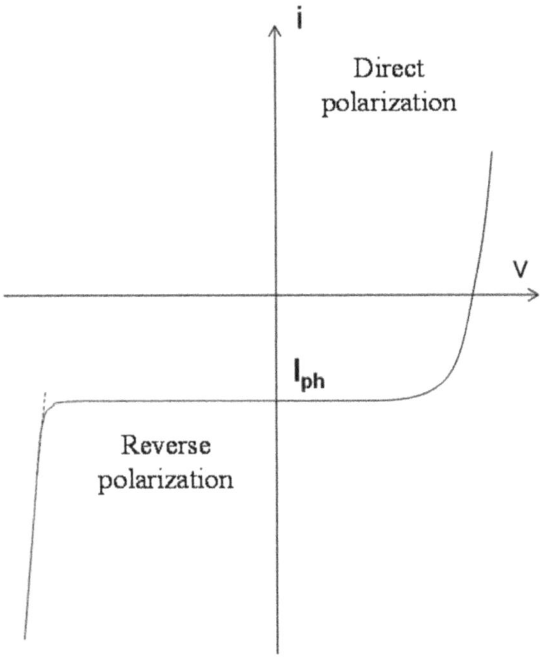

Figure 2.18. *Characteristic of a photovoltaic cell excited by a luminous radiation, in receiver convention. For a color version of this figure, see www.iste.co.uk/robyns/electricity.zip*

As a PV cell is used in generator mode, the characteristic I(V) of a cell is represented in generator convention. To do this, the current has to be reversed (Figure 2.19). If we do not take an interest in its parasitic operating modes, only the first quadrant is kept (positive i and v) (Figure 2.20). Note that the cell is, for the most part, equivalent to a current source (starting from a short-circuit operation) and also partially equivalent to a voltage source (starting from open circuit operation). The shape of this characteristic varies slightly depending on the cell technology (Figure 2.21). For a given technology, it varies evidently with illumination (Figure 2.22), but also with the temperature (in practice, the current fed by the cell slightly increases with the temperature, while the no-load voltage decreases) (Figure 2.23). On the whole, the maximum power decreases when the temperature increases. In Figure 2.21, "FF" corresponds to the fill factor [2.1]. A higher FF value improves the cell efficiency [2.8]. In Figure 2.22, the curve connecting the characteristics is the position of the power maximums (section 2.4).

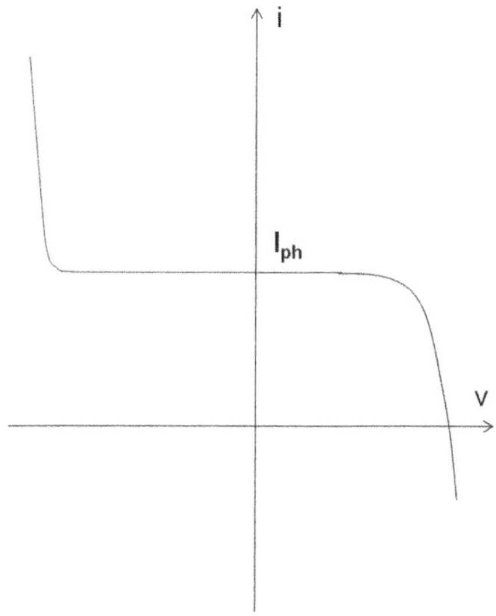

Figure 2.19. *Characteristic of a photovoltaic cell in (classic) generator convention. For a color version of this figure, see www.iste.co.uk/robyns/electricity.zip*

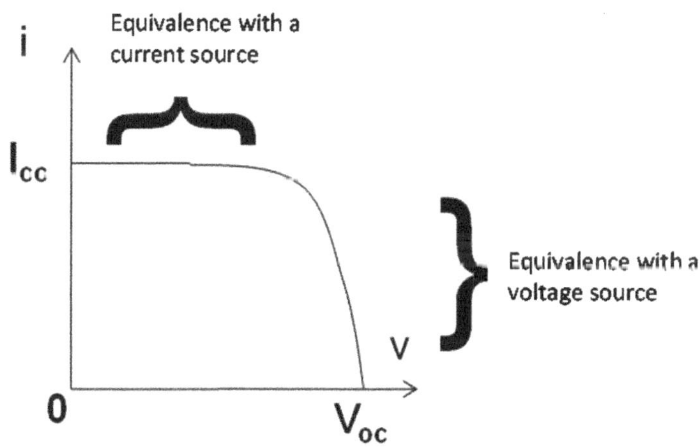

Figure 2.20. *Characteristic I(V) of a photovoltaic cell*

In Figure 2.20, the I_{cc} current corresponds to a current which appears when the cell is short-circuited (null voltage at its terminals) under standard sunshine. In practice, this current is very close to the photocurrent I_{ph}. Similarly, the voltage V_{oc} or V_{cc} corresponds to the no-load voltage under the same conditions.

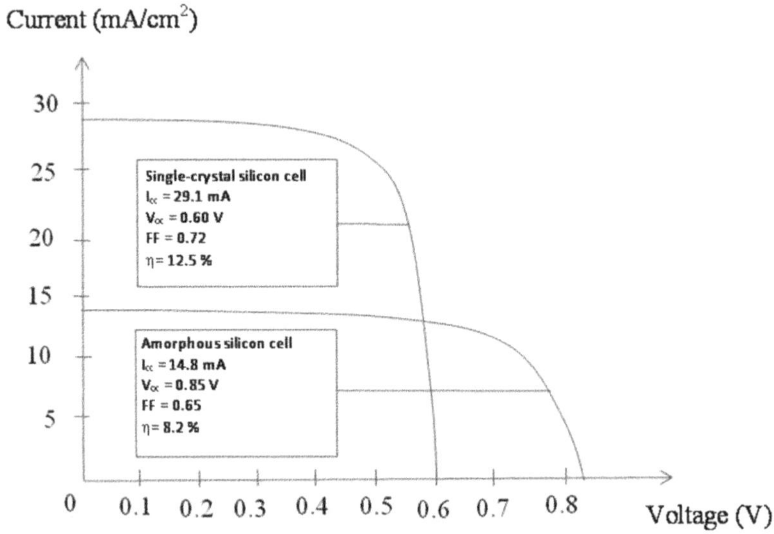

Figure 2.21. *Characteristic I(V) of the two 1 cm^2 cells of different technologies [RIC 05]*

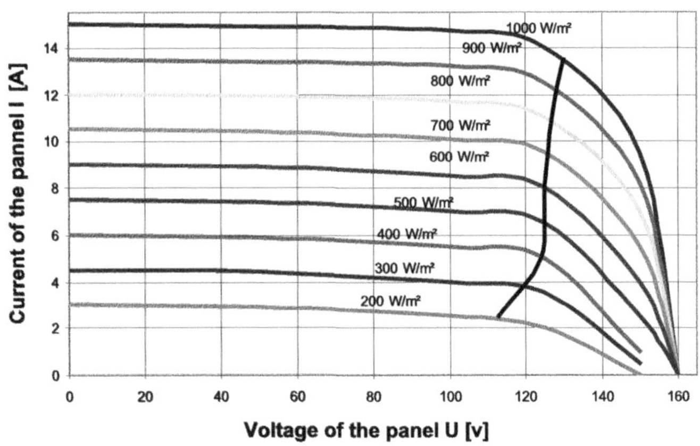

Figure 2.22. *Characteristic I(V) depending on illumination at a constant temperature [PAN 04]. For a color version of this figure, see www.iste.co.uk/robyns/electricity.zip*

In Figure 2.22, we can notice that the current fed by the panel depends on illumination. On the other hand, the open circuit voltage does not depend on illumination, but on the quality of the material and on the type of the considered junction (Figure 2.21).

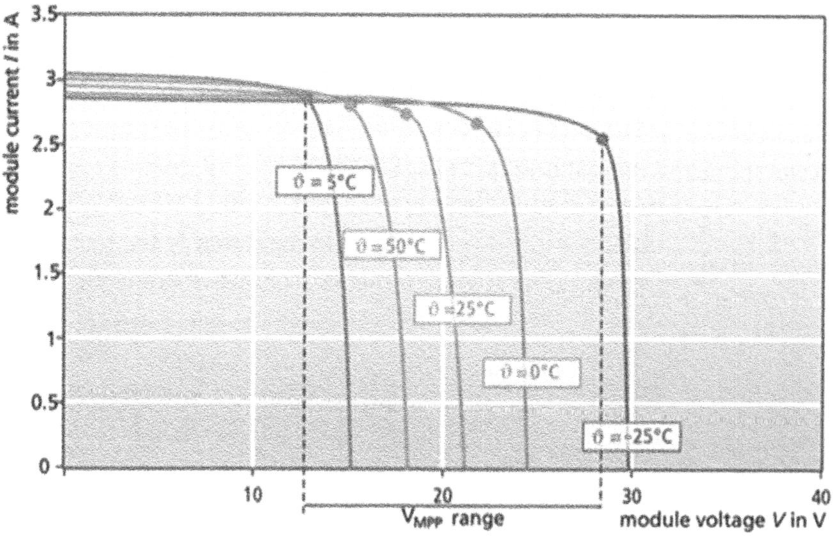

Figure 2.23. *Characteristic I(V) according to the temperature at constant solar radiation [GER 08]. For a color version of this figure, see www.iste.co.uk/robyns/electricity.zip*

We can notice the current–voltage characteristic I = f(U) of a cell, by using the assembly of Figure 2.24, in which the load resistor enables us to scan all the operating points of the cell at a given illumination.

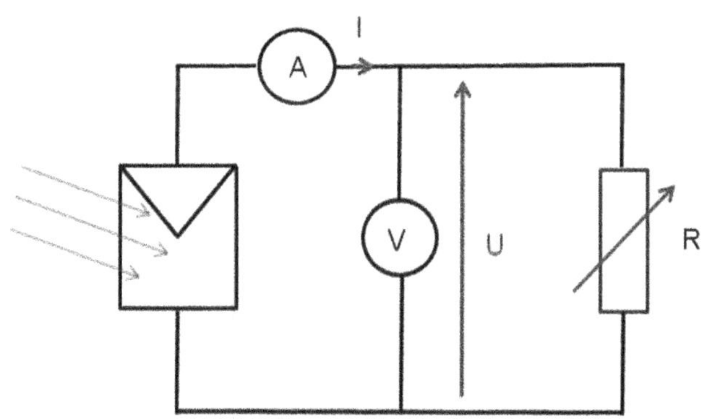

Figure 2.24. *Assembly to measure the characteristic I(V) of a cell. For a color version of this figure, see www.iste.co.uk/robyns/electricity.zip*

By establishing the characteristic in power P = f(U) for given illumination and temperature conditions, we can underline a point of maximum or (peak) power, called P_c, which is visible in Figure 2.25 and corresponds to optimum use [AST 08a, PRO 97]. We can introduce the concept of the FF, which corresponds to the ratio of the power $P_c = V_c.I_c$ on the power $P_m = V_{oc}.I_{cc}$, in relation to the rectangle, and which indicates the degree of ideality of the characteristic.

$$FF = \frac{P_c}{P_m} = \frac{V_c.I_c}{V_{oc}.I_{CC}} \quad [2.1]$$

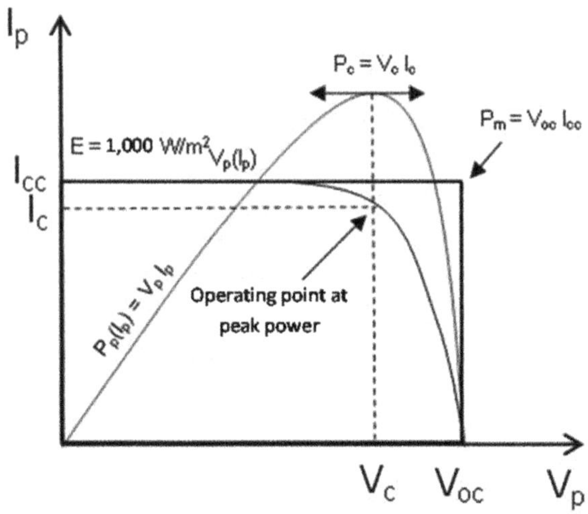

Figure 2.25. *Maximum power point of a cell. For a color version of this figure, see www.iste.co.uk/robyns/electricity.zip*

In Figure 2.21, note that the single crystal structure helps to obtain a higher maximum power (FF = 0.72). However, the no-load voltage V_{oc} is higher for an amorphous structure.

2.3.3.3. *Equivalent model [AST 08a, PRO 97, MUL 07, GER 02]*

An equivalent diagram of a cell can be established by taking into account the current source I_{ph}, the P-N junction and the various Joule losses, which are represented by resistors (Figure 2.26). Although research work generates more complicated models they are more precise at modeling behavior.

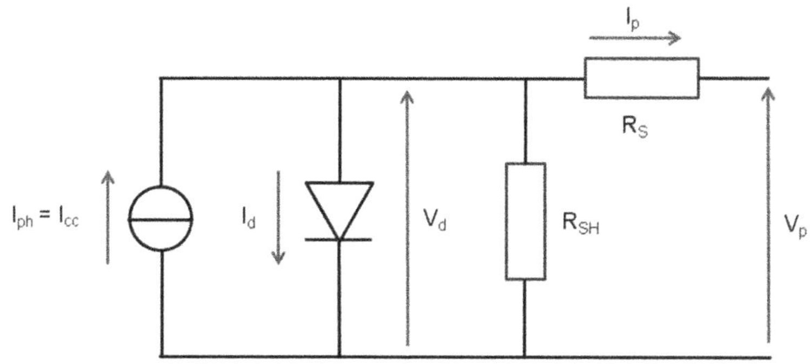

Figure 2.26. *Equivalent diagram of an electrical model of a cell*

The diode models the behavior of the cell in darkness. The current generator models the current I_{ph}, which is generated by an illumination. Finally, the two resistors model the internal losses:

– series resistor R_s represents the material ohmic losses;

– shunt resistor R_{sh} represents the parasitic currents crossing the cell.

The mathematical model associated with a cell can be obtained from that of a conventional PN junction. The current I_{ph} is then added. It is proportional to illumination (or to a luminous flux) and to the surface S of the junction, which is subjected to solar radiation. We also add a term modeling the internal phenomena. Figure 2.26 offers an equivalent diagram including resistors, which enables us to better comprehend the actual observed characteristic. The current I_{cell} coming from the cell can then be written:

$$I_{cell} = I_{ph} - I_d - \frac{V_{cell} + R_S.I_{cell}}{R_{sh}} \qquad [2.2]$$

with:

$$I_d = I_{0d}.(e^{\frac{q.(V_{cell}+R_S.I_{cell})}{k.T}} - 1) \qquad [2.3]$$

Consequently, we obtain the following expression for the current fed by the cell:

$$I_{cell} = I_{ph} - I_{0d}.(e^{\frac{q.(V_{cell}+R_S.I_{cell})}{k.T}} - 1) - \frac{V_{cell} + R_S.I_{cell}}{R_{sh}} \qquad [2.4]$$

with:

– I_{ph}: photocurrent, or current generated by illumination (A);

– I_{0d}: diode saturation current (A);

– R_s: series resistor (Ω);

– R_{sh}: shunt resistor (Ω);

– k: Boltzmann constant ($k = 1.38 \times 10^{-23}$);

– q: electron charge ($q = 1.602 \times 10^{-19}$ C);

– T: cell temperature (K).

We can possibly simplify this model by neglecting the voltage drop $R_s.I_{cell}$ in front of V_{cell} (the value of R_s is lower than 1 Ω):

$$I_{cell} = I_{ph} - I_{0d}.(e^{\frac{q.V_{cell}}{k.T}} - 1) - \frac{V_{cell}}{R_{sh}} \quad [2.5]$$

The shunt resistor has a high value (higher than 10 kΩ). If we neglect its contribution, we obtain:

$$I_{cell} = I_{ph} - I_{0d}.(e^{\frac{q.V_{cell}}{k.T}} - 1) \quad [2.6]$$

The corresponding equivalent diagram of Figure 2.27 is that of a first level cell.

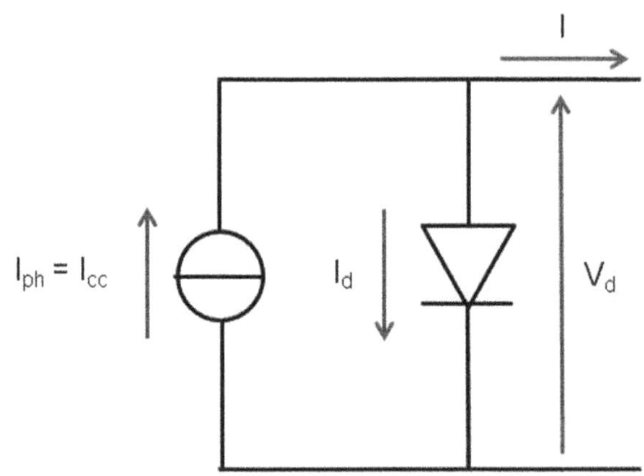

Figure 2.27. *Equivalent diagram of a first level cell*

2.3.3.4. *Maximum power of a cell [MUL 07]*

The power of a cell, a module (assembly of cells in series and/or in parallel) or a photovoltaic system is measured in Watt-peak (W_p). The "peak power" of a PV cell represents the maximum electric power supplied under the following ideal sunshine conditions, the so-called standard conditions, which are noted STC (standard test conditions):

– solar illumination of 1,000 W/m^2;

– cell temperature equal to 25°C;

– an AM 1.5 spectrum corresponding to a solar radiation that has passed through the atmosphere thickness (Sun not at its zenith).

The electromagnetic spectrum on the ground level varies with the position of the Sun and atmospheric conditions. This explains why it is necessary to define a reference spectrum, which helps to compare several panels under conditions that are representative of the real applications. A application of the "peak power" characteristic is the comparison of the efficiency and the price of the photovoltaic production. The peak power is about 50 W_p to 200 W_p per m^2, depending on the technology. The surface of the panels (cell assemblies) varies on average from 0.5 to 2.5 m^2.

The quantity of solar energy which reaches the surface of the Earth depends on the thickness of the atmosphere to be crossed. The latter is characterized by the air mass (AM) number. When the position of the sun in the sky shifts from the zenith to lower points, the light must then pass through a greater thicknesses of air which causes it to lose more energy (Figure 2.28):

– For AM0, outside the atmosphere, at high altitude, an average of 1,367 watts reaches each square meter of the outer edge of the Earth's atmosphere. This is the solar constant, which is equal to 1,367 W/m^2.

– For AM1, with the Sun at the zenith (at 90°), an average of 1,000 W/m^2 reaches the sea level at noon. Since the Sun is only at its zenith for a little while, the air mass is permanently higher and the available energy is thus lower than 1,000 W/m^2.

– For AM1.5, with the Sun at 48°, an average of 833 W/m^2 reaches the ground level.

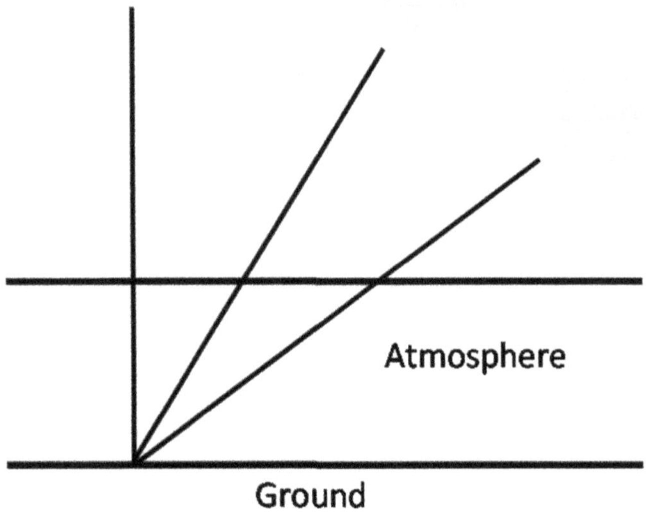

Figure 2.28. *Air thickness crossed by the light, depending on the Sun's position*

In Figure 2.29, we can see the radiation of the reference spectrum AM1.5. At the ordinate we can see the irradiance (flux density) expressed in W/m² for each wavelength. The incident power, called illumination E, is given by the integration of the irradiance. Its value is:

$$\int_0^{\infty} M\lambda d\lambda = 1000\, W/m^2 \qquad [2.7]$$

We can see, in Figure 2.29, that the spectrum spreads from 0.3 µm to 2 µm and that most of the energy is concentrated on small wavelengths.

The different families of photovoltaic cells are not excited by the whole solar spectrum (Figure 2.30).

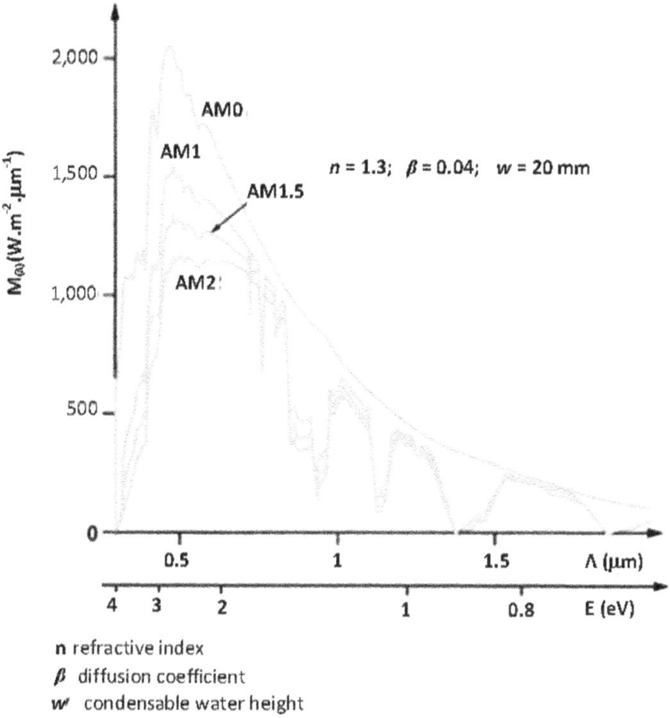

Figure 2.29. *Solar reference spectra [AST 08a]*

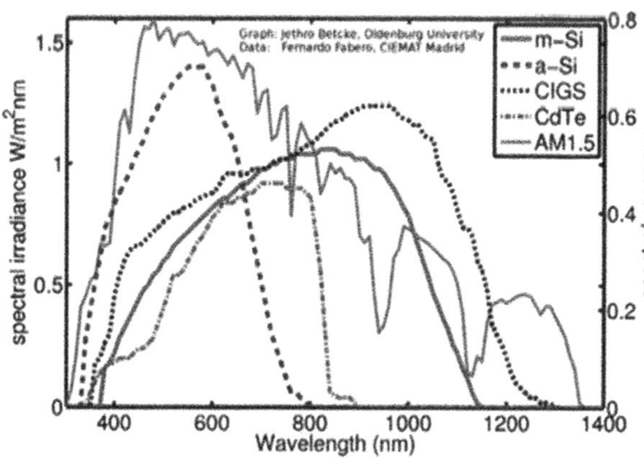

Figure 2.30. *Useful domain of the solar spectrum for different cell types (source: Jethro Betcke, Oldenburg University, Germany and Fernardo Fabero, Ciemat Madrid, Spain). For a color version of this figure, see www.iste.co.uk/robyns/electricity.zip*

2.3.3.5. *Efficiency*

We define the energy efficiency of a cell by the ratio of the maximum power and the incident power:

$$\eta = \frac{P_c}{E.S} = \frac{FF.I_{cc}.V_{oc}}{E.S} \qquad [2.8]$$

with:

– E: incident illumination (W/m^2) at right angles to the cell plane;

– S: active surface of the panels (m^2).

For the calculation of the nominal efficiency, P_c is the maximum power measured under STC (standard test conditions), i.e. under a spectrum AM1.5, at a temperature of 25°C, and with an illumination of 1,000 W/m^2.

The efficiency of a photovoltaic cell is generally quite low (about 10 to 20% for marketed cells). However, we should not lose sight of the fact that the direction of conversion efficiency during the exploitation of renewable resources is very different from the direction of an efficiency, which corresponds to the conversion of non-renewable resources, which are irreparably lost after use. A low photovoltaic efficiency consists, foremost, of the necessity of a larger capture surface, but not necessarily of a low eco-balance. Technologies with the highest efficiencies are not generally those with the best efficiency factor according to the definition given in Chapter 1.

The efficiency depends partly on illumination, but this is not specified by manufacturers, and also partly on temperature (a few tenths of a %/°C). Temperature is an important factor, since the cells are submissive to solar radiation, which is likely to overheat them. A fraction of the absorbed radiation is not converted into electric energy. It is dissipated in the form of heat. Therefore, the temperature of a cell is always higher than the ambient temperature. The colder it is, the more efficient it is. To estimate the cell temperature T_c from the ambient temperature T_a, we can use the following empirical formula:

$$T_c = T_a + \frac{E_m}{800}(TUC - 20) \qquad [2.9]$$

with:

– E_m: average illumination (in W/m^2);

– TUC: temperature of use of the cell (°C).

Let us note that this formula does not take into account ventilation conditions (effect of the wind, natural ventilation, for example from the back). The cell temperature during hot days can go up to 70°C or higher [MUL 07]. Depending on the technology, each warming degree causes an efficiency loss of about 0.1 to 0.5% (Figure 2.31).

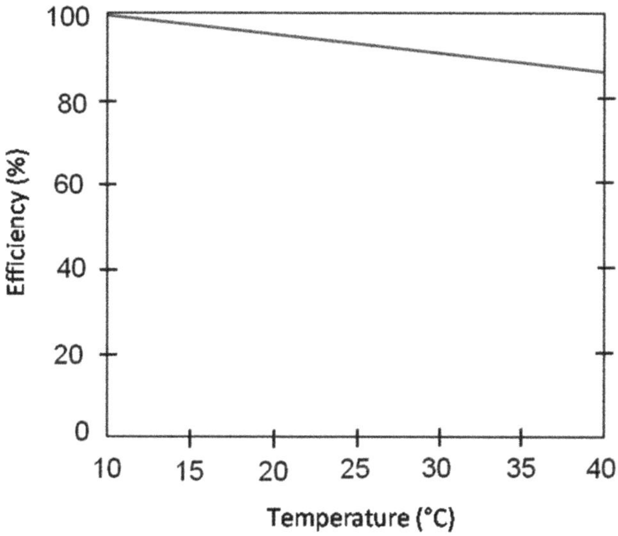

Figure 2.31. *Evolution of the nominal efficiency of a single crystal silicon cell according to the temperature*

Figure 2.32 shows the impact on yield and therefore on the power delivered by a photovoltaic plant on the roof of a residence on a sunny day and on a sunny day with clouds for the same week in early April. The temperature of the day with clouds allows the plant to reach a peak power output that is greater than that of the sunny day.

Figure 2.33 shows the change in the temperature of a cell over a day and the difference with the ambient temperature of a photovoltaic power plant connected to the grid at the *Centre de Développement des Energies Renouvelables* (Renewable Energy Development Center) or CDER in Algiers.

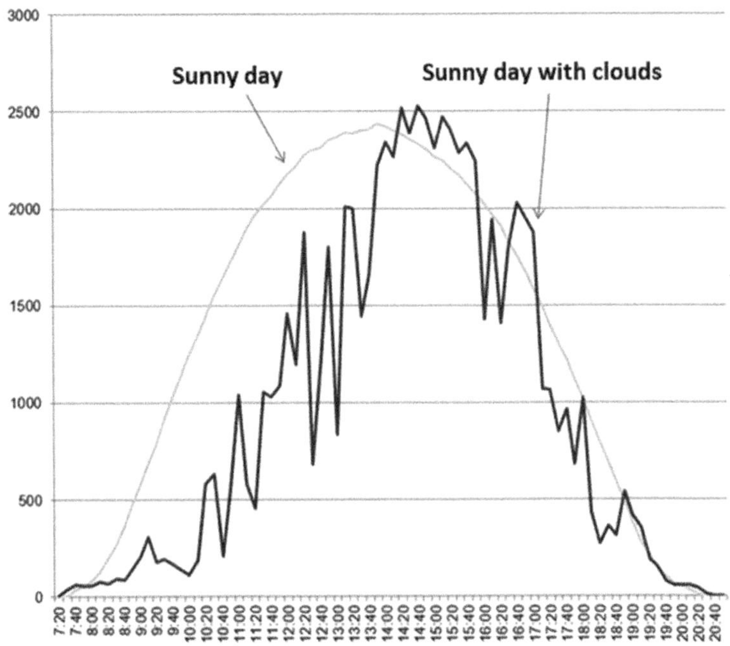

Figure 2.32. *Evolution of the maximum power of a monocrystalline silicon photovoltaic plant as a function of temperature (source: François Gionco, HEI). For a color version of this figure, see www.iste.co.uk/robyns/electricity.zip*

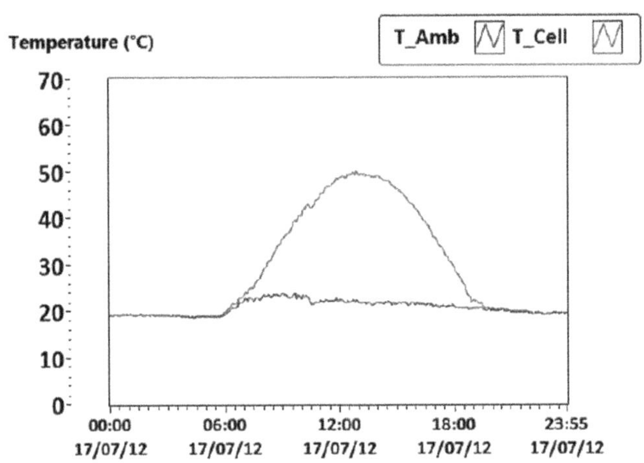

Figure 2.33. *Evolution of the temperature of a cell and the associated ambient temperature [TAG 13]. For a color version of this figure, see www.iste.co.uk/robyns/electricity.zip*

The incidence angle of the Sun's rays on the photovoltaic cell also has to be taken into account for the global efficiency of the cell (Figure 2.34). Indeed, the efficiency of the panels depends on the following relationship:

$$R = 100 \times \sin \beta \quad [2.10]$$

where β is the angle between the incident ray and the panel.

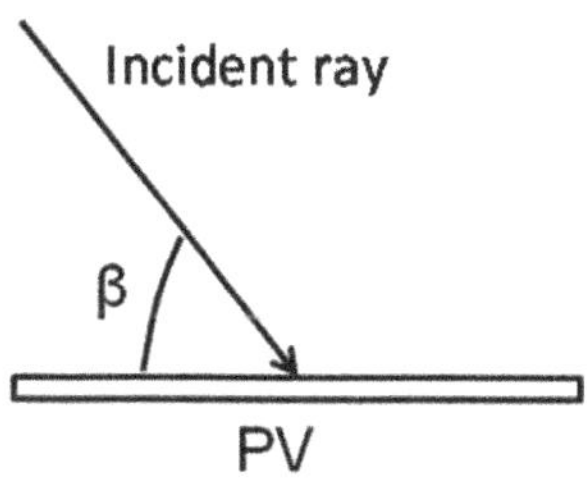

Figure 2.34. *Incidence angle of the Sun rays on a cell*

The global productivity throughout a year will vary according to the slope and orientation of the modules (Figures 2.5 and 2.6). As sunshine is higher in summer, capture is favored in summer, if the objective is to supply (and sell) a maximum of electricity to the network yearly. However, to satisfy needs throughout the year in an isolated site, it is sometimes better to favor capture in winter if the consumption is higher and knowing that the Sun is lower. The optimum slope angle is then increased (in comparison to the horizontal angle).

To increase the illumination of the cells, we advise directing them so that sunbeams hit them perpendicularly. To do so, we can use panels with a fixed orientation. However, panels with a variable slope are even more efficient. If panels are accessible, we may consider manual seasonal actions, but we can also use a single-axis (slope) mechanical tracking system, also called mechanical tracker. In winter, a panel positioned horizontally is two times less efficient than a panel positioned diagonally to face the Sun. We can also use panels that follow the path of the Sun. However, the system is then more expensive (dual axis tracker). These various parameters have to be taken into account, in order to install photovoltaic panels. In an ideal scenario, the photovoltaic cell should always be perpendicular to the Sun beams, in order to receive the maximum quantity of light possible. The slope of the panel will thus have to be adapted to the place. By assuming there is only one orientation (facing directly south in the Northern Hemisphere and facing directly north in the Southern Hemisphere), the optimum slope changes every day and is given by the relationship:

$$slope = (latitude\, of\ the\ place) - \arcsin(0.4 \sin(N.\frac{360}{365})) \quad [2.11]$$

N is the number of days between the vernal equinox (20th or 21st March in the Northern Hemisphere and 22nd or 23rd September in the Southern Hemisphere) and the considered day. It is a negative sign in cold weather.

If we are seeking a fixed slope, it can be increased or decreased, depending on the latitude, in order to find a compromise between winter and summer, to optimize energy production [MUL 07]. If we only want to maximize the annual production, the rule is then to tilt the panels to an angle that is equal to the latitude of the place (for example: about 45° in France, which is a good compromise between summer and winter production).

The installation of a photovoltaic panel is thus decisive: the orientation and slope have to be well considered to maximize the efficiency (Figures 2.5, 2.6, 2.34 and 2.35). However, building specifications will often determine the slope. Finally, as has already been mentioned in section 2.3.3.1, the efficiency also decreases as the panel ages.

Orientation	Inclination from the horizontal (°)						
	0	15	25	35	50	70	90
East	88%	87%	85%	83%	77%	65%	50%
South East	88%	93%	95%	95%	92%	81%	64%
South	88%	96%	99%	max 100%	98%	87%	68%
South West	88%	93%	95%	95%	92%	81%	64%
West	88%	87%	85%	82%	76%	65%	50%

Figure 2.35. *Example of annual efficiency depending on the slope and orientation of the modules in Belgium (about 50° from the north latitude). For a color version of this figure, see www.iste.co.uk/robyns/electricity.zip*

2.3.4. *Cell association*

Under STC, the maximal power for a silicon cell of 10 cm^2 is about 1.25 W (depending on the technology, the produced power is 1 to 3 W under a maximum voltage of less than 1 V). The elementary photovoltaic cell thus constitutes a low-power electric generator, for most domestic and industrial applications. Therefore, photovoltaic generators are carried out by association, in series and/or in parallel with a large number of elementary cells. These groupings are called modules and then panels.

This association must be carried out by respecting specific criteria because of the lack of balance, which can occur in an operating PV cell network. Indeed, although they are supposed to be identical, the numerous cells constituting the generator have different characteristics: because of the inevitable building dispersions, but also due to non-uniform illumination and temperature on the whole cell network (shade, for example).

Series connections of several cells increase the voltage for a similar current, while parallel connections increase the current for a similar voltage. Cells are then assembled to form modules. Figure 2.36 represents the equivalent diagram of a module. The triangle represents the direction of the diodes. For 12 V applications, the marketed modules are often made up of 36 cells in crystal silicon. These cells are connected in series.

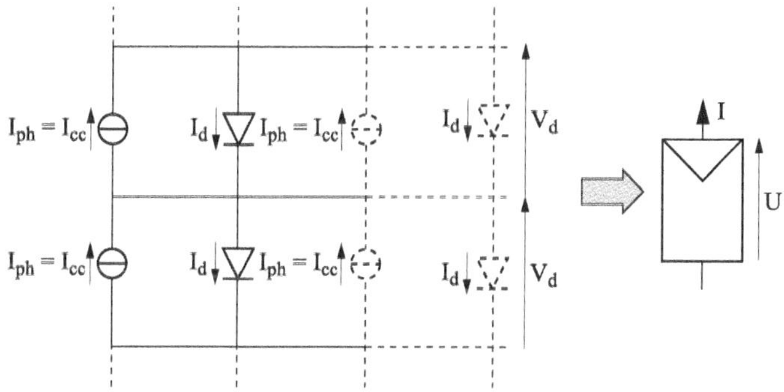

Figure 2.36. *Equivalent diagram of a module*

In a series grouping, modules are crossed by the same current and the resulting characteristic of the series grouping is obtained by the addition of voltages to a given current (Figure 2.37).

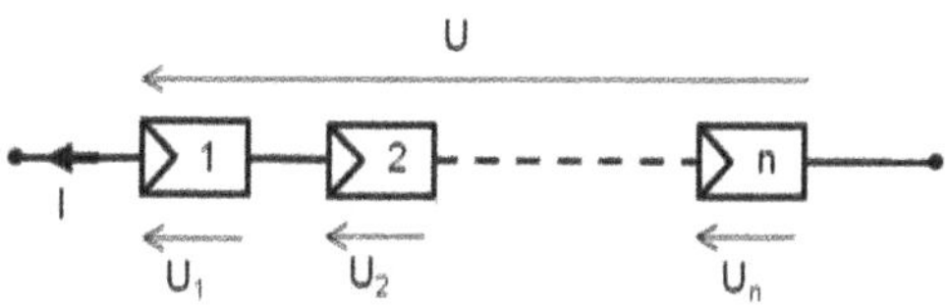

Figure 2.37. *Equivalent diagram of a module*

In a parallel grouping, the modules are subjected to the same voltage and thus intensities are added up: the resulting characteristic is obtained by adding currents at a given voltage (Figure 2.38).

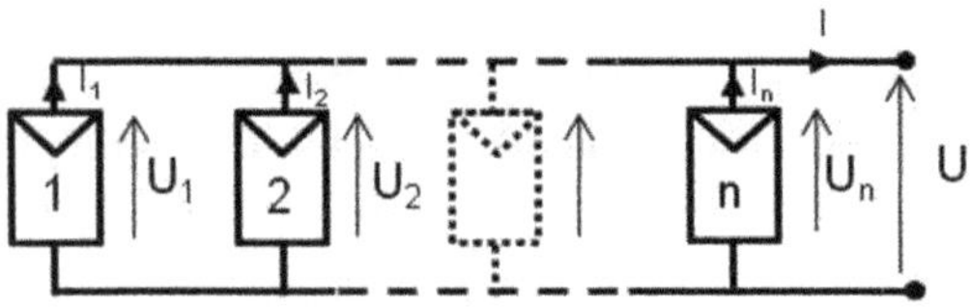

Figure 2.38. *Principle of the series module association*

Finally, to form a panel, modules have to be connected in series, in order to obtain a sufficiently high voltage to be exploitable by inverters, which enable the connection to the networks. Parallel grouping of modules also have to be connected to increase the power (Figure 2.39).

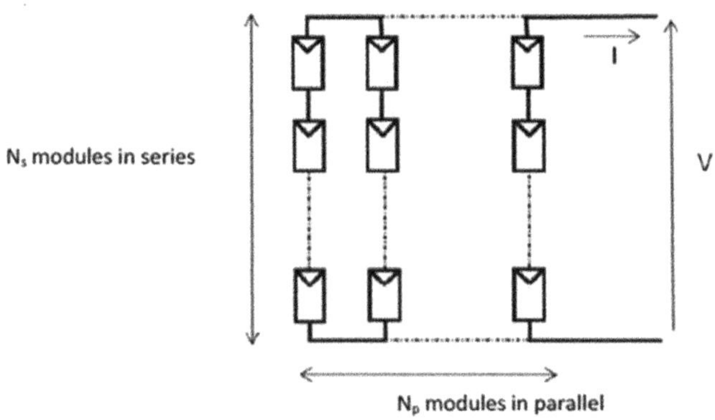

Figure 2.39. *Block diagram of a PV generator made up of a series/parallel module association*

The voltages to be used according to the installed peak power are given in Table 2.3. This data concerns applications in direct current and/or with an electrochemical cell.

Installed power	$0 - 500\ W_p$	$500\ W_p - 2\ kW_p$	$2\ kW_p - 10\ kW_p$	$> 10\ kW_p$
Recommended continuous voltage	12 V	24 V	48 V	>48V

Table 2.3. *Continuous voltages depending on the peak power*

Diodes are inserted within the panels in order to protect the cells [AST 08b]. There are bypass diodes, which are assembled in parallel with a group of cells and blocking diodes, which are assembled in series with the cells (Figure 2.40).

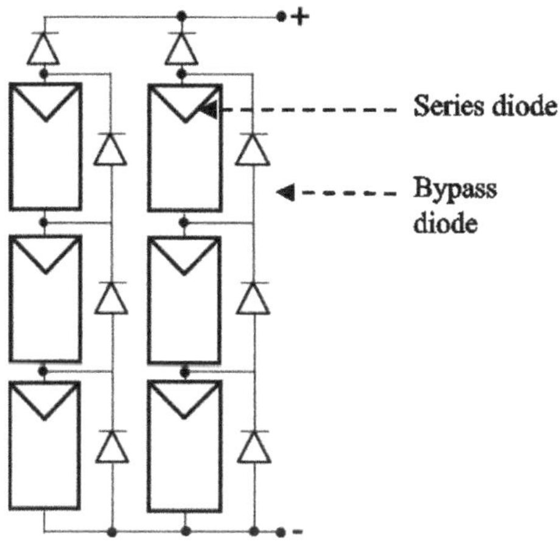

Figure 2.40. *Series–parallel association with protection diodes (bypass and series)*

Modules are sometimes not uniformly exposed to light. Cells are wired in series, and thus the total current is leveled out (the lowest cell imposing its current on the others). Therefore, when a cell no longer feeds, because it is no longer exposed to radiation, the current of the entire chain tends towards zero (Figure 2.41).

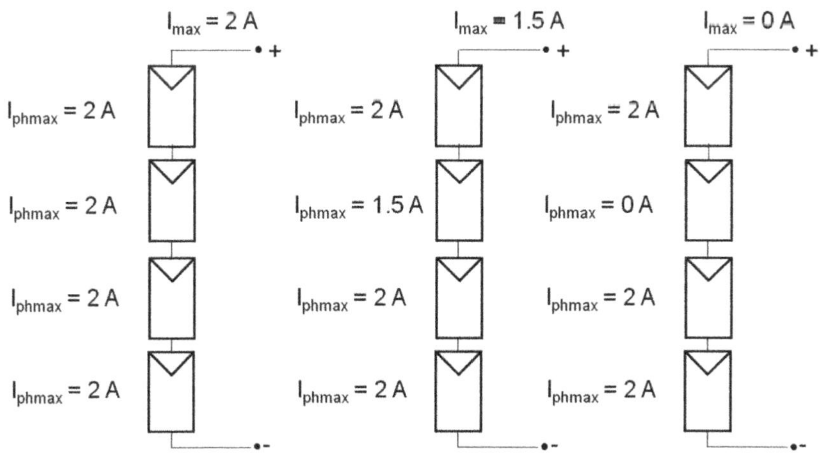

Figure 2.41. *Effect of not exposing a cell to sunshine*

However, it is also possible that the cell, thus masked, becomes the receiver for all the others of the series. It then receives, in reverse voltage, the sum of all their voltages (Figure 2.42). It thus starts to overheat, which can cause destruction and even a fire. To prevent this phenomenon from occurring, we can wire in parallel, one diode for one group of cells (the ideal situation would be to have one diode per cell) (Figure 2.43).

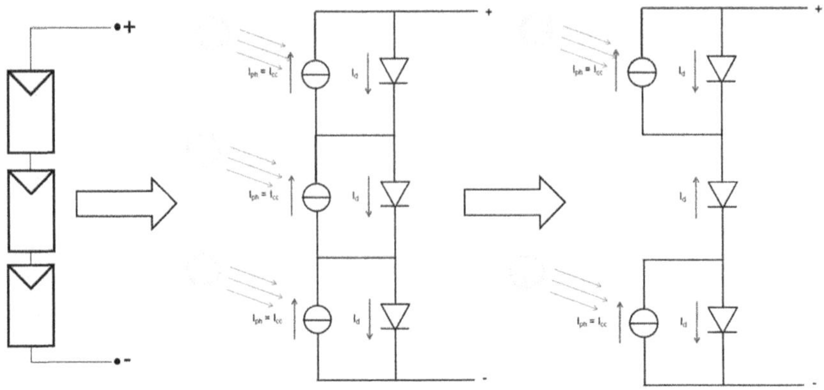

Figure 2.42. *Effect of the non-uniform exposure of cells to sunshine*

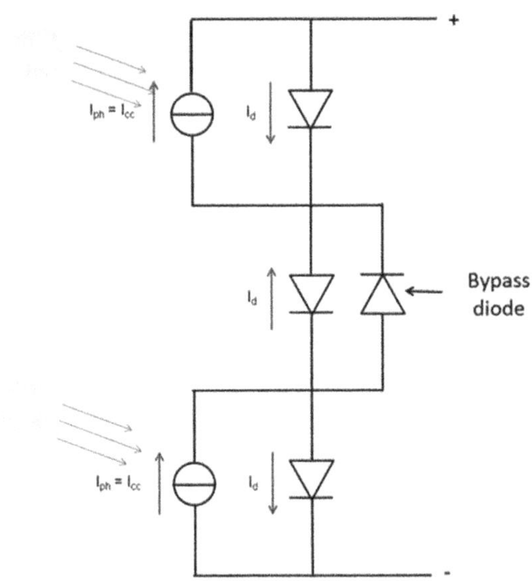

Figure 2.43. *Effect of the bypass diode*

As soon as a reverse voltage occurs at the terminals of the groupings, the spontaneous initiation of this parallel diode limits the voltage to the value V_d of the direct forward voltage of the chosen diode, and the dissipated energy to $V_d.I_s$. By placing one bypass diode per group of cells in series, we maintain the reverse voltage applied to the shaded cell at less than 10 V. This generates a limited overheating (generally lower than 60°C), which is quite well supported by the current modules. Moreover, a blocking diode must protect buses in parallel to reverse currents.

The global electrical current/voltage characteristic of a photovoltaic generator is thus theoretically deduced from the combination of the characteristics of the elementary modules N_s N_p, which are supposed to be identical. The generator is made up by these modules via two affinities of ratio N_s in parallel to the voltage axis and ratio N_p in parallel to the current axis, as shown in Figure 2.39; taking into account that N_s and N_p are, respectively, the total numbers of series and parallel modules. The action of the bypass diodes tends to modify this characteristic (Figure 2.44) [AST 08b].

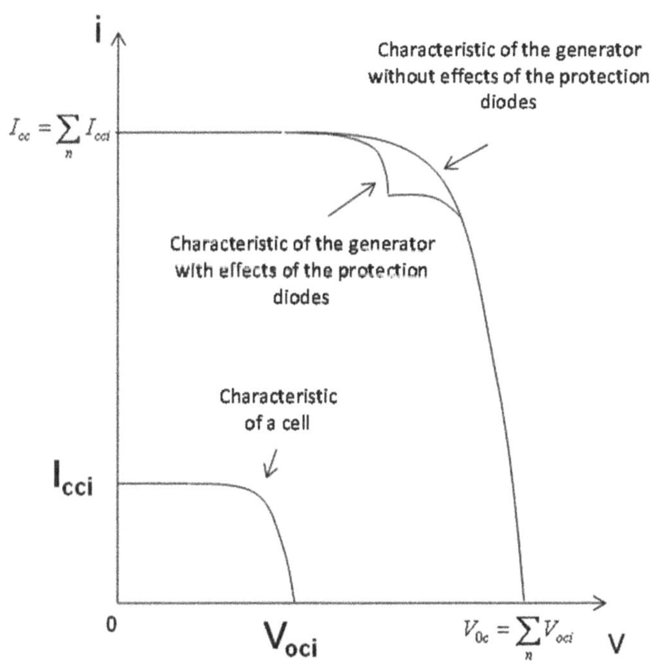

Figure 2.44. *Characteristic of a photovoltaic generator [AST 08b]*

2.4. Maximum electric power extraction [AST 08b, PRO 97, GER 02]

From an experimental point of view, photovoltaic cells (PV) show many variations in their electric power depending on weather conditions. Moreover, when they are connected to a load, the voltage applied to their terminals is not controlled and the power transferred to the load rarely corresponds to the maximum power supplied by the PV generator. The MPP (maximum power point) constantly varies due to variance of the outside conditions, such as cell irradiance and temperature (Figure 2.45)

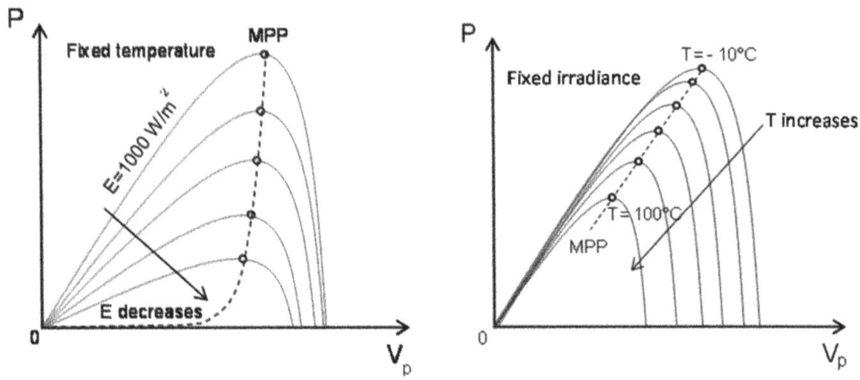

Figure 2.45. *Variation of the MPP with the irradiance and temperature*

In order to extract the maximum available power from a photovoltaic generator and supply it to any load, it would be interesting to use a switching converter associated with a Maximum Power Point Tracking (MPPT) method as an interface between the generator and the load. This method has been used, at least since 1968. These types of controllers are particularly adapted to piloting a non-linear source and forcing the generator to work at its MPP. It thus causes a global improvement of the efficiency of the electric conversion system, on the condition that the efficiency of the conversion stage is also high enough.

When a photovoltaic source is connected directly to a load, the operating point is determined by taking the intersection of the electric characteristics I-V with that of the load (Figures 2.46 and 2.47). This operating point varies because the energy source, or the load, can vary at any time. This is why it is almost impossible to operate in all situations producing maximum power.

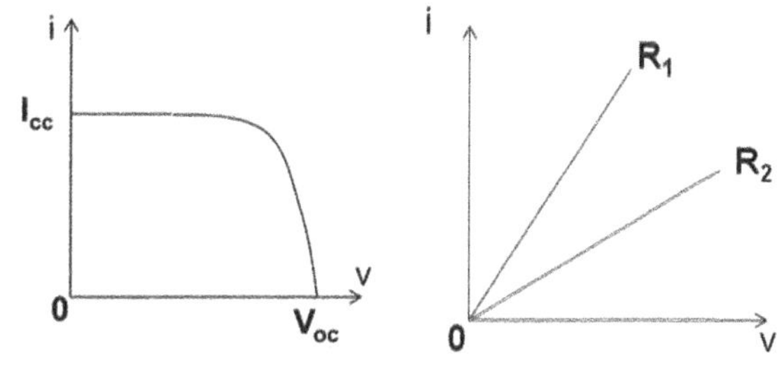

Figure 2.46. *Characteristic I(v) of a cell (a) and of a load resistor (b)*

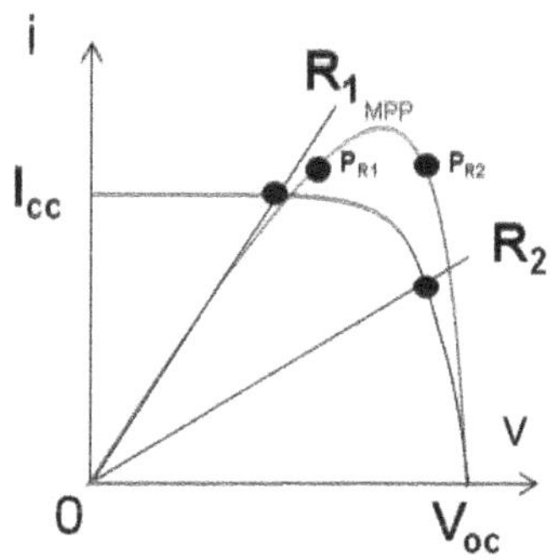

Figure 2.47. *Non-correlation between the characteristic I(v) of a cell and a load resistor. For a color version of this figure, see www.iste.co.uk/robyns/electricity.zip*

The principle of a MPPT controller is to track (as its name lets us assume) the maximum power point of a non-linear electric generator, on the condition that the load accepts its power. This is the case for an electrochemical cell that has not reached its full load (autonomous applications), or when it is connected to the network. These tracking systems can also be associated with wind generators. This controller thus enables the piloting of a static power converter connecting the load (a battery, for example) and the photovoltaic panel, in order to permanently supply the maximum power at the load (Figure 2.48). This converter enables us to adapt the impedance to the optimum value, which enables us to obtain the maximum supplied power.

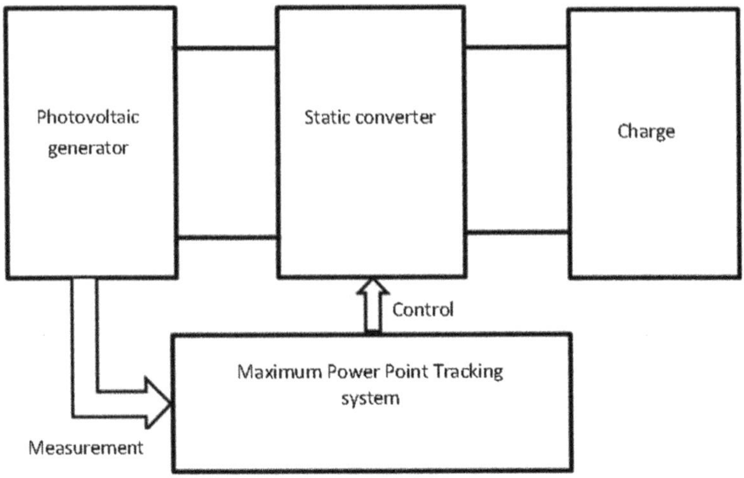

Figure 2.48. *Architecture of a maximum power point tracking system*

In Figure 2.22, which represents the characteristic I(V) as a function of the illumination at constant temperature and on which the curve of current/voltage points makes it possible to extract the maximum power delivered by the panel has been plotted, a controller MPPT will adapt the voltage level at the terminals of the cell in order to follow this curve.

There are different types of MPPT controllers. Ordinarily, each controller has been created for a specific application. Their accuracy and robustness depends on certain criteria:

– The global efficiency of the system sought after by the manufacturer.

– The type of power converter enables the adaptation and connection to a load (DC/DC, DC/AC) or an electrical network.

– The desired application (autonomous, connected to the network or spatial systems).

– The characteristics of the tracking method, depending on its speed and quality.

– The chosen type of implementation (analog, digital, both at the same time).

The principles of these controllers are often based on the research of the "elbow" of the characteristic P-V, which is more or less a trial-and-error method, as seen in Figure 2.49.

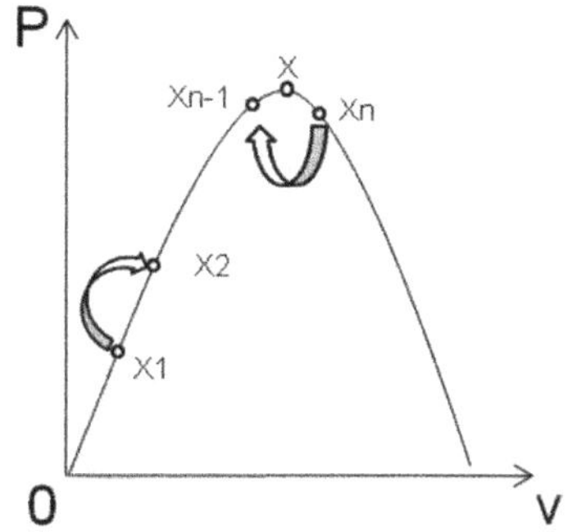

Figure 2.49. *Tracking principle of a controller*

Let us take a point on the curve (X_1) by imposing a voltage V_1. After a certain period, a voltage $V_2 = V_1+\Delta V$ is imposed and we check if the value of the power is higher, or not, than the previous point. If it is, we can go on to the following points, up to the moment when the power at the point (X_n) is lower than that of the previous point (X_{n-1}). At this moment, we take a value range between each lower point and we start again from (X_{n-1}), until we obtain the MPP (X). Therefore, the system permanently adapts the voltage to the terminals of the photovoltaic generator in order to get closer to the MPP, without ever precisely reaching it.

This principle seems easy to carry out under these conditions. However, it becomes less accessible when illumination intervenes. Indeed, when the intensity of the illumination varies, we go to a value E2<E1, and the characteristic P-V changes. The point X, which was until now the maximum point, becomes an operating point under the new conditions, as shown in Figure 2.50. We can see the emergence of a new operating point, called here X'.

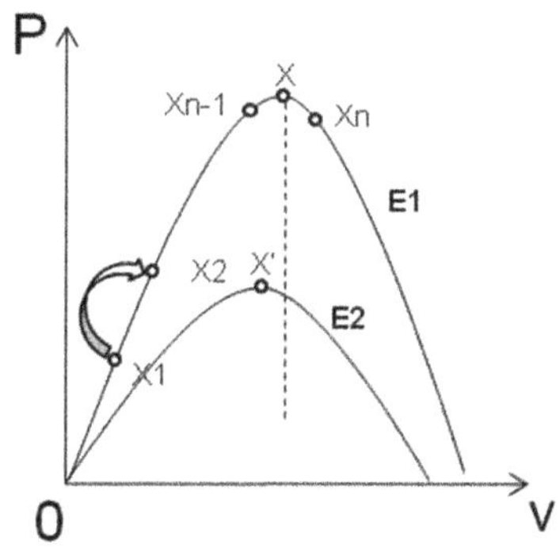

Figure 2.50. *The consequence, for the MPP research, of a change of illumination*

There are, however, several limits:

– The P-V characteristic of the generator can have more than one extremum (local maximums). This is notably the case when cell masking phenomena and operation phenomena of the bypass diode occur (Figure 2.44).

– Dramatic variations can occur at the level of illumination or the load. If the controller does not have good dynamics, the maximum point can be lost. During the time needed to re-find this point, new production losses will occur.

– There are some oscillations around the maximum point during the search for this point. This introduces production losses, because there are no energy losses in the sense of dissipation.

We have to notice that the converters associated with the MPPT controllers work at a high quench frequency (several tens of kHz depending on the power levels). The advantage of high frequency quenching is that it allows a reduction of the size of the magnetic components and the capacitors. The efficiency (in 2020) of the MPPT controller/power converter assembly is around 90% to 95%.

2.5. Power converters

2.5.1. *Introduction*

Power electronics enables us to carry out, with the help of semiconductor elements operating as switches, the interface between two electrical "sources". For instance, a photovoltaic generator made up of a cell assembly and a load, possibly including buffer storage or else an alternative source, in the case of an output on the electrical network, by ensuring (Figure 2.51):

– adjustment of some magnitudes (amplitude, frequency, phase);

– and/or a change of form of this source (alternative continuous).

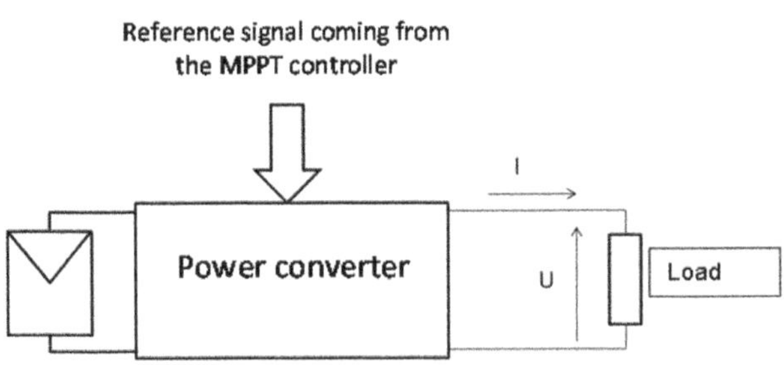

Figure 2.51. *Power converter typology*

In power electronic converters, semi-conductors operate in commutation (on-states and off-states), in order to obtain fundamentally high efficiencies. They are thus associated with filter elements (storage on the scale of the quenching period). In the following section we will present the main conversion structures used in the field of photovoltaic generation.

2.5.2. *Structure of the photovoltaic conversion chains [PAN 04]*

There are structures of inverters which are directly connected to the network and others which are magnetically decoupled from the network with the help of a transformer. In general, inverters have an average lifespan of 10 to 12 years. Generally, the average lifespan for a central inverter of a few kW (a grouping of several panels on the latter which minimizes the cost of the inverter to the detriment of electrical energy production if the inverter is lost) is 10 years, and 20 years for a micro-inverter (one small inverter per panel which allows the production of electrical energy to be increased at the expense of cost). It will take from several hundred to several thousand Euros for an inverter depending on its power, and from a few tens of Euros to hundreds of Euros for a micro inverter. In 2018, the average efficiency of inverters for photovoltaics was 98%. We can expect a return close to 99% in the years to come.

2.5.2.1. *Without an isolation transformer*

The structure of Figure 2.52 is made up of an input capacity, only one converter (thus ensuring the MPPT functions and the DC/AC conversion by itself), and an output filer, which is made up of induction coils.

The advantage of this structure is its minimum number of components, which reduces losses. However, a significant number of PV panels have to be connected in series to obtain a voltage that is at least equal to that of the network peak voltage.

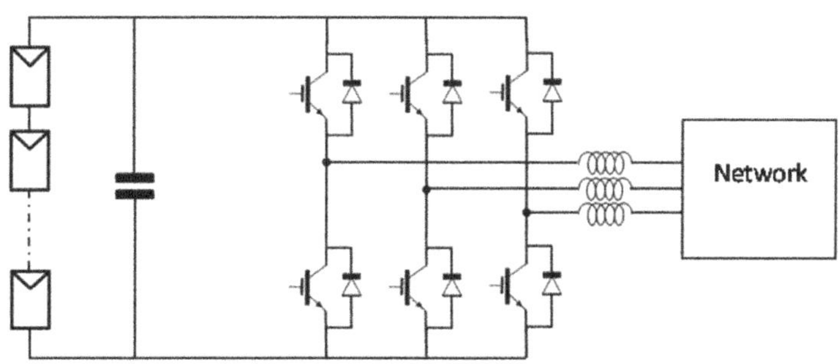

Figure 2.52. *Structure without an isolation transformer*

2.5.2.2. *With an isolation transformer*

2.5.2.2.1. Low frequency transformer

In Figure 2.53 we can see a similar structure to the previous case, except for the fact that a low voltage AC bus and frequency of 50 Hz is created where the transformer is found. Here, the transformer has a dual function: first, it must raise the voltage of the AC bus to the level of the network and ensure the inverter/network isolation. Here the advantage is limiting the number of PV panels to be put in series.

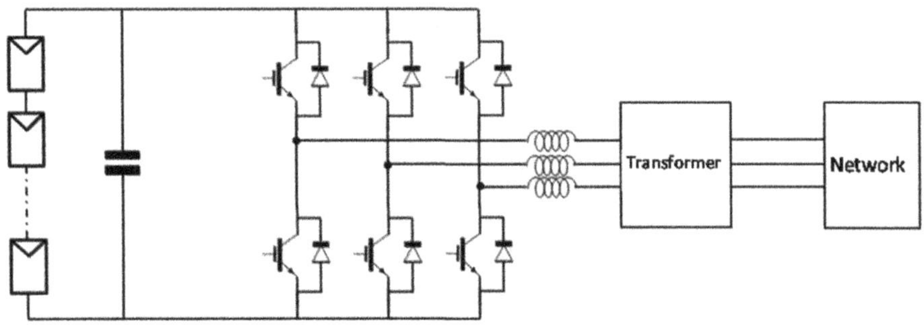

Figure 2.53. *Structure with a low frequency isolation transformer*

In Figure 2.54, we have a similar structure for the AC part to that shown in Figure 2.53, but this converter has an intermediate stage with a continuous bus, which is formed by a DC/DC converter. This additional converter is a boost chopper, which can raise the voltage coming from the PV panel according to the cyclic ratio α.

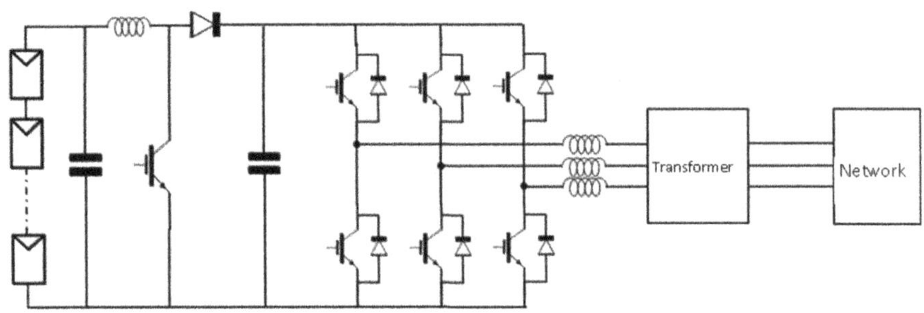

Figure 2.54. *Structure with a low frequency isolation transformer*

2.5.2.2.2. High frequency transformer

The advantage of supplying the transformer with a high frequency voltage is being able to reduce the volume of its magnetic circuit and therefore its dimension. The output voltage of the photovoltaic module is raised to the desired level with the help of a flyback converter (Figure 2.55). It also ensures the galvanic isolation between the photovoltaic module and the inverter.

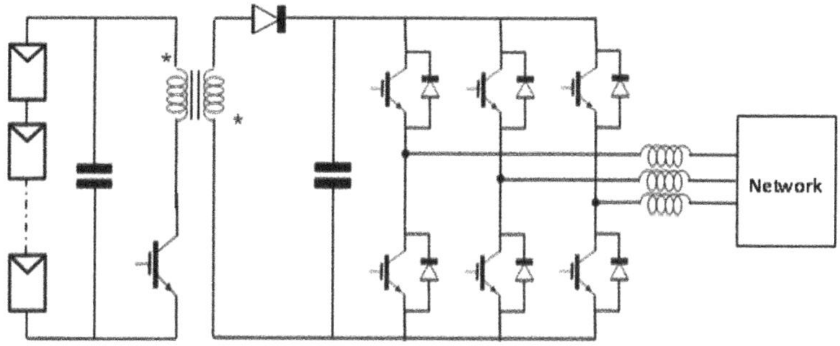

Figure 2.55. *Structure with a low frequency isolation transformer and a DC/DC converter*

2.5.3. *Choppers [SEG 85, BER 02]*

2.5.3.1. *Introduction*

Choppers are used to condition the electric energy between DC generators and receivers. The source feeding the chopper supplies a DC voltage. The chopper enables us to have an adjustable DC voltage. Turning-off the controlled semi-conductors forming the chopper switches can thus not be ensured by the source. These semiconductors must thus be completely controllable, at ignition and turn-off (GTO (gate turn off) or power transistors).

2.5.3.2. *Two switch choppers*

2.5.3.2.1. Series choppers

A series chopper is made up of a switch K, which is controllable and unidirectional and a freewheel diode K', which helps to ensure the continuity of the current flow, when the load is inductive (Figure 2.56). These two components are complementary. One of them is in an on-state while the other is in an off-state and vice versa. An inductance L enables us to smooth the current i_R. During a time located between 0 and αT, the switch K is closed and K' is open: we apply the voltage E to the terminals of the load (Figure 2.57). From αT to T, the switch K is open and K' is closed: we apply a null voltage to the terminals of the load

(Figure 2.58). Figure 2.59 shows the timing diagrams of the various magnitudes. The coefficient α is the ratio of the closing time of the switch K on the period T and is called the cyclic ratio. This is adjustable. It helps to vary the average value of the voltage U:

$$U_{ave} = \frac{1}{T}\int_0^{\alpha T} E.dt = \frac{\alpha T}{T} E = \alpha E \qquad [2.12]$$

The series chopper is a buck converter. The average value cannot be higher than E.

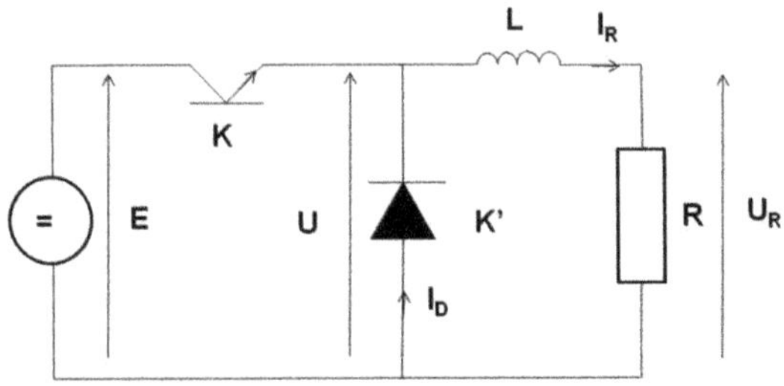

Figure 2.56. *Structure of a series chopper*

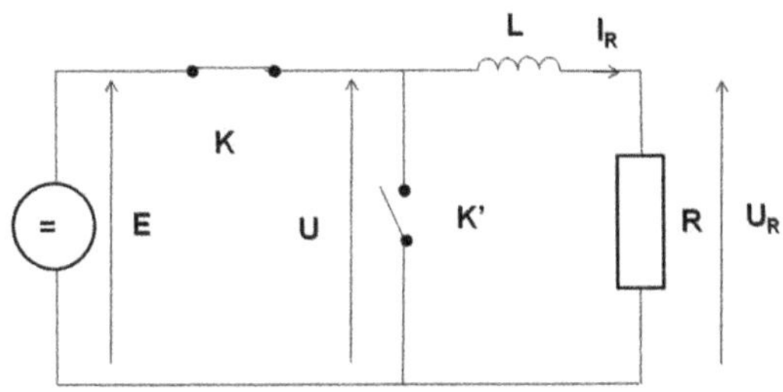

Figure 2.57. *Equivalent diagram of the series chopper of 0 and αT*

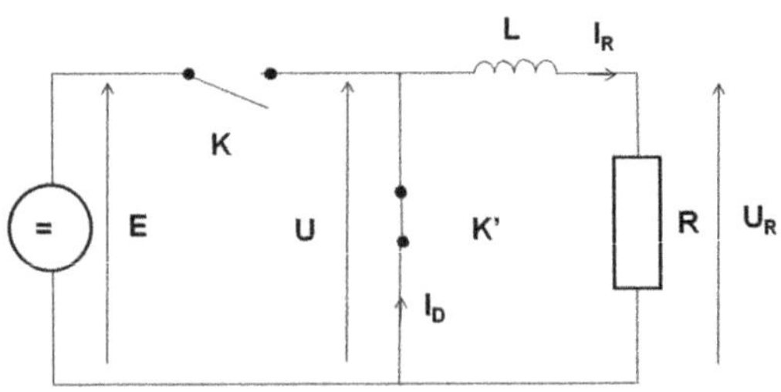

Figure 2.58. *Equivalent diagram of the series chopper from αT to T*

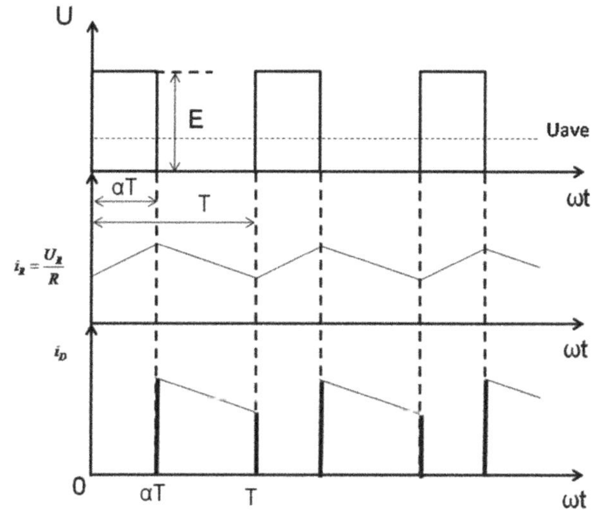

Figure 2.59. *Series chopper operation timing diagrams*

2.5.3.2.2. Parallel choppers

A parallel chopper is made up of a switch K, which is controllable and unidirectional; and a diode K', which enables us to ensure the continuity of the current flow when the load is inductive (Figure 2.60). These two components are complementary. One of them is in an on-state while the other is in an off-state and vice versa. During a time ranging between 0 and αT, the switch K is closed and K' is open: we apply a null voltage at the terminals of the load (Figure 2.61). From αT to T, the switch K is open and K' is closed: we apply a voltage E to the terminals of the load (Figure 2.62). Figure 2.63 shows the timing diagrams of the various magnitudes. The coefficient α is the ratio of the closing time of the switch K on the period of time T. This is called the cyclic ratio, which is adjustable and helps to vary the average value of the voltage U:

$$U_{ave} = \frac{1}{T}\int_{\alpha T}^{T} U_c.dt = (1-\alpha)U_c \quad [2.13]$$

and

$$U_{ave} = E = (1-\alpha)U_c \Leftrightarrow U_c = \frac{E}{1-\alpha} > E \quad [2.14]$$

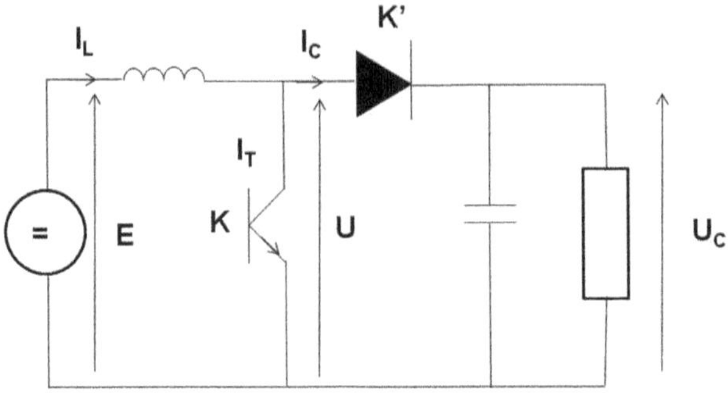

Figure 2.60. *Structure of a parallel chopper*

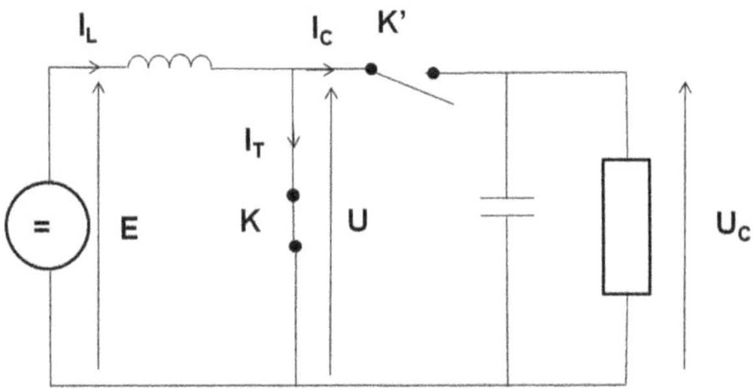

Figure 2.61. *Equivalent diagram of a parallel chopper for 0 and αT*

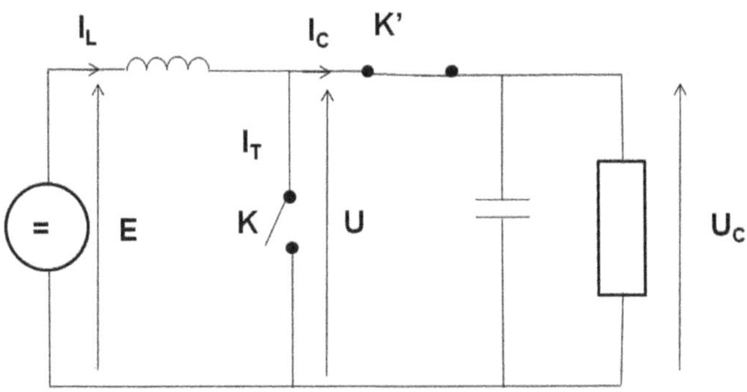

Figure 2.62. *Equivalent diagram of the parallel chopper from αT to T*

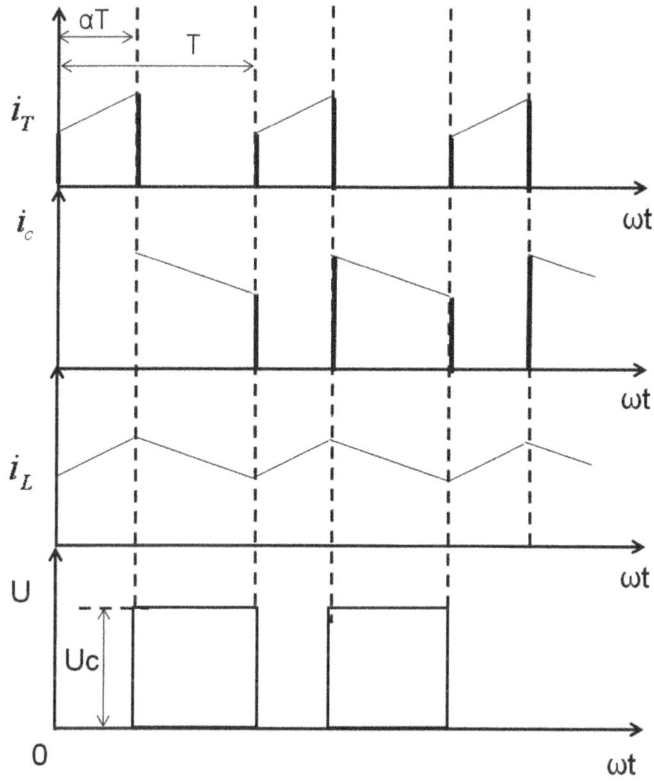

Figure 2.63. *Parallel chopper operation timing diagrams*

The series chopper is a step-up voltage regulator. The average value can be higher than E.

2.5.4. *Inverters [FOC 98]*

2.5.4.1. *Introduction*

An inverter is a static converter ensuring the conversion from DC to AC (Figure 2.64). Depending on the semiconductors carrying out the switching functions, this converter can be reversible in voltage and current and is thus able to carry out a transformation from DC to AC.

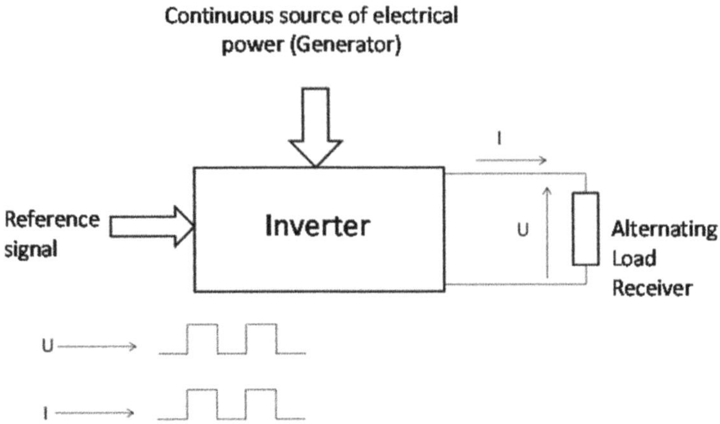

Figure 2.64. *Typology of an inverter*

2.5.4.2. *Single-phase inverters*

A single-phase inverter is reversible in current and voltage. The current in the switches can be positive or negative. Therefore, a diode has been placed in reverse parallel on each transistor. It is made up of 4 commutation elements (Figure 2.65). During a time ranging between 0 and T/2, the switches K1 and K4 are closed, whereas K2 and K3 are open: a voltage E is applied to the load (Figure 2.66). From T/2 to T, the switches K2 and K3 are closed, whereas K1 and K4 are open: a voltage (–E) is applied (Figure 2.66). It is also possible to have the 4 switches open: no voltage is applied. The switches K1 and K3 or K2 and K4 must never be closed at the same time or a short-circuit will occur in order to avoid short-circuiting the voltage generator E. This inverter is a full-wave inverter and frequency control is carried out (Figure 2.67) by adjusting the T period by opening and closing the switches. Pulse width modulation (PWM) enables the simultaneous control of the amplitude and frequency (Figure 2.68).

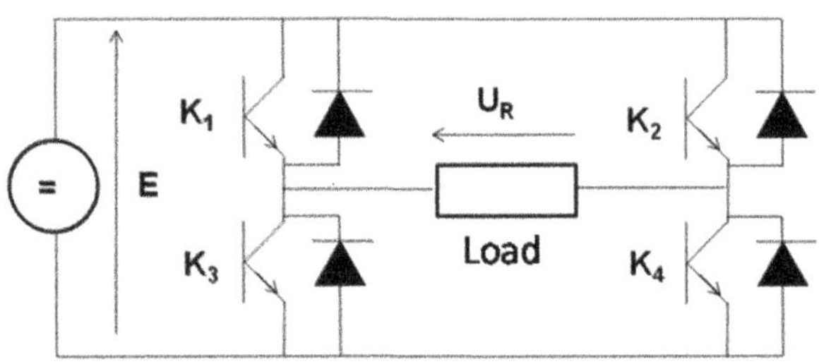

Figure 2.65. *Structure of a bridge chopper*

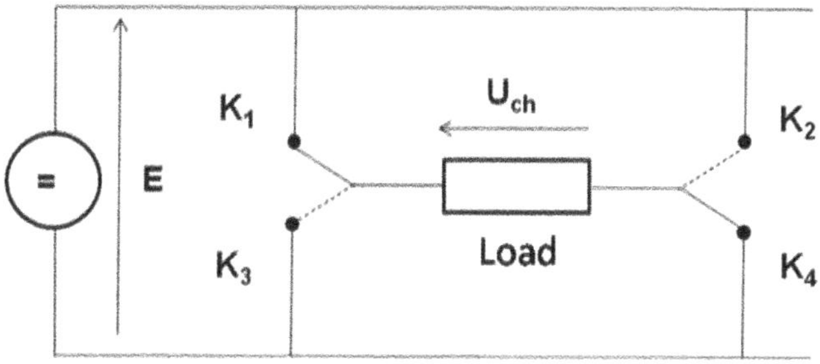

Figure 2.66. *Operation of the single-phase inverter*

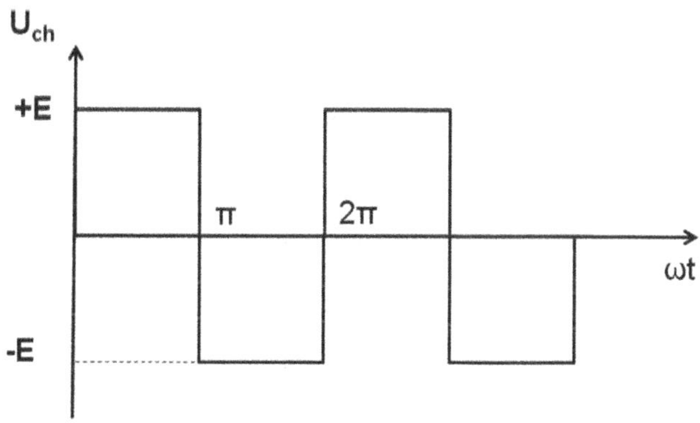

Figure 2.67. *Full-wave single-phase inverter*

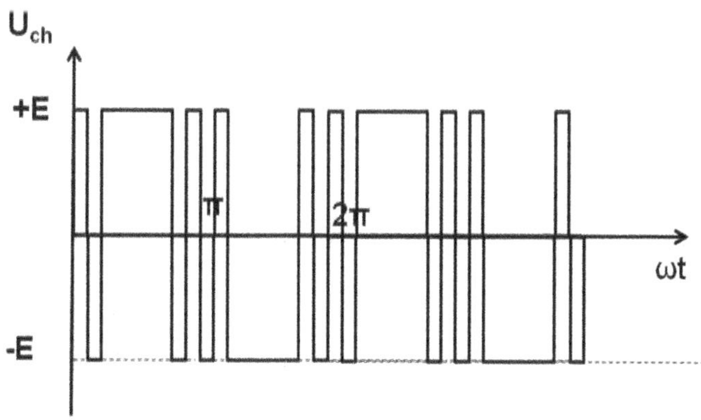

Figure 2.68. *PWM inverter*

2.5.4.3. *Three-phase inverters*

A three-phase inverter is made up of 6 commutation elements (Figure 2.69). In a "full-wave inverter operation", each branch is controlled in a complementary way: in Figure 2.70, the switch is in position $K_{1,\ 2}$ or $_3$ for a half-period and in position $K'_{1,\ 2}$ or $_3$ for the other half-period. The three branches are controlled with ⅓ of a period gap. The phase-to-phase voltages (between phases) are represented in Figure 2.71. A phase to ground voltage is represented in Figure 2.72.

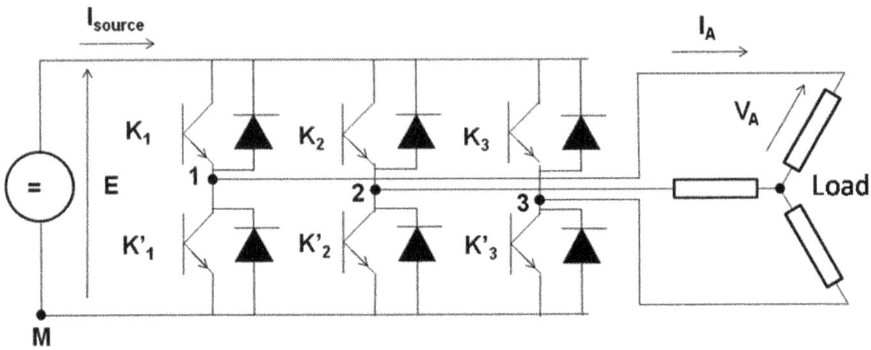

Figure 2.69. *Structure of a three-phase inverter*

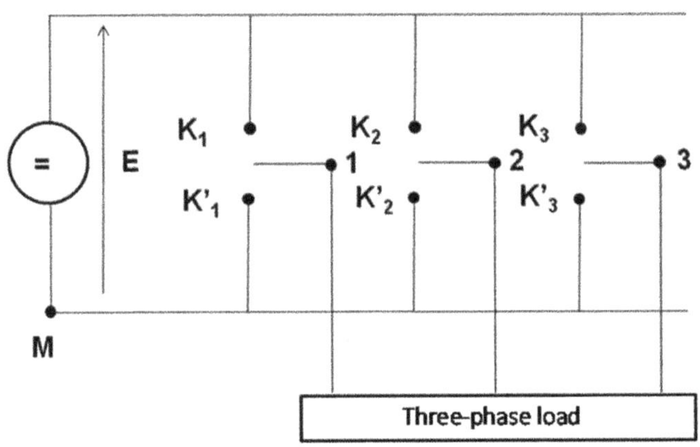

Figure 2.70. *Operation of a three-phase inverter*

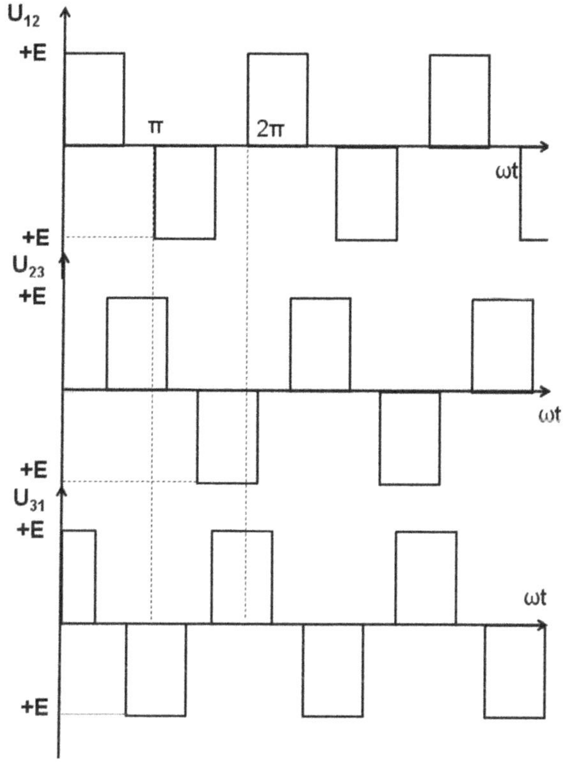

Figure 2.71. *Timing diagram of the phase-to-phase voltages applied between 1, 2 or 3*

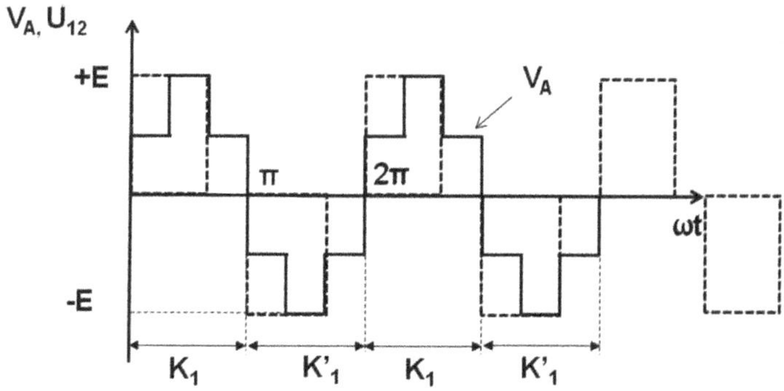

Figure 2.72. *Timing diagram of a phase to ground voltage*

Full-wave inverters enable us to adjust the frequency of the generated voltages. In order to simultaneously adjust the frequency and the amplitude, the switches must be controlled in PWM, as for single-phase inverters (section 2.5.4.2) [FOC 00, SEG 85].

2.6. Adjustment of the active and reactive power

With the help of power converters, it is possible to adjust the phase displacement φ between the voltage and the current (Figure 2.73), in addition to the frequency and the amplitude of the alternative waves.

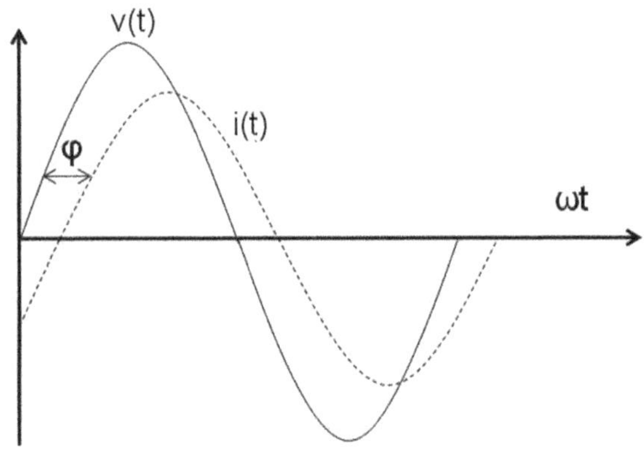

Figure 2.73. *Phase displacement φ between the voltage and current*

Active and reactive powers depend on this phase displacement. It is thus possible to adjust them by controlling the latter. Their expressions in single-phase are as follows:

$$P = VI\cos\varphi \qquad [2.15a]$$

$$Q = VI\sin\varphi \qquad [2.15b]$$

In three-phase: they take the following forms depending on whether we are considering the phase to ground voltages V (of the phase) or the phase-to-phase (of the line) voltages U:

$$P = 3VI\cos\varphi = \sqrt{3}UI\cos\varphi \qquad [2.16a]$$

$$Q = 3VI \sin \varphi = \sqrt{3}UI \sin \varphi \qquad [2.16b]$$

A third power, called the apparent power, is defined by the relationship:

$$S = \sqrt{P^2 + Q^2} \qquad [2.17]$$

We thus define the power triangle, linking P, Q, S and φ (Figure 2.74).

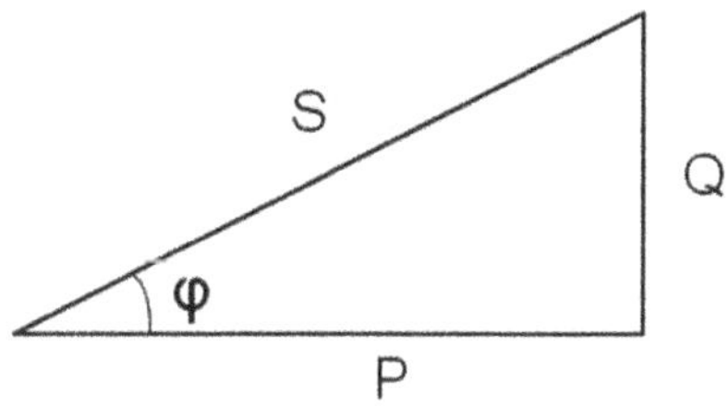

Figure 2.74. *Power triangle*

2.7. Solar power stations [PRO 97, AST 08b, SAB 06]

2.7.1. *Introduction*

Solar power stations can be classified into two categories in relation to the electrical network:

– autonomous power stations (not connected to the network);

– power stations connected to the network.

2.7.2. *Autonomous power stations*

These stations are used in zones where there is no possibility of an electric power supply via a distribution network or when the cost of electrification would be too high. They generally supply mountain chalets, watering systems, road sign systems, parking meters, markers, GSM antennas and so on, or mobile systems: boats, motor homes, satellites, etc.

Their structure is the same as those seen in section 2.5, and they may or may not be associated with energy storage systems, which are often storage batteries (Figure 2.75). They thus often have regulators to control the battery charges and discharges in order to extend their lifespan. These allow the devices to be supplied with direct or alternating voltage, as required, when consumption is not in line with production: in cases where instant production differs from instant consumption or when the power consumed is greater than the power produced. The technology of the batteries used depends on the usage demands made of the stand-alone plant, its coupling with other electrical energy production systems (generator, wind turbine, etc.) and cost factors. Battery storage options are open lead, Absorbed Glass Mat (AGM), gel or lithium-ion technologies [ROB 04, DOU 09, NGO 16, ROB 15]. They will have a lifespan of 7 to 10 years depending on the battery technology. The most efficient, lithium-ion batteries, have an efficiency of 90%. The least efficient ones, the Lead Acid variety, have a low efficiency (~70%–75%), a short lifespan, a negative environmental impact (lead and sulfuric acid) and a significant embodied energy. The efficiency of a battery is defined as the ratio between the energy that the battery gives back when discharging and the energy it takes when recharging. Embodied energy is the total sum of energy consumed during the lifecycle of a material or product, without taking into account the energy consumed during its use. For example, to make a light bulb within the circular economy, one takes into account the energy consumed from the extraction of raw materials, their transport, material transformation, manufacture of the bulb and its eventual recycling at the product's end of life. On the other hand, its consumption during its use to illuminate an area will not be counted.

It will cost, for example, €6,000 to €10,000 TTC for 10 kWh of lithium-ion battery.

Back-up power systems can also be coupled to these power stations (other source of electric energy production).

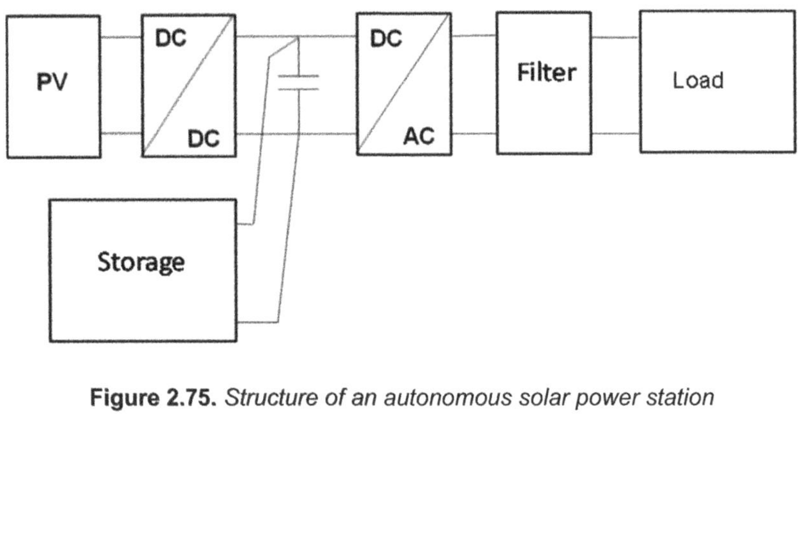

Figure 2.75. *Structure of an autonomous solar power station*

2.7.3. *Power stations connected to the network*

These power stations are photovoltaic installations, which are connected to the electrical network without (Figure 2.76) or with (Figure 2.77) their own storage system.

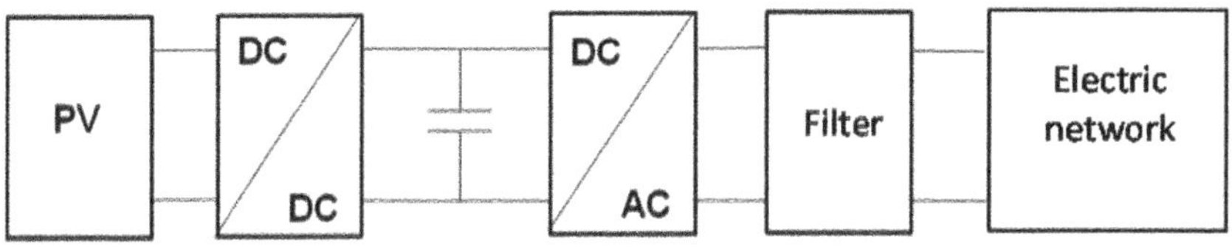

Figure 2.76. *Structure of a solar power station "following the path of the Sun", which is connected to the electrical network*

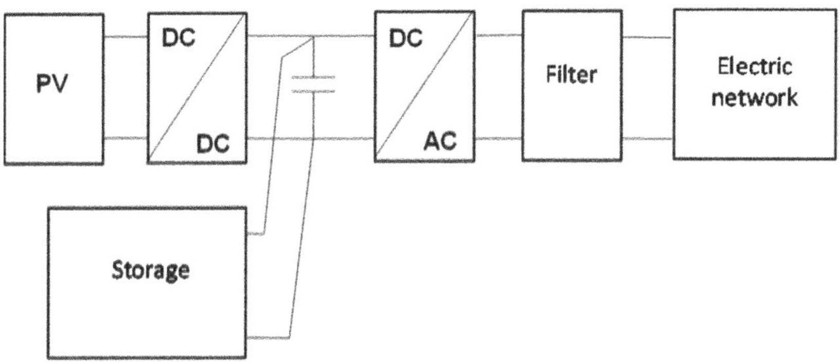

Figure 2.77. *Structure of a solar power station with storage, which is connected to the electrical network*

Without storage, it operates by "following the path of the Sun", i.e. it injects into the electricity network:

– all of the energy that can be captured by the panels if the plant is connected directly to the power grid;

– the difference between the energy captured by the panels and the energy consumed by the receivers, a technique called individual self-consumption with sale of surplus, if the plant is connected at the building level.

In both cases, the producer who owns the photovoltaic installation must pay a contribution to the financing of the electricity networks in two forms:

– the connection cost, which is paid when the installation is commissioned [ENE 20a];

– the TURPE or *Tarif d'Utilisation des Réseaux Publics d'Electricité* (Tariff for the Use of Public Electricity Networks) [ENE 20b, PHO 20] which is payable during the operation of the installation. TURPE is made up of:

- a fixed fee for management (which depends on the type of installation: self-consumption or not) which covers the costs assumed by the managers of the public distribution networks for their administrative part (management of user files, physical and telephone reception, invoicing and collection),

- a fixed metering component that covers rental and maintenance, reading, monitoring and profiling fees (reconstitution of flows), coupled with consumption in the event of surplus injection,

- an injection part for connections on voltage levels other than Low Voltage and a reactive energy part for installations with a power greater than 36 kVA, which effectively absorbs it in order to cover the financing of electrical network equipment.

In the 2000s, there was significant development of primary configurations (without a storage system) which, thanks to law number 2000-108 of February 10 2000 relating to the modernization and development of the public service, a total resale of the electrical energy (with financial support from the public authorities for the installation) to the historical electricity supplier EDF. A preferential kWh repurchase rate was offered for a 20-year contract. It reached a level of 58c €/kWh until March 2011 for PV plants that were fully integrated into the building and with a power less than or equal to 9 kW. This financial support from the public authorities, in the form of repurchase price, made it possible to promote the development of photovoltaic installations on roofs in France in order to reduce the costs of PV installations. After March 2011, public authorities' financial support for the photovoltaic sector decreased for several reasons:

– the return on investment of photovoltaic installations is faster because the price of the panels has decreased year by year while yields have increased;

– the photovoltaic sector is deemed mature and now attracts investors;

– support for the sector represents a significant cost in the state budget and therefore in the tax burden of the French.

The feed-in tariff has therefore gradually decreased (as well as the installation premiums) until reaching a price of 18.44c €/kWh in September 2020 for installations less than or equal to 3 kW integrated into the building and 15.67c €/kWh for installations between 3 and 9 kW that were integrated into the frame (10c €/kWh for the sale of surplus from power installations from 0 to 9 kW). This gradual drop in the feed-in tariff put an end to the marked development of primary configurations, that is to say, a total resale of electrical energy to move towards a strong emergence of the second configuration, i.e. self-consumption, which is defined by consuming the energy that we produce, for individuals, businesses and communities. For this secondary configuration, we will find installations on the roofs of buildings with or without storage systems [ROB 19]. The presence of a storage system can ensure continuity of the power supply as a stand-alone plant in the event of a network outage by adapting the system control. This operating mode is called "islanding". However, in most countries, the power plant is not allowed to operate when the power grid is lost. Here the electricity network can be considered as a back-up.

Due to the timing mismatch between the production of electrical energy and the consumption of electricity, there is a need to modify usages or to control the system. There are different energy management strategies between the photovoltaic panels, the storage system, the receivers, the electricity network and the objectives set: reduce the energy bill to the network, reduce the building's CO_2 footprint [ROB 19]. The facilities are generally sized according to two indicators: the self-consumption rate and the production rate [ROB 19]. These rates can be increased by using a storage system. This technique will become more and more viable when the price per kWh coming from the grid is higher than the cost price per kWh of a photovoltaic installation.

With the increase in the cost of the kWh of the electricity network each year and a cost price of photovoltaic kWh that decreases each year, we are moving towards grid parity [HOR 20], that is to say, a cost of the kWh produced with self-consumption photovoltaics that is equal to the cost of the kWh purchased from the electricity network (generally for photovoltaic installations with a lifespan of 30 years). In a few years, we should even have a cost of photovoltaic kWh that is lower than the cost of peak kWh bought from the grid, and later, even lower for the basic version (Figure 2.78).

Finally, at the end of the 2010s and beginning of the 2020s, it is the photovoltaic installations of more than 300 kWp on the ground or on the roof that are supported thanks to calls for tenders from the *Commission de Régulation de l'Energie* (Energy Regulatory Commission), or CRE, according to specifications aimed at pursuing the objectives of the *Programmation Pluriannuelle de l'énergie* (Pluriannual Energy Programming) or PPE [CRE 20, MIN 20].

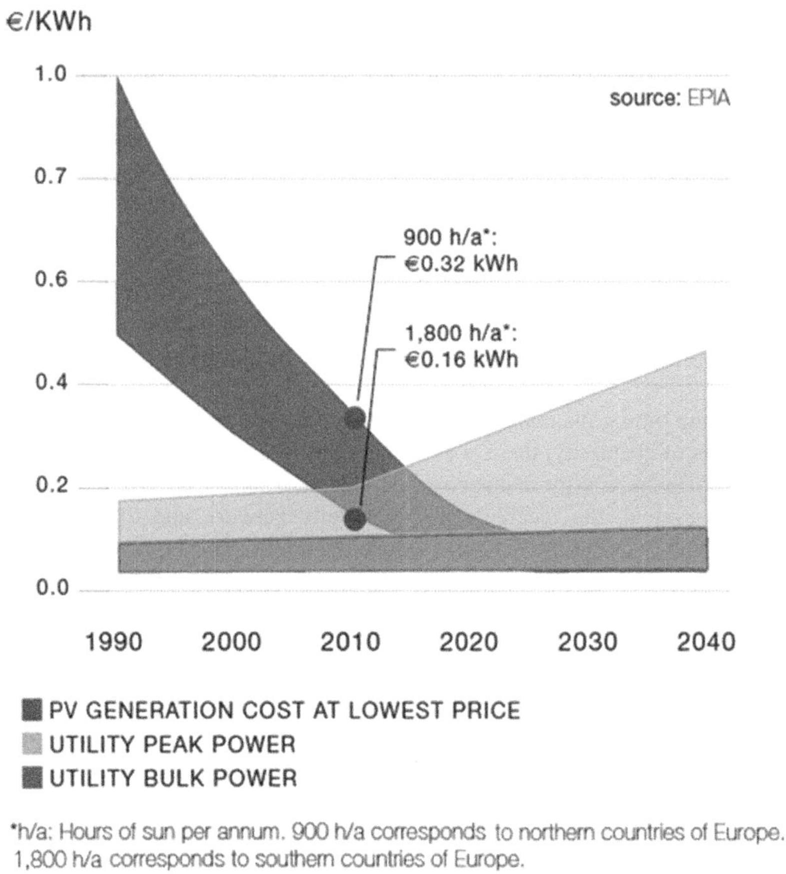

Figure 2.78. *Evolution of the cost of photovoltaic kWh and the electricity market [GRE 11]. For a color version of this figure, see www.iste.co.uk/robyns/electricity.zip*

2.8. Exercises

2.8.1. *Characteristics of a photovoltaic panel*

A photovoltaic panel is formed of two modules which are connected in parallel. Each of these modules is made up of 24 cells which are connected in series. The characteristic of a cell is given in Figure 2.79.

a) Draw the current–voltage characteristic (i-v) of such a generator.

b) What is the short-circuit current of the panel?

c) What is the open-circuit voltage of the panel?

d) What is the peak power, knowing that the form factor is 0.725?

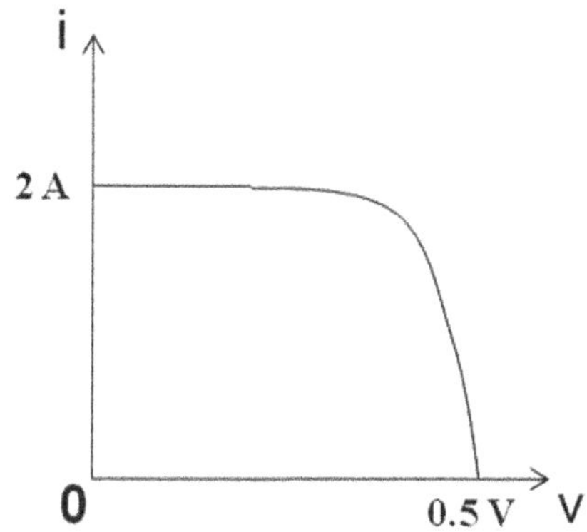

Figure 2.79. *Characteristic of the cell*

Answers

a) A module is made up of 24 cells in series. The voltage of a module is thus the sum of the cell voltages and the drain current is the maximum current delivered by a cell. We thus have an equivalent generator of 12 V – 2 A.

A panel is made up of two modules which are connected in parallel. The voltage of a panel is the voltage of a module and the current is the sum of the maximum module currents. We thus have an equivalent generator of 12 V – 4 A (Figure 2.80).

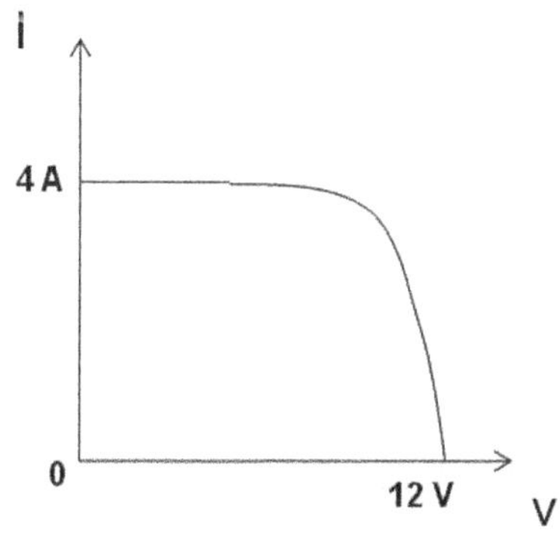

Figure 2.80. *Characteristic of the panel*

b) The sort-circuit current I_{cc} of the panel is 4 A (Figure 2.80).

c) The open-circuit voltage V_{oc} of the panel is 12 V (Figure 2.80).

d) The form factor is 0.725 and the power $V_{co}.I_{cc}$ is 4 x 12 = 48 W.

The peak power Pc is therefore worth:

$P_c = 48.0.725 = 34.8$ W

2.8.2. *Sizing an autonomous photovoltaic installation*

Let us assume we have a small electrical installation (Figure 2.81) made up of:

– a 40 W LCD television;

– two compact fluorescent lamps of 20 W each;

– a 10 W radio receiver;

– The television works 2.5 hours/day;

– The two lamps work 3 hours/day;

– The radio works 1 hour/day.

This installation is supplied by photovoltaic panels. It is associated with a lead-acid battery. The average efficiency of the set MPPT+ DC/DC converter is 92%. The average efficiency of the set batteries + regulator is 70%. The average efficiency of the DC/AC converter is 95%. The energy produced by W_c, which was installed in December for an average solar radiation of the place, is 1.12 Wh/day. Consider that December is the most critical month (least sunshine and maximum consumption).

a) Calculate the total energy consumed by all the devices in a day.

b) Calculate the total energy consumed by all the devices in a day on the DC side of the DC/AC converter.

c) Calculate the nominal capacity (in Ah) of the battery for a voltage of 48 V to be installed, which is supposed to be constant, if we need an autonomy of 3 days without any sun.

d) Calculate the nominal capacity (in Ah) of the battery (48 V) to be installed, if we want a 3-day autonomy and a maximum discharge of 50% (in order to keep its lifespan).

e) Calculate the necessary nominal power of the photovoltaic generator, i.e. the number of W_p to be installed.

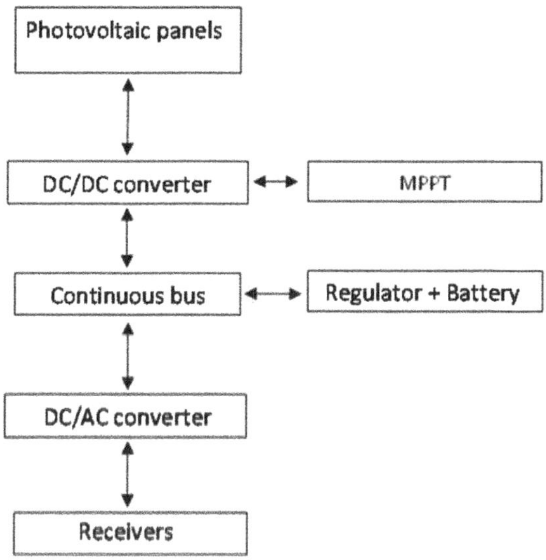

Figure 2.81. *Diagram of the autonomous installation*

Answers

a) The energy consumed by the devices is calculated from the following relationship:

$$E(Wh) = P_{absorbed}(W)\text{operating time (hours)}$$

Power consumed per day per television:

$$40 \times 2.5 = 100\ Wh$$

Power consumed per day by the 2 lamps:

$$2 \times 20 \times 3 = 120\ Wh$$

Power consumed per day by the radio:

$$10 \times 1 = 10\ Wh$$

This give a total consumed power per day of:

$$100 + 120 + 10 = 230\ Wh$$

b) The total power consumed by all the devices in 1 day on the AC/DC side of the converter is thus:

$$\frac{230}{0.95} = 242\ Wh$$

c) If we have 3 days without sun, we have to be able to supply 3 times the energy at the input of the AC/DC converter, i.e.:

$$3 \times 242 = 726\ Wh$$

By taking into account the efficiency of the battery, we thus have to store:

$$\frac{726}{0.7} = 1{,}037\ Wh$$

The capacity of the battery is thus:

$$\frac{1{,}037}{48} \approx 22\ Ah$$

d) If we do not want the battery to discharge at more than 50%, when we have 3 days without sun, we thus have to store twice as much energy, i.e.:

$$2 \times 22 = 44\ Ah$$

e) In this place, a panel of 1 W_p associated with the set MPPT controller + DC/DC converter is able to produce:

$$1.12 \text{ x } 0.92 = 1.03\ Wh/W_p/day$$

To charge the battery at its maximum and, at the same time, supply the necessary energy to supply the devices, we need:

$$\frac{44.8 + 242}{1.03} \approx 2{,}285\, W_p.$$

2.9. References

[AST 08a] ASTIER S., "Conversion photovoltaïque: Du rayonnement solaire à la cellule", *Techniques de l'ingénieur*, article D3935, 2008.

[AST 08b] ASTIER S., "Conversion photovoltaïque: De la cellule aux systèmes", *Techniques de l'ingénieur*, article D3936, 2008.

[BER 00] BERNOT F., "Composants de l'électronique de puissance", *Techniques de l'Ingénieur*, article E3960, 2000.

[BER 02] BERNOT F., "Hacheurs: Fonctionnement", *Techniques de l'ingénieur*, article E3964, 2002.

[BER 04] BERNARD J., *Energie solaire, calculs et optimisation*, Éditions Ellipses, Paris, 2004.

[COU 19] COUSCOUS H., BENCHEKROUN A., ALMAKSOUR K. *et al.*, "Photovoltaic power forecasting using back propagation artificial neural network", *ITISE 2019*, Granada, Spain, September 25–27, 2019.

[CRE 16] CRE, Recommandation de la CRE no 14 du 8 décembre 2016 (R. 2016-14). Available at: https://www.cre.fr/, accessed December 2020.

[CRE 20] CRE. Available at: www.cre.fr, accessed December 2020.

[DES 04] DESTRUEL P., SEGUY I., "Les cellules photovoltaïques organiques", *Techniques de l'ingénieur*, article RE25, 2004.

[DOU 09] DOUBLET M.-L., "Batteries Li-ion, conception théorique", *Techniques de l'ingénieur*, article AF6612 V1, 2009.

[ENE 20a] ENEDIS, Raccorder votre installation de production d'électricité, Enedis. Available at: https://www.enedis.fr/raccorder-votre-installation-de-production-delectricite, accessed December 2020.

[ENE 20b] ENEDIS, Notre financement: Le tarif d'acheminement, Enedis. Available at: https://www.enedis.fr/tarif-acheminement, accessed December 2020.

[EQU 09] EQUER B., "Les Filières photovoltaïques: Progrès récents, recherches actuelles", *Actes de la journée SEE – ISA France "L'énergie Photovoltaïque"*, 29–30 April, 2009.

[FOC 98] FOCH H., FOREST F., MEYNARD T., "Onduleurs de tension: Structures, principes. Applications", *Techniques de l'ingénieur*, article D3176, 1998.

[FOC 00] FOCH H., FOREST F., MEYNARD T., "Onduleurs de tension: Mise en œuvre", *Techniques de l'ingénieur*, article D3177, 2000.

[FRA 20] FRAUNHOFER, PHOTOVOLTAICS REPORT 2020 [Online]. Available at: https://www.ise.fraunhofer.de/content/dam/ise/de/documents/publications/studies/Photovoltaics-Report.pdf, accessed December 2020.

[GER 02] GERGAUD O., Modélisation énergétique et optimisation économique d'un système de production éolien et photovoltaïque couplé au réseau et associé à un accumulateur, PhD thesis, Ecole Normale Supérieure de Cachan, SATIE, December 2002.

[GER 08] GERMAN ENERGY SOCIETY, *Planning & Installing Photovoltaïc Systems*, 2nd ed., Earthscan, London, 2008.

[GHA 17] GHALEHKHONDABI I., ARDJMAND E., WECKMAN G.R. *et al.*, "An overview of energy demand forecasting methods published in 2005–2015", *Energy Systems*, vol. 8, pp. 441–447. Available at: https://link.springer.com/article/10.1007/s12667-016-0203-y, 2017.

[GRE 11] GREENPEACE. Available at: www.greenpeace.org.Final-SolarGeneration-VI-full-report-lr.pdf, accessed December 2020.

[HOR 20] HORSIN MOLINARO H., MULTON B., Energie électrique: Génération photovoltaïque, Culture Sciences de l'Ingénieur, Ecole normale supérieure Paris Saclay, Portail national de ressources – éduscol, Sciences et Techniques Industrielles. Available at: https://eduscol.education.fr/sti/si-ens-paris-saclay, accessed December 2020.

[INM 13] INMAN R.H., PEDRO H.T.C., COIMBRA C.F.M., "Solar forecasting methods for renewable energy integration", *Progress in Energy and Combustion Science*, vol. 39, no. 6, pp. 535–576, 2013.

[LÉG 08] LÉGIFRANCE, Arrêté du 23 avril 2008 relatif aux prescriptions techniques de conception et de fonctionnement pour le raccordement à un réseau public de distribution d'électricité en basse tension ou en moyenne tension d'une installation de production d'énergie électrique [Online]. Available at: https://www.legifrance.gouv.fr, accessed December 2020.

[LÉG 15] LÉGIFRANCE, Décret n° 2015–1697 du 18 décembre 2015 relatif à la programmation pluriannuelle de l'énergie de Corse [Online]. Available at: https://www.legifrance.gouv.fr, accessed December 2020.

[LET 99] LETURCQ P., "Physique des semi-conducteurs de puissance", *Techniques de l'Ingénieur*, article D3102, 1999.

[LIV 20] LIVE TREE, L'Université Catholique de Lille en transition énergétique et sociétale. Available at: https://livetree.fr/, accessed December 2020.

[MAT 17] MAT DAUT M.A., HASSAN M.Y., ABDULLAH H. *et al.*, "Building electrical energy consumption forecasting analysis using conventional and artificial intelligence methods: A review", *Renewable and Sustainable Energy Reviews*, vol. 70, pp. 1108–1118, 2017.

[MIN 20] MINISTÈRE DE LA TRANSITION ÉCOLOGIQUE, Programmation Pluriannuelle de l'Energie (Pluriannual Energy Programming). Available at: https://www.ecologie.gouv.fr/programmations-pluriannuelles-lenergie-ppe and www.ecologie.gouv.fr, accessed December 2020.

[MUL 07] MULLER J.-C., "Electricité photovoltaïques", *Techniques de l'ingénieur*, article BE8578, 2007.

[MUL 11] MULTON B., THIAUX Y., BEN AHMED H., "Consommation d'énergie, ressources énergétiques et place de l'électricité", *Techniques de l'ingénieur*, article D3900v2, 2011.

[NGO 16] NGO C., "Stockage de l'énergie", *Techniques de l'ingénieur*, article BE8090 V1, 2016.

[NRE 20] NREL. Available at: https://www.nrel.gov/pv/cell-efficiency.html, accessed December 2020.

[PAN 04] PANKOW Y., Etude de l'intégration de la production décentralisée dans un réseau Basse Tension. Application au générateur photovoltaïque, PhD thesis, Ecole Nationale Supérieure d'Arts et Métiers, L2EP, December 2004.

[PAT 99] PATEL MUKUND R., *Wind and Solar Power Systems*, CRC Press, Boca Raton, 1999.

[PHO 20] PHOTOVOLTAIQUE. Available at: www.photovoltaique.info/fr, accessed December 2020

[PRO 97] PROTIN L., ASTIER S., "Convertisseurs photovoltaïques", *Techniques de l'ingénieur*, article D3360, 1997.

[PVC 20] PV CYCLE. Available at: https://pvcycle.fr/recyclage/, accessed December 2020.

[RIC 05] RICAUD A., "Modules photovoltaïques – Filières technologiques", *Techniques de l'ingénieur*, article D3940, 2005.

[ROB 04] ROBERT J., ALZIEU J., "Accumulateurs – Accumulateurs au plomb", *Techniques de l'ingénieur*, article D3352 V1, 2004.

[ROB 15] ROBYNS B., FRANÇOIS B., DELILLE G. *et al.*, *Energy Storage in Electric Power Grids*, ISTE Ltd, London, and John Wiley & Sons, New York, 2015.

[ROB 19] ROBYNS B., DAVIGNY A., BARRY H. *et al.*, *Electrical Energy Storage for Buildings in Smart Grids*, ISTE Ltd, London, and John Wiley & Sons, New York, 2019.

[RTE 16] RTE, Bilan électrique. Available at: www.rte-france.com, 2016, accessed December 2020.

[RTE 19] RTE, Bilan électrique. Available at: www.rte-france.com, 2019, accessed December 2020.

[RTE 20] RTE, Prévisions de production solaire en France. Available at: www.rte-france.com, accessed December 2020.

[SAB 06] SABONNADIÈRE J.-C., *Renewable Energy Technologies*, ISTE Ltd, London and John Wiley & Sons, New York, 2009.

[SEG 85] SEGUIER G., *L'électronique de puissance*, Dunod, Paris, 1985.

[SEI 16] EDF, Les référentiels techniques applicables, déconnexion des installations de production mettant en œuvre de l'énergie fatale à caractère aléatoire dans les zones non interconnectées au réseau métropolitain continental. Available at: https://reunion.edf.fr/producteur/se-raccorder/les-referentiels-techniques-applicables-0, 2016, accessed December 2020.

[SLA 19] SLAOUI A, "Électricité photovoltaïque – Matériaux et marchés", *Techniques de l'Ingénieur*, article BE8579 V4, 2019.

[TAG 13] TAGHEZOUIT B., CHOUDER A., HADJARAB A. *et al.*, "Surveillance d'une centrale photovoltaïque connectée au réseau BT utilisant LabVIEW, Centre de Développement des Energies Renouvelables", *16èmes Journées Internationales de Thermique (JITH 2013)*, Marrakech, Morocco, 13–15 November, 2013.

[YAN 14] YAN X., ABBES D., FRANÇOIS B., "Solar radiation forecasting using artificial neural net-work for local power reserve", *2014 International Conference on Electrical Sciences and Technologies in Maghreb (CISTEM)*, pp. 1–6, Tunis, Tunisia, 3–6 November, 2014.

[ZHA 12a] ZHANG H., DAVIGNY A., COLAS F. *et al.*, "Fuzzy logic based energy management strategy for commercial buildings integrating photovoltaic and storage systems", *Revue Elsevier Energy and Buildings*, vol. 54, pp. 196–206, 2012.

[ZHA 12b] ZHAO H., SU S., MI Z. *et al.*, "Short-term solar irradiance forecasting model based on artificial neural network using statistical feature parameters", *Energies*, vol. 5, no. 5, pp. 1355–1370, 2012.

3

Wind Power

3.1. Characteristic of the primary resource

3.1.1. *Variability*

Chapter 1 has shown the extent of the potential for wind power. Wind is present all over the world with very variable characteristics. In addition to this significant geographical variability, we can also note a significant temporal variability, because the wind speed on a specific site varies with the seasons, months, days. It can also vary a lot within a few seconds. Figure 3.1 gives an example of the wind speed evolution measured every 3 seconds on a site in France along the North Sea coast.

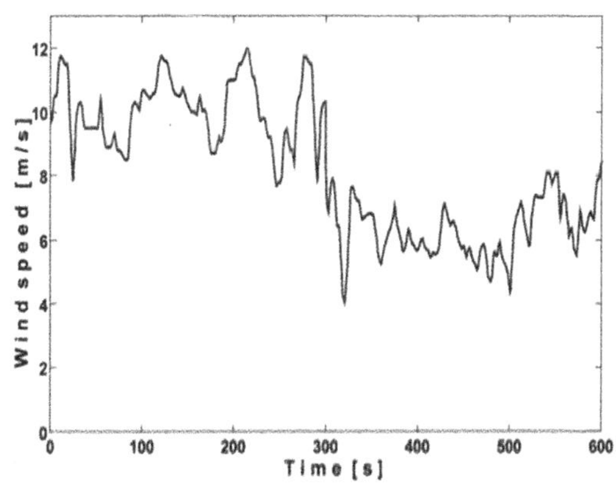

Figure 3.1. *Example of the instantaneous evolution of the wind speed*

Chapter written by Bruno FRANÇOIS and Benoît ROBYNS.

Although these fast variations are difficult to predict, the wind speed can be characterized by using an average of the measurements carried out for a given site per month over a year (Figure 3.2).

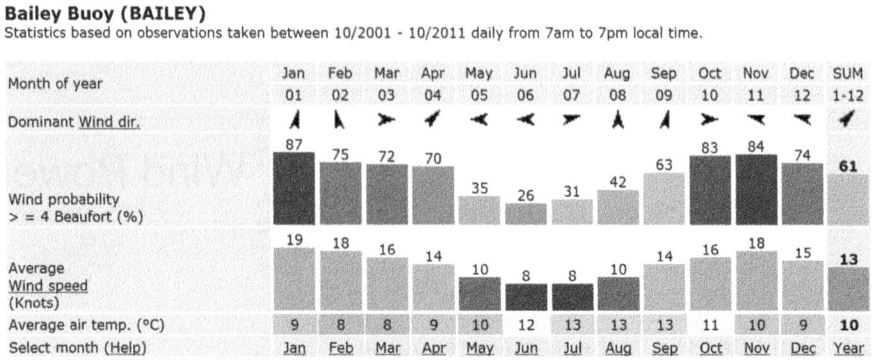

Figure 3.2. *Example of the wind speed evolution on the site of Bailey Buoy (United Kingdom) [WIN 10]. For a color version of this figure, see www.iste.co.uk/robyns/electricity.zip*

These measurements help us to characterize the wind power potential on the considered site. Dispersed over a territory, they help to characterize the most favorable sites for wind power collection (Figure 3.3).

A European Wind Atlas has been developed by the Danish research center RISO with support from the European Commission [TRO 89]. Sites exposed to strong winds are more common in northern Europe. Figure 3.4 shows an example of the wind power potential in France. The windiest zones (zone 5) are located in Languedoc Roussillon and in the Camargue.

3.1.2. *The Weibull distribution*

Knowing the annual average speed is not enough to precisely calculate the available energy for two reasons: because of the wind's kinetic power, which is proportional to the cube of its speed (detailed in section 3.2, equation [3.6]) and because of the productivity characteristic of wind turbines, which is fundamentally non-linear. It is also interesting to consider the frequency of the various occurrences of wind speeds.

Let us note that speed measurements are generally values that are averaged every 10 minutes. After taking measurements on site for a year, the frequency of occurrence of a specific wind speed (i.e. its probability) can be mathematically modeled by a Weibull distribution curve [3.1]:

$$P(v)=1-e^{-\left(\frac{v}{A}\right)^k} \qquad [3.1]$$

The distribution is defined by:

– the form factor k characterizing the distribution asymmetry;

– the scale factor A in m/s.

#	Weather Station	Country	Average Wind Speed (Knots)	Average Air Temp. (°C)
1	Bailey Buoy	United Kingdom	39	9
2	East Dellwood Buoy	Canada	36	7
3	South Brooks Buoy	Canada	36	8
4	W Ireland Buoy	Ireland	35	9
5	Akraberg/Suðuroy	Faroe Islands	35	8
6	West Hebrides	United Kingdom	35	10
7	Sartine Island	Canada	34	7
8	Vagar/Faroer	Faroe Islands	34	8
9	Rockall	United Kingdom	33	11
10	Andoya	Norway	33	7
11	Andoya / Bleik	Norway	33	7
12	South Uist/Hebrides	United Kingdom	33	11
13	Benbecula	United Kingdom	33	11
14	Tórshavn	Faroe Islands	33	8
15	Irish West Coast Buoy	Ireland	33	11
16	Rada Tilly	Argentina	33	15
17	Punta Indio	Argentina	33	15
18	Comodoro Rivadavia	Argentina	33	15
19	Miyakejima	Japan	33	14
20	Solander Island	Canada	33	7
21	Torsvåg Fyr	Norway	33	7
22	Urakawa	Japan	32	5
23	Bø/Vesterålen	Norway	32	6
24	Leknes	Norway	32	7
25	Fjaervoll	Norway	32	6

Figure 3.3. *Ranking of the high wind energy potential sites in relation to their annual wind speed, 1 m/s = 1.943844 knots [WIN 10]. For a color version of this figure, see www.iste.co.uk/robyns/electricity.zip*

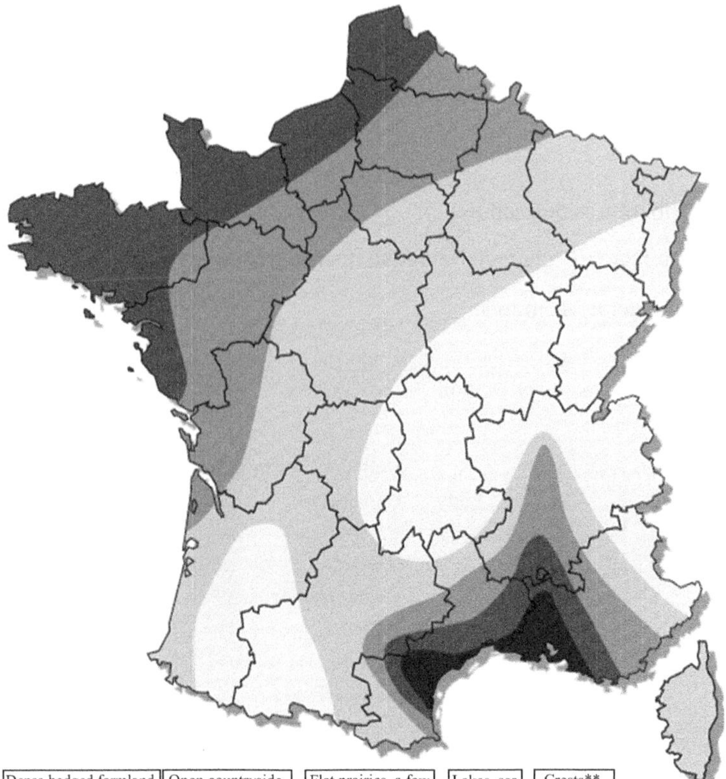

Dense hedged farmland, woods, suburbs	Open countryside, scattered obstacles	Flat prairies, a few bushes	Lakes, sea	Crests**, hills	
<3,5	<4,5	<5,0	<5,5	<7,0	Zone 1
3,5 - 4,5	4,5 - 5,5	5,0 - 6,0	5,5 - 7,0	7,0 - 8,5	Zone 2
4,5 - 5,0	5,5 - 6,5	6,0 - 7,0	7,0 - 8,0	8,5 - 10,0	Zone 3
5,0 - 6,0	6,5 - 7,5	7,0 - 8,5	8,0 - 9,0	10,0 - 11,5	Zone 4
>6,0	>7,5	>8,5	>9,0	>11,5	Zone 5

* Wind speed at 50 meters above the ground according to the topographie.
** Montainous areas require a specific deposit study.

Figure 3.4. *Map of the wind power potential in France quantified in annual average value of the wind speed [ADE 09]. For a color version of this figure, see www.iste.co.uk/robyns/electricity.zip*

The form factor (ranging between 1 and 3) is a very important indicator in the knowledge of the wind climate being studied. Indeed, as this coefficient becomes increasingly higher, the wind speed supplies less energy. The probability distribution of wind speeds varies from one place to another, because it depends on the local weather conditions, the landscape and its surface (more accurately on the roughness characterizing it). The Weibull distribution thus tends to vary, in shape but also in average value. Figure 3.5 presents a typical example of a Weibull distribution for a given site.

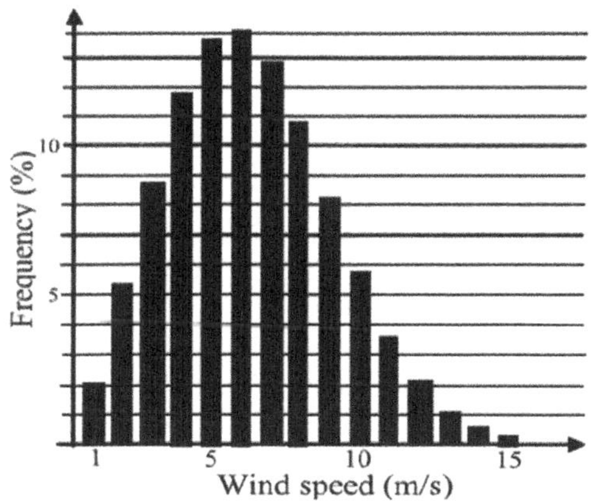

Figure 3.5. *Example of the distribution characteristic of the wind speed on a site, from which we can deduce the Weibull distribution*

The Weibull distribution of a site enables us to determine the energy that could be produced, by multiplying the power generated by the wind turbine (Figure 3.27) for each value of the wind speed with the number of hours that each wind speed occurs (discretized). The Weibull curve also helps to determine the wind speed for which a wind turbine will produce its nominal power. Therefore, it also enables us to optimize the design of the wind power system for the wind speed that supplies the most power. The most interesting sites will thus have a regular wind speed between 6 m/s and 10 m/s.

3.1.3. *The effect of relief*

Typically, the wind speed v varies according to the height at which it is measured and the surrounding terrain. The evolution of the wind speed according to the height h can be estimated with the following law of approximation:

$$\frac{v}{v_o} = \left(\frac{h}{h_o}\right)^{\alpha} \qquad [3.2]$$

v_0 is the average speed in m/s at the height h_0 . α is the Hellman coefficient characterizing the site relief. Indeed, any obstacle, such as mountains, trees, buildings, etc., has an influence on wind propagation in terms of speed and direction. The typical values of α are equal to 0.13 at sea, 0.16 on the shore, 0.2 in plains and 0.3 in cities [MAR 02].

3.1.4. *Loading rate*

It seems obvious that a wind turbine will produce much more energy when it is installed in an especially windy and very high location. Its use is characterized by the loading factor, which is calculated at a given period of time. The annual loading factor is defined by the ratio between the number of hours of operation at nominal power (full power) and the number of hours in a year (8,760 hours/year). A 3 MW wind turbine, which has produced a total of 5,256 MWh in a year, would have produced the same energy if it had operated at its nominal power during 5,256 MWh / 3 MW = 1,752 hours. Knowing that there are 8,760 hours in a year, its annual loading rate (or factor) is thus equal to 1,752 / 8,760 = 20%.

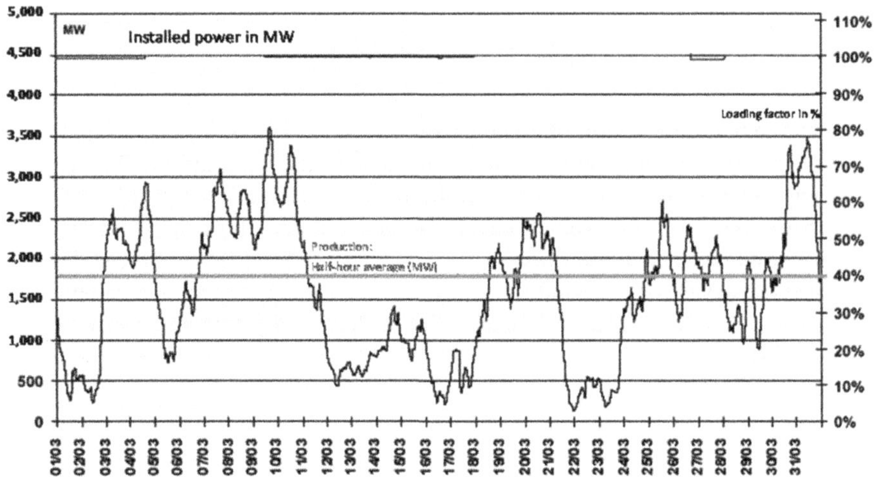

Figure 3.6. *Wind power production in March 2010 (Centre d'information du réseau électrique français (RTE) – Information Center for the French Electrical Network)*

For the entire wind power in France, RTE calculated an annual loading factor of 22% in 2009 [RTE 10]. This loading factor can vary significantly from one region to another and from year to year. The loading factor of offshore winds is higher, between 30 and 50%, thanks to greater regularity of the wind at sea and the absence of significant obstacles (low α coefficient in equation [3.2]).

Looking at the measured powered readings, the global loading factor of a wind turbine can be calculated on different time scales. For example, March 2010 was windy: the average wind power production was close to 1,574 MW (annual loading factor: 35%), with the highest measurements on March 21st with 142 MW (daily loading factor: 3%) and March 9th with 3,611 MW (daily loading factor: 81%) (Figure 3.6).

3.1.5. *Compass card*

A compass card is a polar diagram representing the wind speed frequency depending on its direction when the measurements were made. To give an example, Figure 3.7 shows that a compass card characterizes the wind power potential of a site and that it is quite different for two sites close to each other, even if common trends can be distinguished.

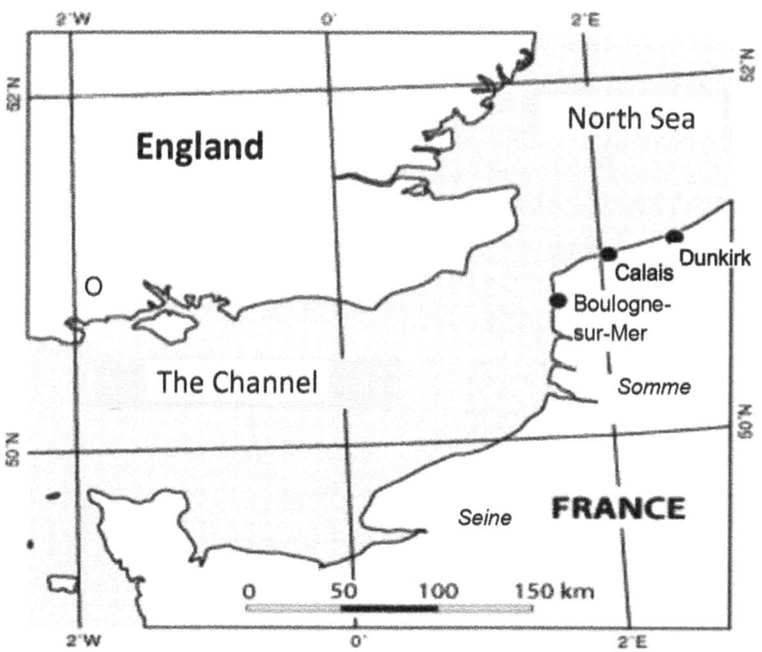

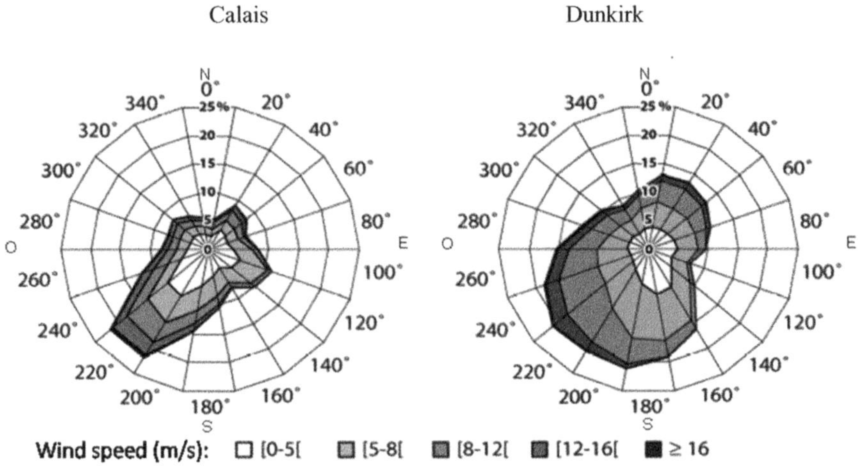

Figure 3.7. *Compass cards made from the tri-zone wind data of Météo-France (meteorological information system) [CHA 08]*

3.2. Kinetic wind energy

A wind tube, at a speed v, contains a certain amount of the kinetic energy that the wind turbine receiving the section of this tube will convert into mechanical energy. To calculate it, we consider an air tube of length $\mathrm{d}l$, section S and density ρ, which is driven by a speed v in accordance with Figure 3.8.

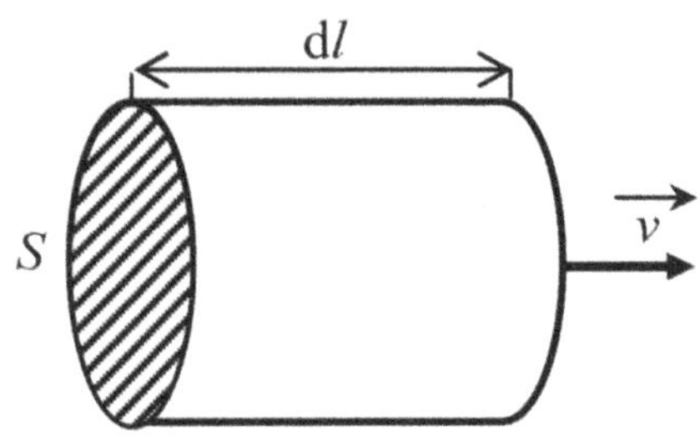

Figure 3.8. *Air tube in motion*

As for any solid in motion, the kinetic energy of this tube is expressed by the following expression:

$$W_c = \frac{1}{2} m v^2 \qquad [3.3]$$

m represents the total mass of the air volume that is contained in the tube. This mass depends on the length of the considered volume, which can be expressed by using the air density (ρ):

$$dm = \rho S dl \quad [3.4]$$

with $dl = vdt$

The air density depends on the pressure p, on the specific ideal gas constant R_a and on the air temperature T_{air} : $\rho = \frac{p}{R_a T_{air}}$ with, for a dry air, $R_a = 287{,}058\ J.kg^{-1}.K^{-1}$ and $p = 101{,}325\ kPa$. For a given location of the wind turbine, temperature is the main parameter fluctuating throughout the year; this leads to density variations. Under normal temperature and pressure conditions, $\rho =$ 1,225 kg/m^3. Given that this mass moves at speed v, the kinetic energy variation of this tube is expressed by:

$$dW_c = \frac{1}{2}v^2 dm = \frac{1}{2}v^2 \rho S v dt \quad [3.5]$$

The available gross power can then be written as follows:

$$P_v = \frac{dW_c}{dt} = \frac{1}{2}\rho S v^3 \quad [3.6]$$

This relationship shows that a small variation of the wind speed results in a significant power variation, because this speed appears at the cube. The available power per m^2 for a wind speed of 36 km/h (10 m/s) is equal to:

$$\frac{P_v}{S} = \frac{1}{2}\rho v^3 = 625 W / m^2 \quad [3.7]$$

Collecting the power in its entirety would presuppose that the wind speed behind the wind turbine is null. However, we do not know how to do this with our current knowledge. The next section will discuss the conversion of wind power into mechanical power. There are two main categories of wind turbines: vertical axis and horizontal axis wind turbines.

3.3. Wind turbines

3.3.1. *Horizontal axis wind turbines*

3.3.1.1. *Introduction*

Most of the currently installed wind turbines have a horizontal axis (Figure 3.9). The kinetic wind energy is converted into mechanical energy with the help of blades. The assembly of several blades on an axis of rotation constitutes the turbine. Horizontal axis wind turbines are comprised of one to three blades, which are aerodynamically faired. Tri-blade wind turbines have frequently been used, since recoverable power depends (to some extent) on the number of blades, and because a tri-blade rotor is an excellent compromise between recoverable power and the machine cost. Moreover, the efforts undertaken by the machine are better balanced and, visually speaking, the rotation of a tri-blade rotor is seen as more harmonious (and thus esthetic) than that of a two-bladed rotor or a single-blade rotor. Figure 3.10 gives an idea of the size of horizontal axis wind turbines depending on the power. In 2019, the largest wind turbine reached 8 MW with a turbine measuring 180 m in diameter. For offshore wind, prospects of up to 20 MW are being studied with a turbine of 250 m in diameter to be installed on a 200 m high mast [DE 18].

Figure 3.11 presents the main components of a conventional horizontal axis wind turbine with a speed multiplier enabling the use of a conventional electric generator and an orientation system for direction when facing the wind.

Figure 3.9. *Horizontal axis wind turbine*

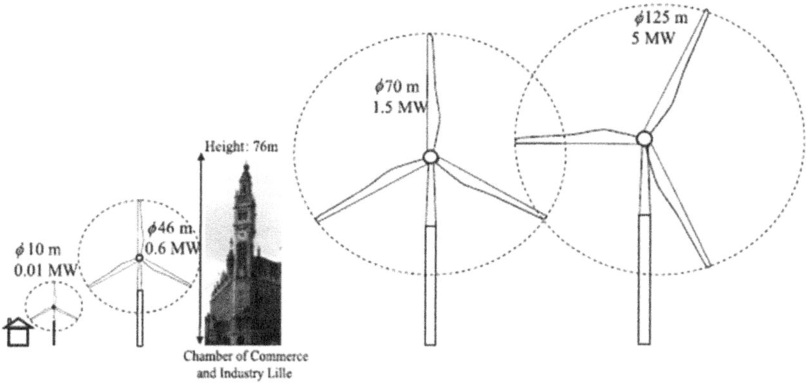

Figure 3.10. *Typical relationship between the wind turbine diameter and power*

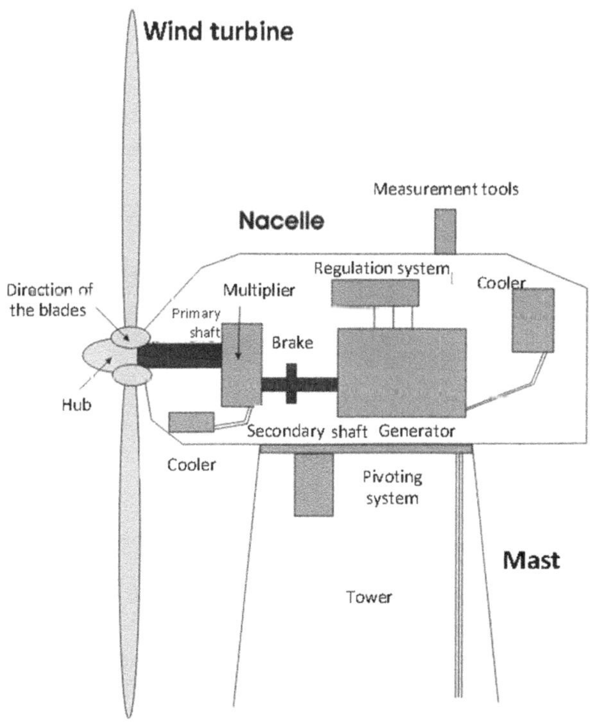

Figure 3.11. *Main components of a wind turbine with a fast generator*

Most of these wind turbines are omnidirectional sensors, which means they have to direct their rotor according to the wind, either passively (rudders) or actively (gears). Conventional wind turbines incorporate certain positioning devices which allow the turbines to face the direction of the wind.

Small wind turbines (lower than 50 kW) mainly use passive control of the direction via a rudder for simplicity reasons. Some rudders are equipped with dampers or other devices, which are designed to reduce the orientation speed of the wind turbine and, more importantly, all direction oscillations. Beyond a certain size (about 300 kW), wind turbines use a motorized, controlled active orientation system. This system helps to keep the turbines more stable when facing the wind. A wind vane placed on top of the nacelle (Figure 3.12) sends signals to a motorized system, which mechanically orientates the nacelle to face the wind.

Figure 3.12. *Anemometers and wind vanes on the back of a wind turbine nacelle (Laborelec)*

3.3.1.2. *Conversion into mechanical power by a lift effect*

Horizontal axis wind turbines are based on the principle of windmills and the use of the lift force on each blade. The turbine is supposed to face the wind (as in Figure 3.9), which has a certain speed, v. This wind speed is measured far in front of the turbine (upstream infinity speed). Indeed, the air flow is disturbed by the presence of the blade well before touching it and it is deflected before reaching the blade's leading edge (Figure 3.13).

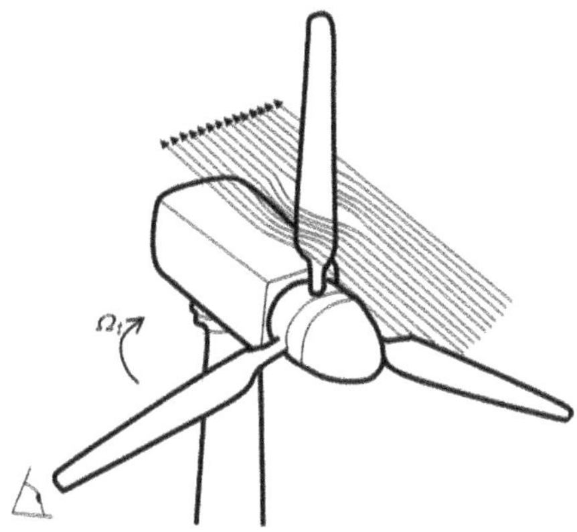

Figure 3.13. *Deflection of a flow by a blade*

For this study, we consider a cross-section view obtained by "cutting" a section of the blade at one point of its wingspan (Figure 3.14). The hatched surface corresponds to the profile, which is comprised of a leading edge, a trailing edge and a chord, which connects these two ends (direction "x") with the camber (direction "y"). The third dimension corresponding to the blade length (wingspan) can be integrated into accurate calculations of the aerodynamic properties, as a superimposition of the 2D behaviors on the blade length. In the case of a complete 3D blade, the chord varies according to the position along the wingspan. Moreover, the profile shape can vary with this distance (Figure 3.19).

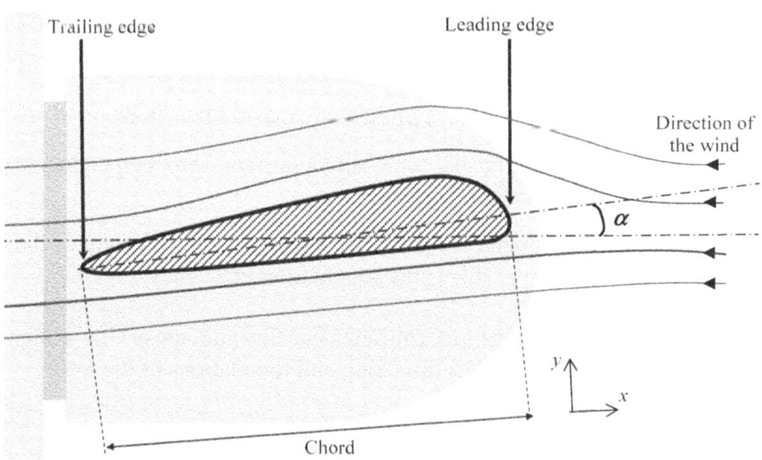

Figure 3.14. *Deflection of a flow by a blade in cross-section view. For a color version of this figure, see www.iste.co.uk/robyns/electricity.zip*

The chord presents an angle of attack (α), with the upstream direction of the air flow. As this angle becomes larger, the air flow is increasingly deflected by the profile and modified around it. The air flow follows the curvature of the blade surface and changes direction; this attachment phenomenon is also called the Coanda effect. The blade and especially its shape deflect the air flow crossing its plane of rotation. There follows a reaction that pushes the blade. Thus, the force, which is deflecting the air flow and is created by the blade, creates another force, which is equal and opposite, depending on Newton's third law of motion.

The shape of the blade is rounded on the upper part. Therefore, the air circulating in this zone has to travel a longer distance (Figure 3.14). There is an acceleration of the flow on the upper part, which is accompanied with a pressure reduction. Similarly, there is a deceleration on the lower part of the blade, which comes with a pressure increase. The difference of pressure above and below the blade causes an upwards force: the resulting aerodynamics can be calculated by using the Bernoulli equation (Figure 3.15). It is even more significant as the angle of attack becomes larger. There is a limit leading to a stall phenomenon. This phenomenon will be presented in section 3.4.2.

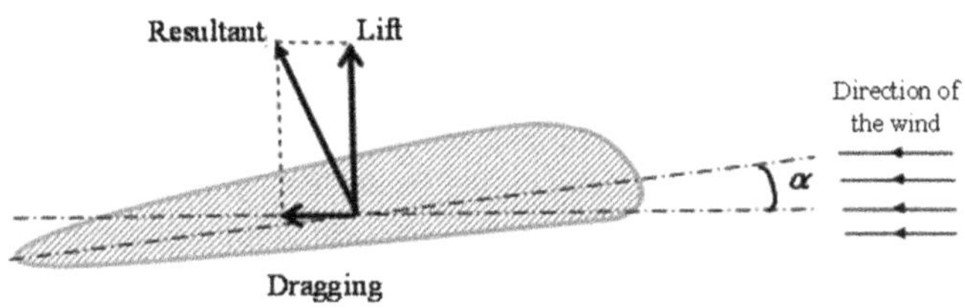

Figure 3.15. *Forces exerted on a blade in cross-section view*

The resulting aerodynamic force is broken down as follows into:

– a component depending on the wind direction (and thus perpendicular to the turbine's plane of rotation), called the drag force;

– a perpendicular component, the lift force, which it is at the origin of the torque enabling the turbine's rotation. This is the sought after efficiency.

The drag force subjects the set (turbine, nacelle and tower) to a significant mechanical thrust, which conditions the sizing and foundations of the tower.

3.3.1.3. *Influence of the rotational speed on the angle of attack*

In reality, the blade's rotational speed will modify the perceived wind speed. If the wind turbine has a rotational speed of Ω_t (rad/second), then the blade tangential speed induced by the rotation, at a certain length, r, ranging between the hub and the blade end is equal to:

$$U = r\Omega_t \quad [3.8]$$

This tangential speed (U) is higher than the wind speed and significantly influences the flow, which will be received by the profile. It must be combined with the wind speed (V) in order to obtain the relative wind speed Vr. From the aerodynamic point of view, it is as if wind arrives at the wind turbine profile with this relative speed and according to its direction (Figure 3.16).

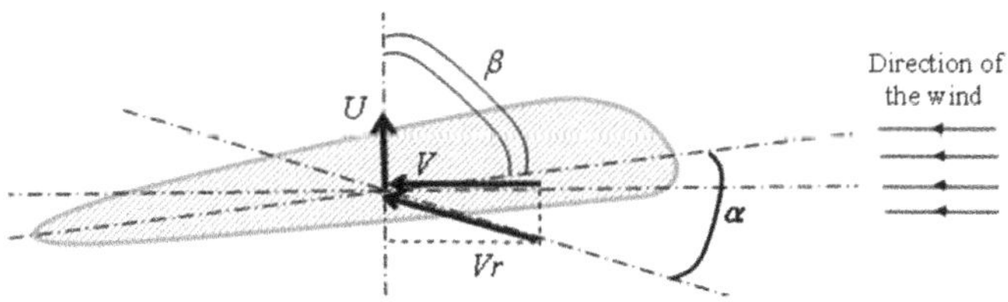

Figure 3.16. *Direction of the speeds of a blade in cross-section view*

The angle of attack is modified and the directions of the drag and lift forces are deduced from the relative speed (Figure 3.17).

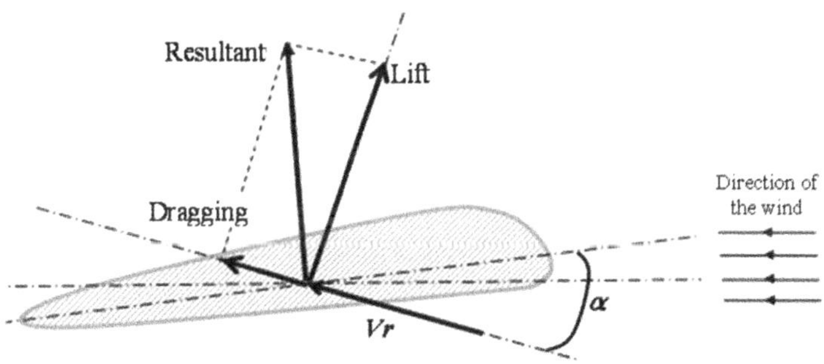

Figure 3.17. *Modification of the forces related to the relative wind speed*

The tangential speed increases proportionally along the blade. The maximum tangential speed is obtained at the tip of the blade (r is then equal to the blade length). As the relative speed, Vr, also increases along the blade, its geometry must be adapted to this speed increase. To do so, the profile parameters are defined by the pitch angle β, which is defined between the plane of rotation and the chord of the profile. Generally, profiles are designed by modifying the pitch angle along the blade, in order to keep a constant angle of attack (α) all along the blade; this gives a twist aspect to the blade (Figure 3.18).

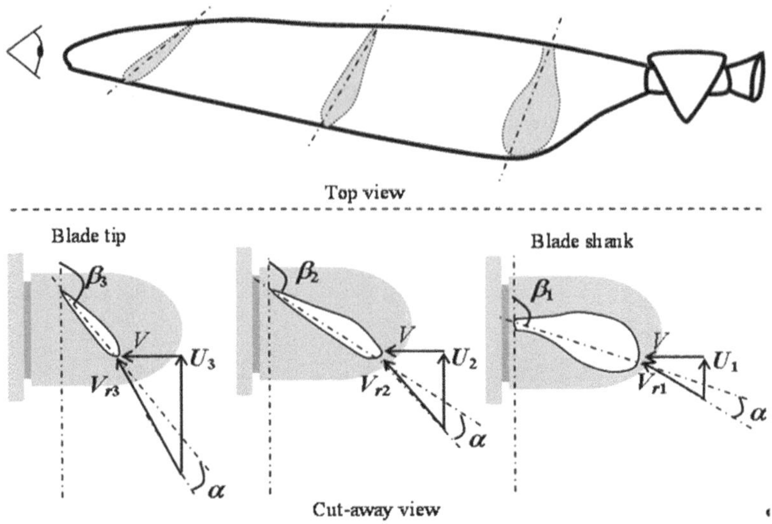

Figure 3.18. *Variation of the pitch angle along the blade (twist)*

Figure 3.19 shows the 3D shape of the blade of a horizontal axis wind turbine in the "flag" position. In this position, the air flow is practically not disturbed (on the contrary to Figure 3.9) and the lift effect is very small. This will contribute to maintaining the turbine at a standstill, in case of excessive wind speeds.

Figure 3.19. *A blade in its initial position, the flag position (photo: Laborelec)*

3.3.1.4. *Efficiency and power coefficient*

The turbine extracts a mechanical power P_m, which is lower than the aerodynamic power P_v (the air mass speed is not null behind the turbine). We then define the turbine power coefficient (or the aerodynamic efficiency) by:

$$C_p = \frac{P_m}{P_v}, \; C_p < 1. \tag{3.9}$$

The power collected by the turbine is then written:

$$P_m = C_p \frac{1}{2} \rho \pi R^2 v^3 \tag{3.10}$$

with R as the blade length.

Betz has demonstrated that the power coefficient cannot exceed a value $C_{P\max}$ [HAU 06], which is called the Betz limit and is equal to $\frac{16}{27} \approx 0.5925$. The value of this power coefficient depends on the tip speed ratio λ :

$$\lambda = \frac{R\Omega_t}{v} \tag{3.11}$$

Ω_t is the rotational speed of the turbine and $R\Omega_t$ is the peripheral linear (tangential) speed at the blade tip.

All the masses (blades and hub) are reported on the axis of rotation and constitute the inertia J. The mechanical energy stored at the rotational speed Ω_t is then equal to:

$$E = \frac{1}{2} J \Omega_t^2 \tag{3.12}$$

3.3.2. *Vertical axis wind turbines*

Vertical axis wind turbines capture the wind speed regardless of its direction; there is thus no need for a rotor orientation device, as in the case of horizontal axis wind turbines. However, winds are quite low near the ground and show strong turbulences, because of the strong impact of relief. The generated power is lower in comparison to a centralized capture at a greater height.

Vertical turbines offer the possibility of accommodating, at their foundation, the entire energy conversion device (generator, multiplier, etc.). While for safety reasons this turbine has to be inaccessible, these components are located at an even lower height than for horizontal axis turbines, thus facilitating maintenance operations.

There are two types of vertical turbines: the Savonius rotor relying on the differential drag principle and the Darrieus rotor relying on cyclic incidence (or lift) variation.

The Savonius rotor is designed to exploit differential dragging. The efforts exerted by wind on each of the faces of a hollow body have varying intensities. Figure 3.20 shows the assembly of two hollow bodies on an axis, subjected to wind. The air flow is modified and the resultant forces will create a resultant force F_1 on the concave face which is higher than that of the convex face F_2. This "differential drag" principle creates a torque, leading to the rotation of the whole.

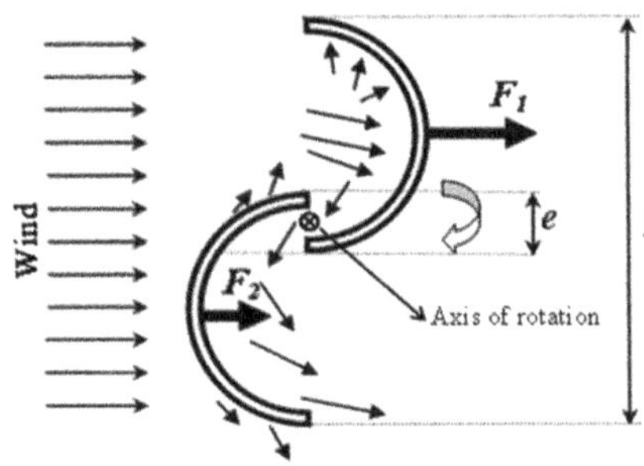

Figure 3.20. *Operating principle of an elementary Savonius principle (top view)*

The Savonius rotor is characterized by a low rotational speed and a high torque. The wind speed that enables the machine startup is quite low: about 2 to 3 m/s. The power coefficient reaches at maximum the value 0.3 for a tip speed ratio close to 1 [FAC 48]. The first models are described as two half-cylinders, whose axes are out of line from one another (distance e in Figure 3.20). They are thus quite easy to produce. Very good performances are obtained by taking the ratio:

$$e / D = 1 / 3 \quad [3.13]$$

Figure 3.21. *Savonius wind turbines for urban applications (project of engineers from the Centrale Lille Institut)*

These turbines have evolved towards devices using several levels of half-cylinders or a twist form. They enable a continuous circulation of the flows inside the turbine and have a starting torque that is independent of the position of the blades in relation to the wind direction (Figure 3.21).

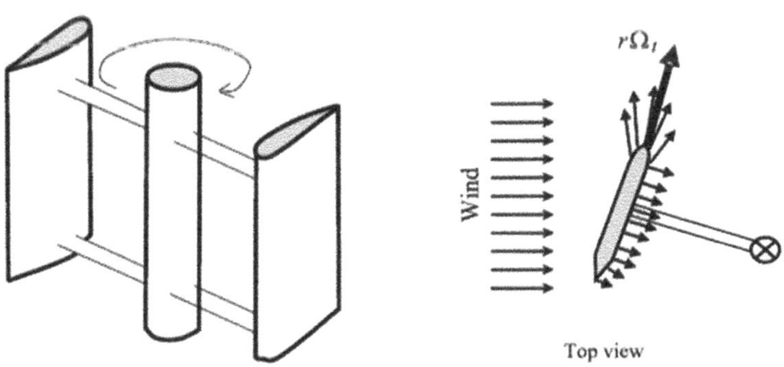

Figure 3.22. *Block diagram of the Darrieus rotor*

Darrieus rotors are based on the principle of cyclic incidence variation. A profile positioned in an air flow according to various angles is subjected to forces of variable intensities and directions. The resultant of these forces then generates a torque leading to rotation of the device (Figure 3.22).

Wind turbines using lift are able to extract more energy from the wind than those using drag. Indeed, they create a low-pressure area, and the blade shape (creating the lift force) crosses the air flow, thus generating a force vector higher than the wind speed (quick wind turbine). The power coefficient can reach the value 0.35 for a tip speed ratio ranging between 4 and 5 [MAR 02]. As an example, the F64-10 wind turbine of the manufacturer FAIRWIND (Figure 3.23(a)) is able to produce 10 kW under a wind speed of 9 m/s. There is a large variety of Darrieus rotors with this principle, such as cylindrical rotors, tapered rotors, dish rotors, H rotors, helical rotors (Figure 3.23(b)), etc.

a) FAIRWIND F64-10 wind turbine b) QUIETREVOLUTION wind turbine

Figure 3.23. *Darrieus wind turbine*

Darrieus vertical axis wind turbines are not widespread, but they are however the best adapted to some sectors, such as integration into buildings, extreme areas (observatories, airfields, etc.). Most of them have a lower efficiency than horizontal axis wind turbines, but they free us from the limits resulting from the blade size and rotational speed. The global bulk is lower and in some cases, where the electric machine is located at the wind turbine foundation, this type of wind turbine is more economical.

This type of solution significantly reduces noise, while still permitting operation, when there are winds higher than 90 km/h and no matter their direction. However, some variants from this category cannot reach the mentioned speeds.

The main defect of this type of wind turbine is their difficulty in starting. Indeed, the rotor weight bears upon its base; a more significant starting torque needs to be overcome because of the friction losses.

3.3.3. *Comparison of the various turbine types*

From the aerodynamic point of view, we can compare the various types of turbines by analyzing their power coefficients in relation to the tip speed ratio λ (Figure 3.24). We can notice that all these curves go through a maximum, which reveals the existence of an optimal rotational speed, maximizing the aerodynamic efficiency at a given wind speed. The higher the optimal value of λ gets, the higher the optimal rotational speed at a given turbine diameter becomes.

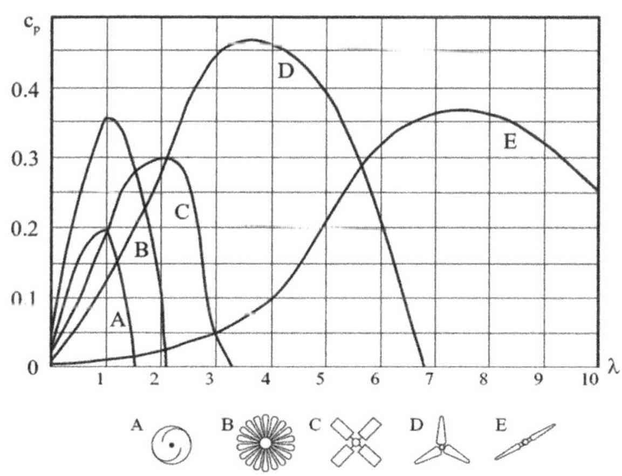

Figure 3.24. *Power coefficients according to the tip speed ratio λ for different types of turbines [FAT 48]*

Turbine A is a Savonius vertical axis turbine. Turbines B and C are horizontal axis turbines, which are respectively multi-blade ("Far West") and windmill turbines. Turbines D and E are modern three- and dual-blade wind turbines. A low value of the tip speed ratio λ corresponds to a low speed turbine. Figure 3.24 shows that with fewer blades, a horizontal axis wind turbine has a higher optimal rotational speed. Therefore, turbine D is a good compromise (for average wind speeds) and brings the best efficiency. For this type of three-blade turbine, the efficiency can reach 85% of the theoretical Betz limit. If reducing the number of blades helps to reduce manufacturing and installation costs, the disadvantage of a single- or dual-blade wind turbine is that it has to spin faster. This makes it louder, reduces the lifespan of the mechanical components and creates a more irregular power. Indeed, when a blade goes into the aerodynamic protected area located just in front of the tower, the instant power captured by the turbine decreases significantly (tower shadow phenomenon).

3.4. Power limitation by varying the power coefficient

High wind speeds enable us to obtain significant powers, but they are not very common. Wind turbines are thus not oversized for these speeds, because this would lead to additional costs. Turbines are generally designed to reach their nominal power at a wind speed of about 15 m/s and are automatically stopped at 25 m/s. To limit the power produced at high wind speeds, although the aerodynamic gross power of the wind keeps on increasing in v^3, there are two categories of power regulation systems:

– the "stall" or the natural aerodynamic stall system, for which blades have a profile that is designed to naturally reduce the power coefficient C_p when the wind speed increases, i.e. when the tip speed ratio λ decreases – this is the stall principle;

– the "pitch" or variable pitch angle system, with mobile blades around their longitudinal axis. This enables us to reduce the lift and therefore the C_p coefficient for significant wind speeds.

3.4.1. *The "pitch" or variable pitch angle system*

A blade is at a variable pitch angle or at a variable step if the orientation of the blade in relation to its initial position can be modified during operation. Blades are equipped with actuators (hydraulic jacks or electric motors), which are responsible for changing their orientation. Given the inertia of a blade and the importance of the forces exerted on it, orientation is slow. For example, the temporal angle variation is about 8°/s for a wind turbine MM82 REPOWER (2 MW, 80 m) and the pitch motor requires an electric power of about 4 to 5 kW.

The value of the aerodynamic lift that is generated on each blade can thus be modified. This will influence the power coefficient C_p of the turbine. Figure 3.25 highlights the pitch angle effect on the power coefficient. A null pitch angle corresponds to a blade facing the wind, and a negative value of C_p means that the turbine is in brake operating mode (excessive rotational speed).

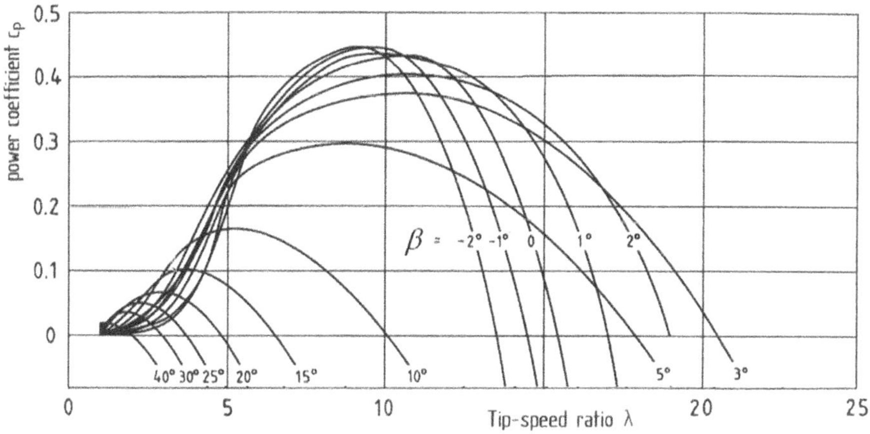

Figure 3.25. *Effect of the pitch angle on the power coefficient C_p of the WKA-60 turbine [HAU 06]*

Figure 3.25 clearly shows the existence of a maximum power coefficient for a particular value of the tip speed ratio and the step. To extract the maximum power (thus at speeds lower than the speed previously mentioned and beyond which we wish to reduce the lift), and because of the highly variable nature of wind, the adjustment of the tip speed ratio requires us to optimize the lift by an action on the step and to adapt the turbine speed in order to maintain the energy conversion efficiency at its maximum.

The regulation system of the pitch angle enables active control of the recovered power, which is mainly used for two tasks: power or rotor speed control and aerodynamic braking for stopping the turbine. Thus, the power can be limited to the nominal power and we can guide the blades in the direction of the wind (also called the "flag position", Figure 3.19) for high speed winds, and thus protect the machine. For the MM82 REPOWER wind turbine, the transition to this mode can be carried out in 11 s. This complex and expensive system took a few years to enter the market, but nowadays it can be found on almost all the modern wind turbines with a power higher than or equal to 1 MW.

3.4.2. *The "stall" or aerodynamic stall system*

If the angle of attack is too large, the air flow no longer manages to follow the trajectory that has been imposed by the (highly sloping) profile. As there is a smaller deflection from the trajectory, there is a less significant acceleration of the flow on the upper side of the blade. Thus, the decrease of pressure and lift are less significant (Figure 3.26).

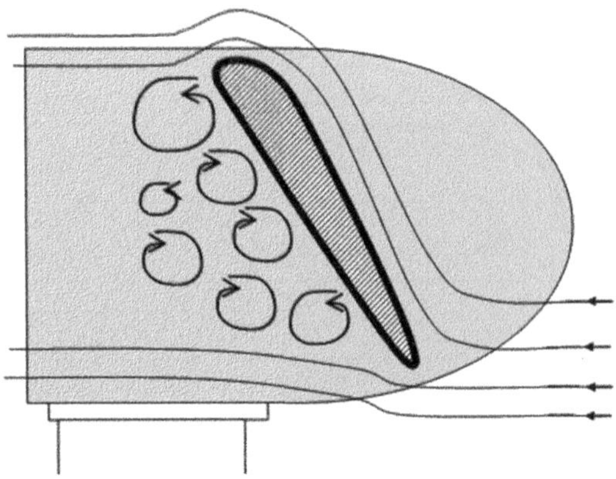

Figure 3.26. *Aerodynamic stall effect at high rotational speeds*

To exploit this phenomenon, blades have particular forms in order to progressively "uncouple" from the nominal wind speed. At high speeds, intrusive turbulences appear and naturally make the power drop, by causing an efficiency loss. This passive regulation system therefore does not require any external control system. It is thus less expensive and more reliable than the variable pitch angle system, since it only has a small number of elements in motion and since it is predominant in the field of low-power wind turbines. However, it is more difficult to optimize the operation of a wind turbine, which is equipped with a "stall system", since we have one less control parameter (the pitch angle) in comparison to the variable pitch angle system. The power curve of such a machine can then be recognized by its wavy (not flat) shape beyond the nominal wind speed and up to the cut out speed (Figure 3.27).

In addition, there is also a more marginal intermediary system, the "active stall", which consists of adjusting the stall effect by a very small variation of the pitch angle. This thus makes the mechanical rotation device of the blades lighter and more robust.

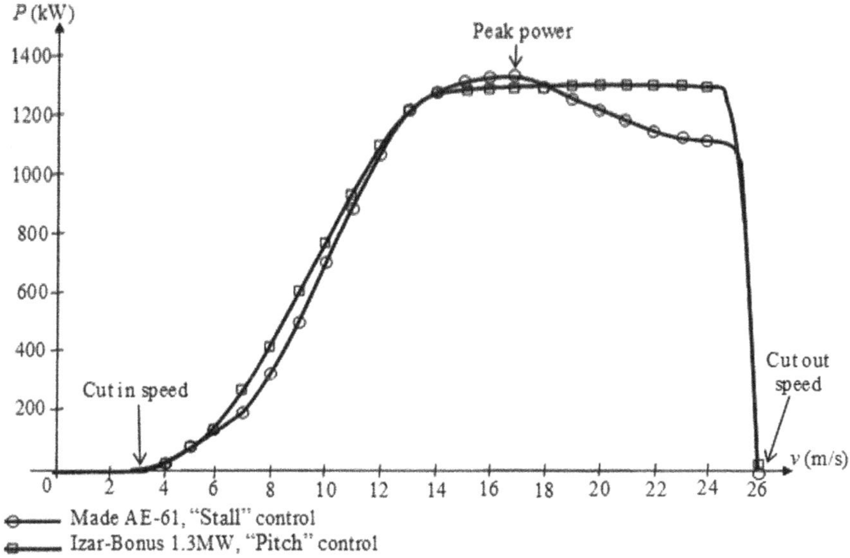

Figure 3.27. *Comparison of the power characteristics according to the wind speed*

3.5. Mechanical couplings between the turbine and the electric generator

3.5.1. *Connection between mechanical speed, synchronous speed and electrical network frequency*

In private homes in France, electrical energy is available in the form of a sine-wave voltage with a frequency of 50 Hz and a root mean square value of 230 V. It is thus expressed as:

$$v = 230\sqrt{2}\sin(2\pi 50t) \qquad [3.14]$$

In fact, this voltage results from a balanced three-phase system of the same frequency, same root mean square value and a $2\pi/3$ phase:

$$
\begin{aligned}
v_1 &= 230\sqrt{2}\sin(2\pi 50t) \\
v_2 &= 230\sqrt{2}\sin(2\pi 50t - \frac{2\pi}{3}) \\
v_3 &= 230\sqrt{2}\sin(2\pi 50t - \frac{4\pi}{3})
\end{aligned}
\quad [3.15]
$$

In order to describe the elementary physical principles of electricity production, we consider three identical coils, which are supposed to be ideal (notably not saturating) and regularly spaced out in a circle of a 120° angle (Figure 3.28(a)). To form an electric generator, the three coils are assembled, in order to form a stator circuit (Figure 3.28(b)). A magnet is placed on the rotor.

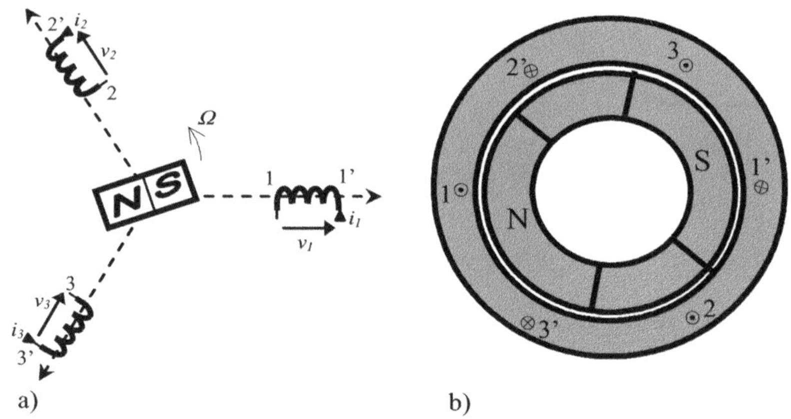

Figure 3.28. *Elementary principle of the rotating magnetic field*

When we supply the coils with these three-phase voltages, each coil creates a pulsed magnetic field. At the center of this system, the vector sum of the three magnetic fields then results in a magnetic field that turns at a speed of 50 turns per second. If we put a magnet at the center, it will turn synchronously with this field and at the same speed. This physical phenomenon is at the origin of the technology of synchronous machines. In this example, the latter operate in motors. To reduce the rotational speed, the number of rotor poles can be increased. For any number of pole pairs (p), we then define the synchronous speed by:

$$N_s = \frac{50}{p} 60 \text{ rpm} \quad [3.16]$$

or

$$\Omega_s = \frac{50}{p} 60 \frac{2\pi}{60} = \frac{100\pi}{p} \text{ rad/s} \qquad [3.17]$$

In this example, the synchronous machine operates as a motor; power is extracted from the electrical network, to then be transformed into mechanical power, which is the origin of the rotational motion. Now if we apply a mechanical force to the axis, in order to accelerate it, the power transfer is reversed and the surplus mechanical power is converted into electric power, which is then sent to the network.

3.5.2. *"Direct drive" wind turbines (without a multiplier)*

In order to generate significant mechanical power, blades have a large size. Because of the optimal value of the tip speed ratio λ, these dimensions result in quite a slow rotational speed for the turbine. The speed never exceeds one turn per second for wind turbines. Ideally, to be able to directly couple the generator to the 50 Hz (or even 60 Hz) network, we would have to use a synchronous machine with a very large number of pole pairs, so that the synchronous speed is slightly lower than the turbine's slow mechanical speed (for the most probable wind speed found on the site, Figure 3.5). Synchronous machines with a very large number of poles thus have a very large diameter and also a more significant mass than that of a fast machine of the same power, with a lower nominal torque. For these reasons, amongst other things, there are no conversion chains which are based on synchronous machines directly coupled to the network.

As an example, the synchronous machine used in the J48 wind turbine and manufactured by Jeumont Industrie is comprised of a turbine with a 48 m diameter for a nominal power of 750 kW with a speed of 9 to 25 rpm. It also relies on an axial field (discoid). However, it was coupled to the network via an electronic power converter operating a change of frequency between that of the machine and that of the network. This multiplier-free solution is particularly advantageous in the case of offshore wind since it reduces the maintenance requirements of wind turbines, access to which is more difficult than if it were onshore.

3.5.3. *Use of a speed multiplier*

Many industrial applications use induction machines with two pole pairs and thus with a synchronous speed of $N_s = 1500$ rpm. Currently, this approach is a good compromise in terms of dimensions, mass and cost. However to use this type of machine in a wind turbine, a speed multiplier is required in order to obtain a speed of about 1,500 rpm on the generator shaft.

The mechanical solution consists of using gear trains, which offer an acceptable cost and mass and the possibility of making the generator revolve more quickly with few losses (about 2% of losses per level), thus obtaining an electromechanical conversion chain, which nowadays has the best mass/cost compromise. But even if the multiplier reliability has significantly improved, it is less significant than that of the direct drive conversion chains. Moreover, they require certain maintenance operations (lubricating oil change and cooling.)

If we do not take the mechanical losses into account, the speed obtained at the multiplier shaft output is expressed by using the multiplication ratio (m):

$$\Omega = m\Omega_t \qquad [3.18]$$

If we assume this mechanical system to be without losses, the conservation of the mechanical power on the slow axis ($c_t \Omega_t$) and on the fast axis ($c\Omega$) leads us to establish the torque expression, which is transmitted on the slow axis (c_t):

$$c_t = mc \qquad [3.19]$$

3.6. Generalities on induction and mechanical electric conversion

The synchronous machine is an electric machine technology that is used to convert mechanical power into electric power. To illustrate the physical principle of synchronous machines, we consider the experiment which consists of making a magnet turn in front of a fixed coil whose terminals are connected to an oscilloscope (Figure 3.29).

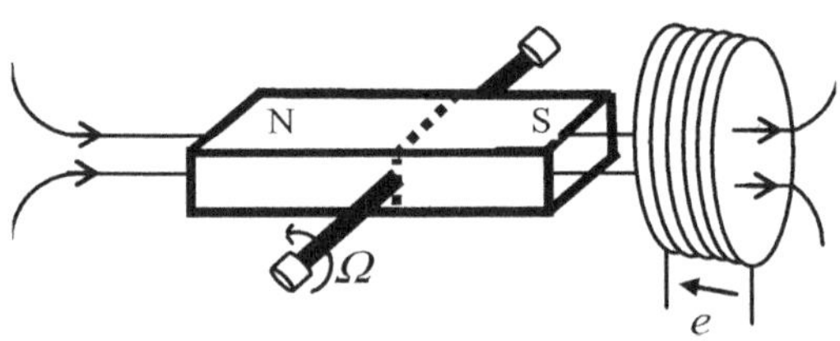

Figure 3.29. *The motion of a magnet in front of a coil*

If the angular speed of the magnet has a constant value Ω, on the screen we observe a sine-wave periodic curve. By definition, the magnetic flux characterizes the intensity and the spatial distribution of the magnetic field through a surface. Figure 3.30 shows that in one complete revolution of the magnet, with a duration $T=\frac{2\pi}{\Omega}$, the total magnetic flux Φ through all the coil spires successively takes:

– at the instant $t = 0$, the maximum value Φ_m;

– at the instant $t = \frac{T}{4}$, a null value;

– at the instant $t = \frac{T}{2}$, the value $-\Phi_m$;

– at the instant $t = \frac{3.T}{4}$, a null value;

– at the instant $t = T$, again the value Φ_m.

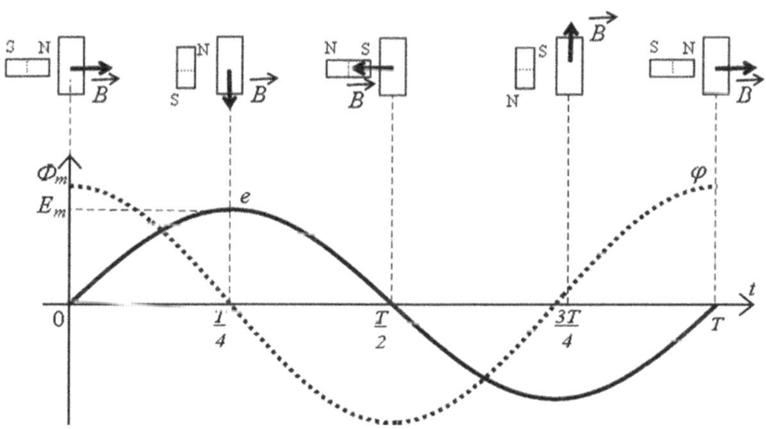

Figure 3.30. *Temporal evolution of the magnetic flux and the voltage at the coil terminals*

This flux is thus periodic, of period T, and its variations can be represented at least approximately by the sinusoidal function:

$$\Phi(t)=\Phi_m \cos(\Omega t) \quad [3.20]$$

Thus, in the induced circuit constituting the coil, the production of an induced e.m.f. with an instant value given by the Lenz-Faraday law is [ROB 12]:

$$e(t) = -\frac{d\Phi}{dt} = \Phi_m\ \Omega\sin(\Omega\ t) = E_m\sin(\Omega\ t) \qquad [3.21]$$

Its angular frequency is also equal to the angular speed Ω of the field magnet. The number of pole pairs is 1. Therefore, this speed is equal to the synchronous speed ($\Omega = \Omega_s$, equation [3.17]). Its amplitude is proportional to the maximum value of the flux crossing the armature and to the rotational speed of the field magnet:

$$E_m = \Phi_m \Omega \qquad [3.22]$$

The technological production of the synchronous machine thus relies on a stator, which includes an induced e.m.f and a rotor at the origin of a magnetic field. The stator is made up of three windings, which are coupled in a star configuration. It is also at the origin of the magnetic field revolving in the machine. Depending on the rotor design, two types of machines can be distinguished.

If magnets are used, then a "permanent magnet machine" is obtained.

The magnetic field at the rotor can also be produced by using coils, which then must be externally supplied. This circuit is still called an inductor.

3.7. "Fixed speed" wind turbines based on induction machines

3.7.1. *Physical principle*

By again taking the example of Figure 3.28, and if we replace the magnet with a metal part, the latter will also revolve when stator windings are supplied by three-phase voltages. However, the rotational speed will be different from the synchronous speed, hence the designation of the term "induction machine" [LE 09]. The difference between the mechanical speed (Ω) and the synchronous speed (Ω_s, imposed by the network frequency), is characterized by a slip:

$$g = \frac{\Omega_s - \Omega}{\Omega_s} \qquad [3.23]$$

A unit slip corresponds to the machine at standstill and a null slip corresponds to the synchronous speed (the operation is similar to a synchronous machine).

3.7.2. *Constitution of induction machines*

The stator of an induction machine is made up of three windings, which are coupled in a star or delta configuration. Depending on the rotor design, two machine technologies can be distinguished.

For the first technology, industrial productions generally use a rotor. The latter is made up of copper or aluminum bars, which are conductive and short-circuited by a conductive ring at each end. This looks like a squirrel-cage, hence the name "squirrel-cage induction machines". This operation enables the circulation of the induced currents (in the rotor), which, since they are revolving, cause a second magnetic field. The second magnetic field even plays the same role as that of the magnet in Figure 3.31. The interaction of this field with the field created by stator windings (supplied by network voltages) is at the origin of the couple of forces applied on the axis. We can show that this squirrel-cage rotor behaves as a wound rotor, whose windings at the rotor are short-circuited on themselves.

For the second technology (in wind power, these machines are doubly-fed induction machines), the rotor is really wound and includes a three-phase winding. The latter is similar to that of the stator connected in a star configuration. Moreover, the free end of each winding is connected to a revolving ring with the rotor shaft. Three brushes establish connections on this ring and enable the external connection of the windings to the rotor. This connection can be a link to external resistors (in order to modify the machine characteristic in some operation zones) or an external supply, which helps to carry out control of the rotor sizes. This second technology is still known under the name "doubly-fed machine".

The rotor cage induction machine (Figure 3.31) is very often used in industrial applications. Indeed, because of its design, its cost is relatively low in comparison to other machines, its robustness is important at an electromechanical level and there is good standardization between manufacturers. The good robustness of this machine and its lower cost compared to other types of electrical machines have resulted in its use in generator operation at almost constant speed in the first high power wind turbines that were developed on a large scale. However, the relative simplicity of this machine design hides quite a complicated operation mode.

Figure 3.31. *Rotor and stator of a squirrel-cage induction machine [ROB 12]*

3.7.3. *Modeling*

3.7.3.1. *Equivalent single-phase diagram*

Induction machines are often modeled with the help of a simple model based on an equivalent single-phase diagram. As long as a study is carried out in permanent regime (a gauging study for example), the three-phase machine components are identical and the magnitudes are expressed in root mean square values (Figure 3.32).

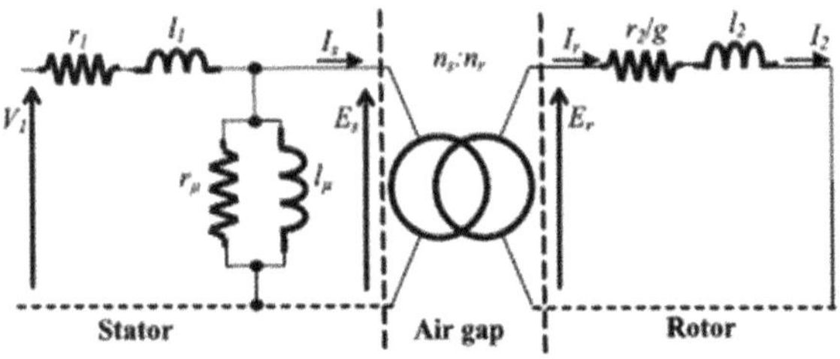

Figure 3.32. *Equivalent single-phase diagram of the squirrel-cage induction machine*

The left part of the diagram in Figure 3.32 corresponds to the stator circuit and is comprised of the resistance of conductors r_1 and an inductance l_1 representing the magnetic flux leakages. The right part corresponds to the rotor circuit and is comprised of the resistance of the conductors r_2 and an inductance l_2, representing the magnetic flux leakages. The vertical circuit models the magnetization fixing the flux in the machine, and comprises a resistance r_μ which models magnetic losses and a magnetizing inductance l_μ. V_1 is the phase to ground voltage applied to one of the stator phases.

The magnetic field generated by the stator creates e.m.f in the rotor winding of root mean square value E_r at standstill. These e.m.f. are at the origin of currents in rotor circuits and thus of a magnetic field, which is generated by the rotor. This rotor magnetic field creates e.m.f in the stator of root mean square value E_s.

The rotor resistance plays a fundamental part in the electromechanical conversion of energy. The diagram in Figure 3.32 is modified to highlight this role by distinguishing it from the modeling of the losses by Joule effect at the rotor:

$$\frac{r_2}{g} = r_2 + r_2 \frac{(1-g)}{g} \qquad [3.24]$$

We then obtain the diagram in Figure 3.33, which enables us to determine the various losses and powers exchanged within the machine.

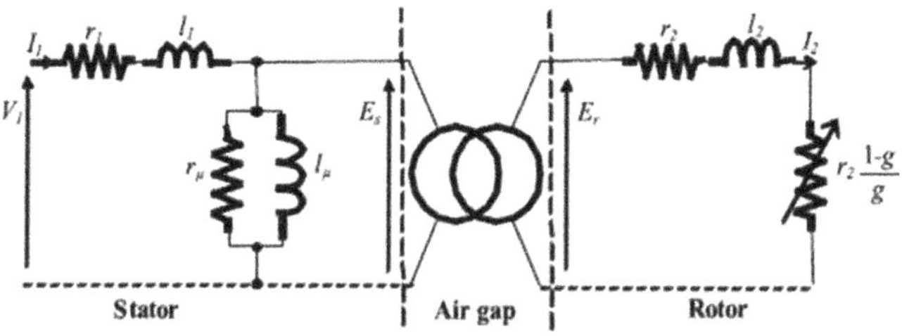

Figure 3.33. *Modified equivalent single-phase diagram of a squirrel-cage induction machine*

All the voltages and currents are expressed in root mean square value. As I_1 is the stator current and I_2 is the rotor current, the various losses have the following expressions:

– for losses by Joule effect at the stator

$$P_{js} = 3\, r_1\, I_1^{\,2}, \qquad [3.25]$$

– for magnetic losses

$$P_\mu = 3\,\frac{E_s^{\,2}}{r_\mu}, \qquad [3.26]$$

– for losses by the Joule effect at the rotor

$$P_{jr} = 3\, r_2\, I_2^{\,2}. \qquad [3.27]$$

NOTE.– The single phase diagram only models one phase of the machine; hence a multiplication by 3 is required to obtain the powers in the three-phase machine.

The electric power transformed into mechanical power is expressed by:

$$P_{em} = 3 r_2\,\frac{1-g}{g}\, I_2^{\,2} \qquad [3.28]$$

Part of this power is dissipated in mechanical losses P_{pm}, which are not represented in the wiring diagram in Figure 3.33. The remaining power is the useful mechanical power P_m on the machine shaft:

$$P_{em} = P_{pm} + P_m \qquad [3.29]$$

We can thus deduce from the equivalent diagram, the balance of powers linking the electric power P_e supplied or absorbed by the machine and the mechanical power P_m on the machine shaft. By using the motor convention (the most conventional one), the electric power consumed corresponds to the sum of the losses and of the converted power:

$$P_e = P_{js} + P_{\mu} + P_{jr} + P_{pm} + P_m \qquad [3.30]$$

The consumed reactive power corresponds to:

$$Q = \sqrt{(3V_1 I_1)^2 - P_e^2} \qquad [3.31]$$

3.7.3.2. *Static characteristics*

If the power supplied to the machine is mechanical, the machine will operate as a generator. If the power supplied is electric, the machine will operate as a motor. To characterize the machine operation, we have to determine the expression of the electromagnetic torque. According to [3.28]:

$$P_{em} = C_{em}\Omega = 3r_2 \frac{1-g}{g} I_2^2 \qquad [3.32]$$

From the equivalent single-phase diagram of Figure 3.33, we obtain the rotor current expression:

$$I_2 = \frac{E_r}{z_2} = \frac{n.E_s}{z_2} \quad \text{with } |z_2| = \sqrt{\left(\frac{r_2}{g}\right)^2 + (l_2\omega)^2} \qquad [3.33]$$

$n = n_r / n_s$ is the ratio of the number of spires between the stator windings and the equivalent rotor windings. By using this expression of the current, the torque is written as:

$$C_{em} = 3r_2 \frac{1-g}{g} \frac{n^2 E_s^2}{\left(\frac{r_2}{g}\right)^2 + (l_2\omega)^2} \frac{1}{\Omega} \qquad [3.34]$$

An approximate expression of this torque can easily be obtained by disregarding the stator resistance r_1 and the stator leakage inductance l_1, whose values, in practice, are low. The induced voltage drops are insignificant compared to the voltage V_1, which is imposed by the network to which the machine is connected. The single-phase diagram in Figure 3.33 is then reduced to that in Figure 3.34.

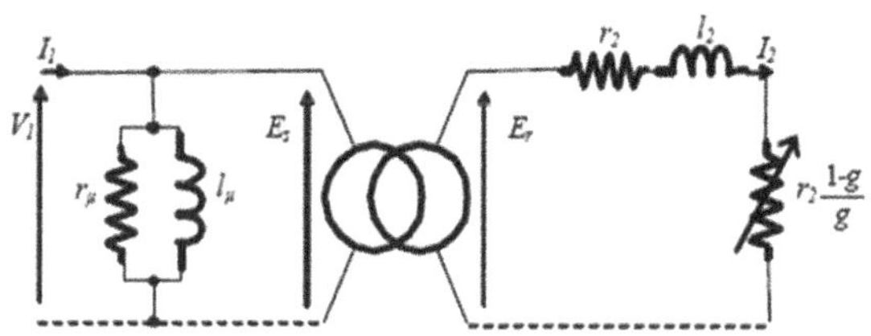

Figure 3.34. *Simplified equivalent single-phase diagram of a squirrel-cage induction machine*

In that case we obtain: $V_1 = E_s$. By using the slip expression, the mechanical speed can still be expressed in the following form:

$$\Omega = \Omega_s(1-g) \qquad [3.35]$$

To simplify how the torque is written, we define the angular frequency of the electric magnitudes at the stator:

$$\omega = p\Omega_s = p\frac{\Omega}{1-g} \qquad [3.36]$$

with p as the number of pole pairs.

The torque expression then depends on the voltage applied to the stator, on the angular frequency of the magnitudes at the stator, the slip and the resistance r_2:

$$C_{em}=3\left(nV_1\right)^2\frac{p}{\omega}\frac{\frac{r_2}{g}}{\left(\frac{r_2}{g}\right)^2+\left(l_2\omega\right)^2} \qquad [3.37]$$

Figure 3.35 shows the evolution of the electromagnetic torque according to the slip.

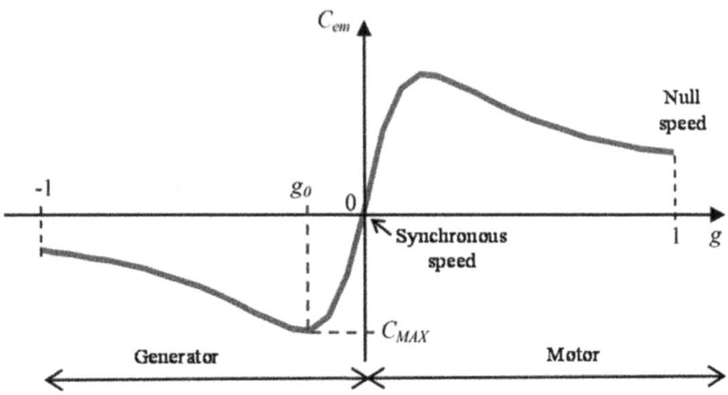

Figure 3.35. *Torque–slip characteristic*

With the classically chosen motor convention, the machine will operate as a generator, if the mechanical power and the electromagnetic torque are both negative. On the contrary, if those two magnitudes are positive, the machine will operate as a motor. From expressions [3.37] and [3.32], we can determine that the machine will operate as a motor when the slip ranges between 0 and 1, and as a generator when the slip is negative. This latter operating mode is implemented in wind turbines. Let us note that a slip that is higher than 1 corresponds to an uninteresting operating mode, because the machine then simultaneously receives electric and mechanical power that it cannot dissipate.

3.7.4. *Conversion system*

As for the previous wind power system, a speed multiplier is required to drive the induction machine at a higher speed than that of the turbine. The stator circuit of this machine is directly coupled to the electrical network, as shown in Figure 3.36. Therefore, the root mean square value of the voltage and the angular frequency are imposed by the network, the resistance r_2 is constant and the torque (relationship [3.37]) thus depends on the slip value and thus on the speed.

Often, the rotor is made up of short-circuited windings, which are designed to create magnetic poles. This solution is chosen because there is then the possibility of adapting the synchronous speed (equation [3.17]) according to the mechanical speed and thus to the available energy (equation [3.12]). For example, two pole pairs and thus a synchronous speed of 1,500 rpm can be chosen for high rotational speeds. For low rotational speeds, four poles can be used to obtain a lower synchronous speed of 750 rpm. Therefore, we can also find wind turbines based on induction machines, whose number of pole pairs is automatically commuted depending on the rotational speed.

When it starts, some reactive power is consumed, in order to install the magnetic field into the machine. This is why a capacitor bank is used to carry out an electric compensation of this power. For the connection of three capacitors of capacity C in a star configuration (Figure 3.36), the available reactive power is worth:

$$Q = 3\frac{1}{2}CV_1^2 \qquad [3.38]$$

V_1 is the root mean square value of the phase to ground voltage of the three-phase network.

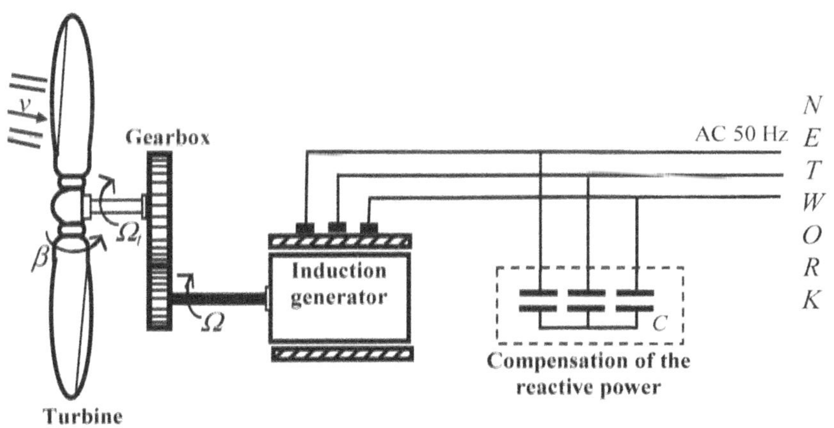

Figure 3.36. *"Fixed speed" squirrel-cage induction generator wind turbine*

3.7.5. *Operational characteristics*

For a wind turbine of 300 kW connected to an electrical network, the theoretical characteristic of the generated power according to the wind speed is represented with a continuous line in Figure 3.37. The relatively slow dynamics of the pitch control (several dozens of seconds) and the fast variations (turbulences) of the wind speed bring this type of wind turbine to approximately follow the adjustment characteristic, as shown by the points in Figure 3.37, which are measured on an effective wind turbine.

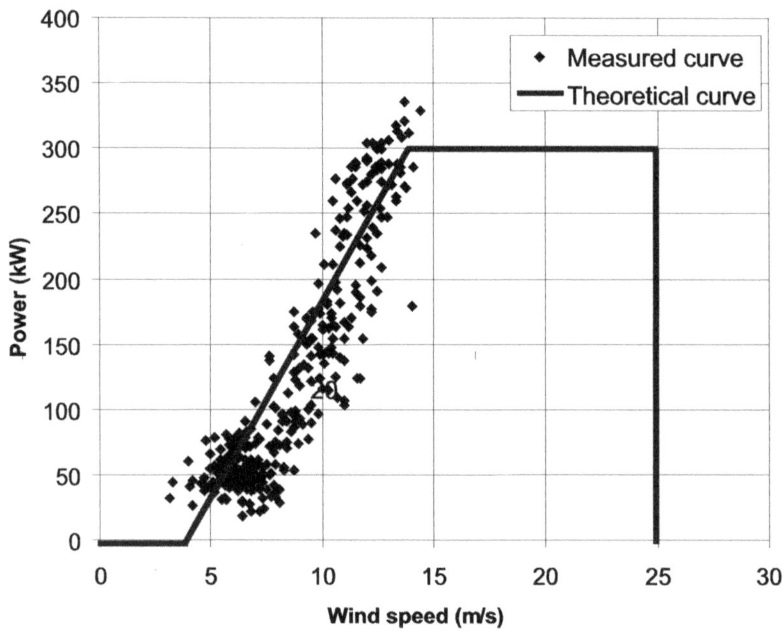

Figure 3.37. *Characteristic of a 300 kW wind turbine [ROB 06]*

The recording in Figure 3.38 illustrates the greatly fluctuating nature of the power generated by this type of wind turbine, and shows that this power can be subjected to variations of more than 100 kW in 3 seconds and that the nominal power can be exceeded by more than 10%. This type of wind turbine thus has limited adjustment dynamics, which is less adapted to overly turbulent winds. Moreover, let us note that frequent operation of the pitch system leads to fatigue, which can then increase the failure rate.

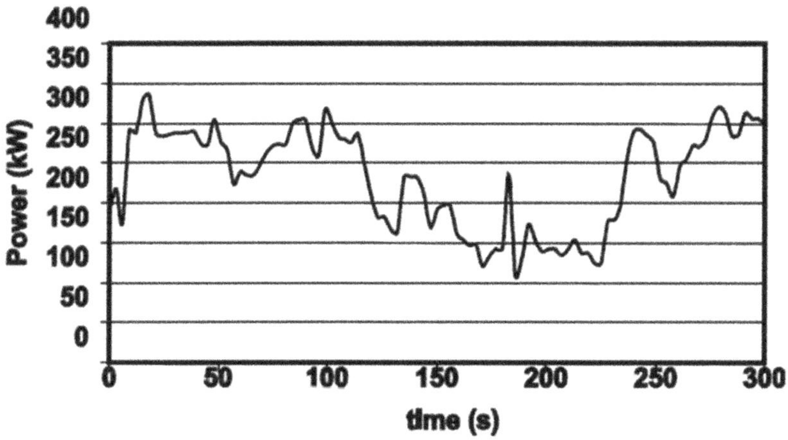

Figure 3.38. *Example of power generated by a fixed speed wind turbine of 300 kW*

3.8. Variable speed wind turbine

3.8.1. *Issues*

The typical power characteristic of a turbine according to the wind speed and the speed ratio (λ) is presented in Figure 3.39.

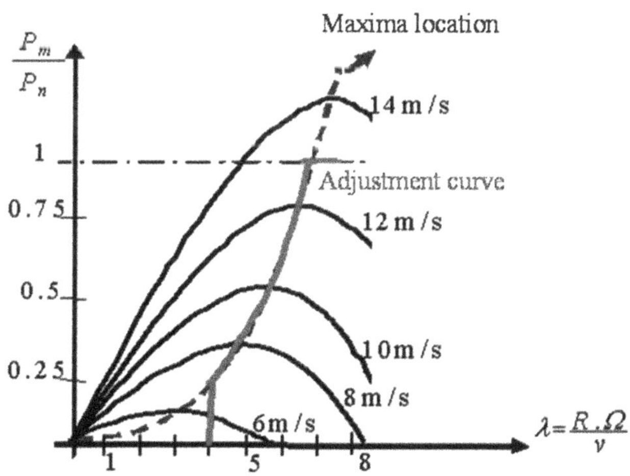

Figure 3.39. *Example of the speed adjustment characteristic. For a color version of this figure, see www.iste.co.uk/robyns/electricity.zip*

The place of the point representing the maximum converted power (represented by the curve in dotted lines) can be followed by adapting the turbine speed (thick line). Thus, in order to maximize the converted power, the turbine speed must be adapted in relation to the wind speed. This is why the market share of the high power variable speed wind turbines is in constant development and tends towards 100%. The main advantages of the variable speed wind turbines in comparison to the fixed speed generators are as follows:

– They increase the recovery range, notably for small wind speeds, where the maximum recovery efficiency can be obtained. Thus, indirectly, the availability and the generated power of the system are increased.

– The system of orientation of the blades is only used to deteriorate the turbine efficiency in the case of high-speed winds. Indeed, the possibility of controlling the generator speed via the electromagnetic torque helps to reduce the role of the orientation system of the blades, which will then mainly act to limit the turbine powers to high wind speeds. Consequently, for low wind speeds, the orientation angle of the blades becomes fixed.

– They reduce the mechanical constraints, due to the fact that, during wind variations, the turbine speed is adapted. The resulting "elasticity" reduces the impact of the wind blasts on the generated power for this operating range.

– They reduce the acoustic noise during low power operations, because the speed is then slow and some manufacturers have even proposed the possibility of a little production deterioration in order to reduce noise (in residential areas) and to find a better compromise.

– They enable better integration of the wind turbine into the electrical network.

3.8.2. *Classification of the structures according to machine technologies*

Speed is variable and therefore so is the electrical magnitude frequency of the (synchronous or induction) machine. The variable speed operation is made possible by the use of power electronic converters. The latter helps to adapt the variable frequency of the generator's electrical magnitudes with the fixed frequency of the network (50 Hz in Europe). Three categories of variable speed wind turbine technologies have been developed:

– with doubly-fed induction machines, in which the power electronic converter is connected to the wound rotor;

– with synchronous machines (with wound rotors or magnets), which are coupled to the turbine (direct drive) or via a multiplier;

– with squirrel-cage induction machines, where the stator is fed by a power electronic converter of the same type as that of synchronous machines. These conversion chains are quite rare.

The first technology is based on a wound rotor induction machine, whose rotor circuits are connected to the network via two back to back AC/DC power electronic converters, which are separated by a continuous bus (Figure 3.40). In modern wind turbines, these converters are equipped with IGBT transistors and are entirely reversible in power. Given the fact that only one part of the power crosses the rotor circuit (this characteristic will be detailed), electronic converters are sized for one part of the generator nominal power (25 to 30%). This is what makes this technology interesting and the most frequently used for high power wind turbines (>1 MW). Electronic converters will help to control the rotor currents of circuits and thus the torque and the rotational speed (equation [3.32]).

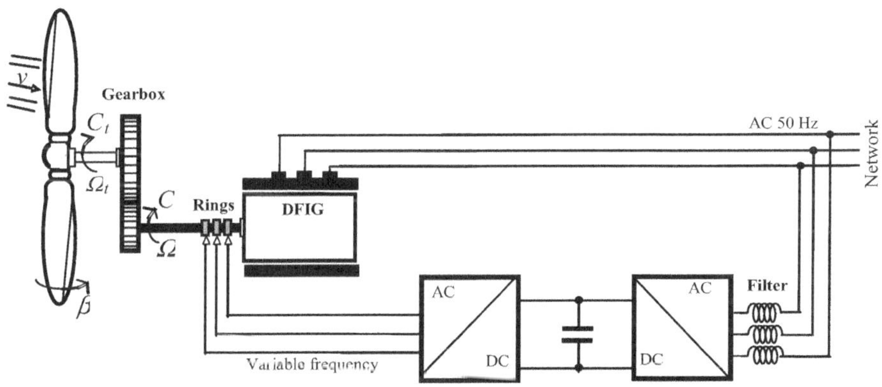

Figure 3.40. *Variable speed wind turbine based on a doubly-fed induction machine*

The second category of conversion chains is based on a synchronous generator (with a wound rotor or a permanent magnet synchronous machine), whose stator is connected to the network via two back-to-back DC/AC converters, which are entirely reversible and separated by a DC bus, which is sized to the generator's nominal power (Figure 3.41). The root mean square value of the stator voltage and the angular frequency might then be adjustable (by controlling the converters), in order to adjust the torque (equation [3.37]). Synchronous machines enable us to avoid a speed multiplier, insofar as it is possible to design machines with a large number of pole pairs (Figure 3.42).

In the third category, the generator is a squirrel-cage induction machine (Figure 3.41) and it exploits the same power electronic structure as synchronous machines. However, on the contrary to synchronous machines, it is not possible to obtain satisfactory performances with a large number of poles. Therefore, these machines require a speed multiplier.

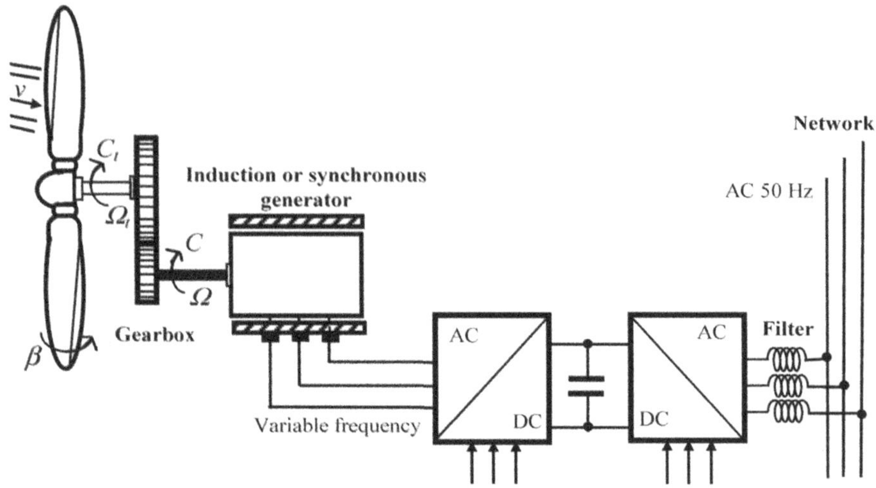

Figure 3.41. *Variable speed wind turbine based on a squirrel-cage induction machine or else on a synchronous machine*

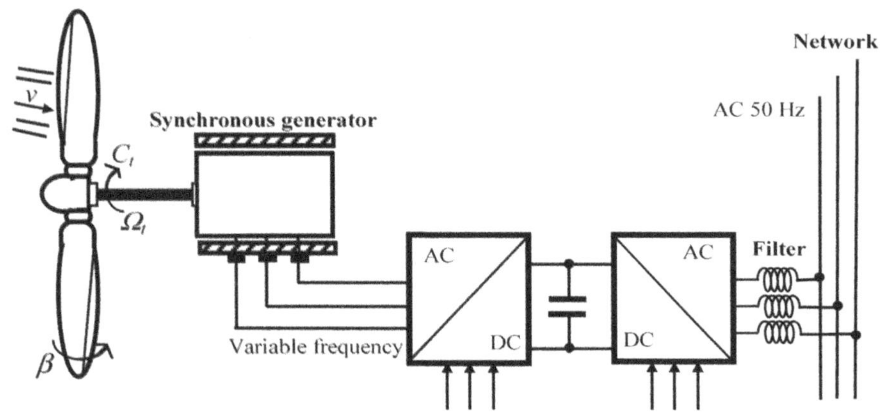

Figure 3.42. *Variable speed wind turbine with direct drive based on a synchronous machine with a large number of pole pairs*

3.8.3. *Principle of element sizing*

To illustrate the main sizing stages of the wind turbine conversion chain, we consider the structure of a variable speed wind turbine, which is based on a squirrel-cage induction machine with a multiplier. The wind turbine has a blade orientation system, limiting the power of the turbine to the value of the generator's nominal power (P_N). To determine the multiplication ratio to optimally operate the electrical machine, the turbine and generator speed power characteristics are represented on a single graph (Figure 3.43). In practice, the multiplier ratio (m) should be chosen so that the parallel of the generator characteristic is the closest to the maximum power curve. Therefore, the turbine speed for which the generator starts to produce electric power (Ω_{t_dem}) is determined, as well as the multiplication ratio:

$$m = \frac{\Omega_s}{\Omega_{t\_dem}} \qquad [3.39]$$

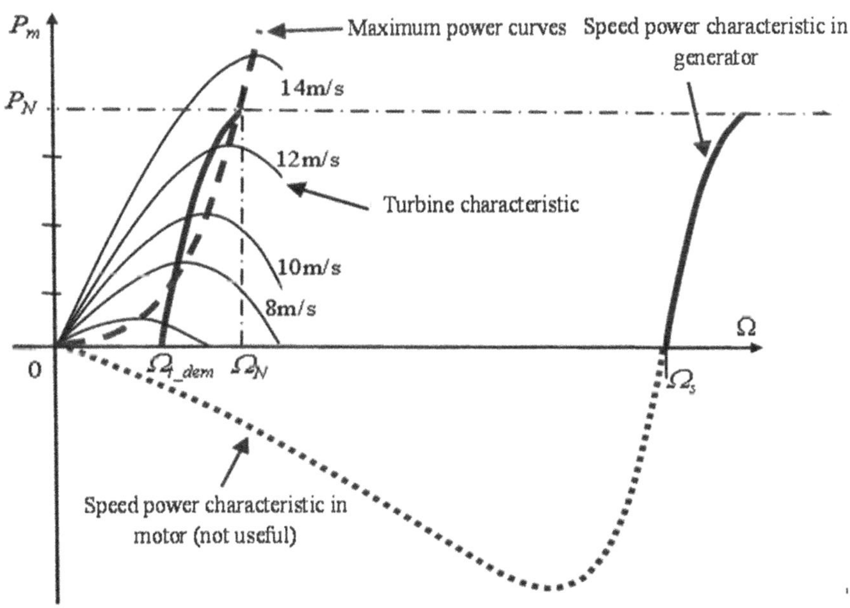

Figure 3.43. *Variable speed wind turbine with direct drive based on an asynchronous machine. For a color version of this figure, see www.iste.co.uk/robyns/electricity.zip*

Using the turbine characteristic $C_p(\lambda)$ (Figure 3.24), the maximal value of the power coefficient is identified ($C_{p\_\max}$), as well as the corresponding value of the tip speed ratio λ_{opt}.

By fixing a tip speed ratio, which is constant and equal to this value, the wind speed for which the turbine will start producing power (v_{dem}) is calculated by generalizing equation [3.11]:

$$\lambda = \lambda_{opt} = R\frac{\Omega_N}{v_N} = R\frac{\Omega_{t\_dem}}{v_{dem}} \Rightarrow v_{dem} = \Omega_{t\_dem}\frac{v_N}{\Omega_N} \qquad [3.40]$$

From equation [3.10], the mechanical power available on the axis of the generator for this wind speed is thus worth:

$$P_{m\_dem} = C_{p\_\max}\frac{1}{2}\rho\pi R^2 v_{dem}^3 \qquad [3.41]$$

$\eta_{multiplier}$ corresponds to the multiplier efficiency. The blade length is calculated so that the starting power is equal to the minimum power of the electrical machine drive ($P_{\min}$).

$$\boldsymbol{R} = \sqrt{\frac{\boldsymbol{P}_{m\_dem}}{\boldsymbol{C}_{p\_\max}\boldsymbol{v}_{dem}^3}\frac{2}{\rho\pi}} \qquad [3.42]$$

The nominal electrical power produced for the nominal wind speed must be higher than the generator's nominal power. This constraint must be verified:

$$C_{p\_\max}\frac{1}{2}\rho\pi R^2 v_N^3 > P_N \qquad [3.43]$$

3.8.4. *Adjustment of active and reactive powers*

The theoretical principle of controlling powers exchanged between the inverter and the network was presented in Chapter 2. As an example, we will consider a variable speed wind turbine relying on a permanent magnet synchronous machine, whose turbine is emulated with the help of a 3 kW test bench [ROB 06]. Three-phase currents and voltages are measured in a three-phase frame (a, b, c) and are transformed in a synchronous orthogonal frame (d,q) with a network voltage (thus revolving at 50 rev/sec) [ROB 07]. These currents are controlled ($\widehat{i}_{td}, \widehat{i}_{tq}$). This helps to set them equal to reference currents ($\widehat{i}_{td\_ref}, \widehat{i}_{tq\_ref}$) resulting from active and reactive power references (P_{t_ref}, Q_{t_ref}) (Figure 3.44). The DC bus voltage (v_{c_ref}) is regulated at a constant value (v_{c_ref}) by an equalizer calculating the active power reference (P_{t_ref}) that needs to be exchanged with the network.

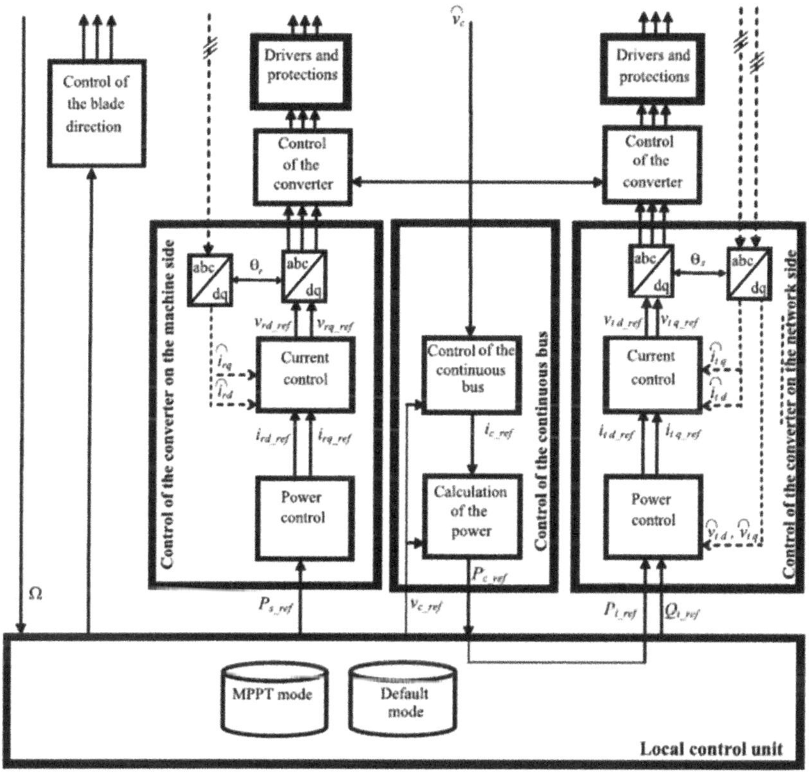

Figure 3.44. *Organization of the control system*

Figures 3.45(a), (b) and (c) show the highly fluctuating wind speed to which the emulated wind turbine is subjected, the active power sent back to the network by extracting the maximum energy from the wind (a negative power represents a generated power), as well as the reactive power, which is imposed as null at the wind turbine connection point.

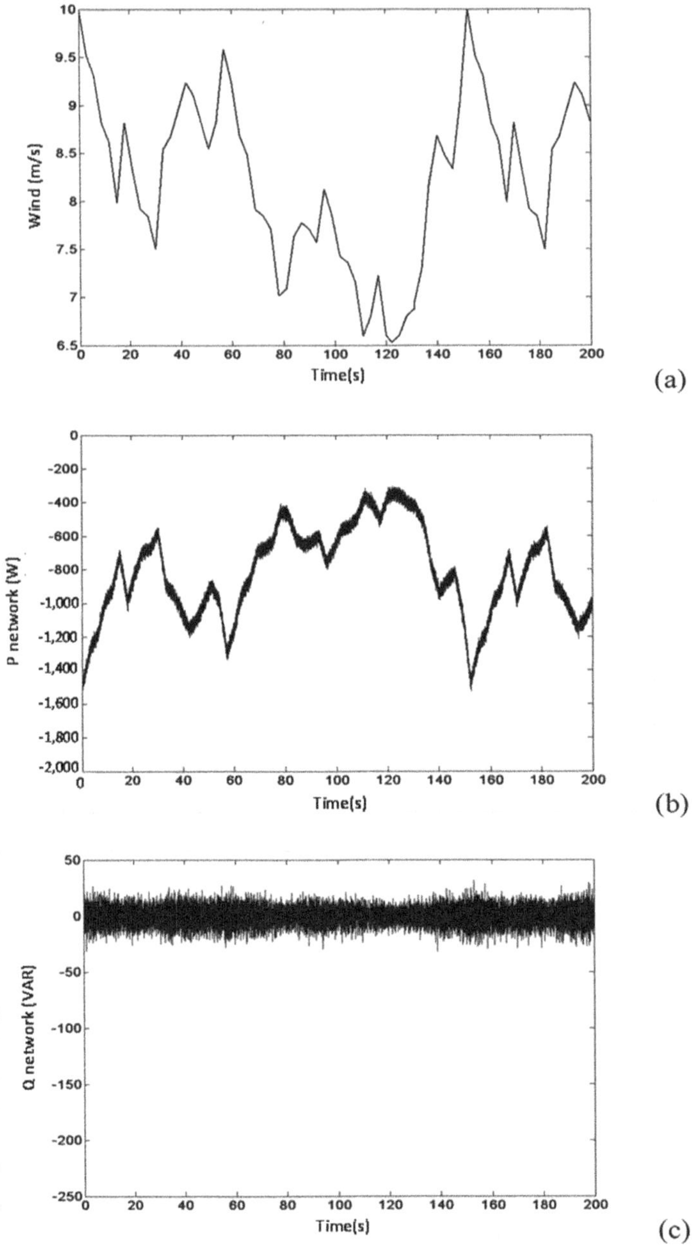

Figure 3.45. *Variable speed wind turbine, which is emulated on a 3 kW test bench: (a) wind speed; (b) active power in MPPT mode; (c) reactive power*

The active electrical power generated can be smoothed by controlling the generator torque, in order to impose the turbine rotational speed, to follow a constant power set point and to no longer extract the maximum power. This technique amounts to degrading the power coefficient C_p and obviously depends on the available wind. Figure 3.46(a) shows the active power sent to the network when its reference value is imposed at 1 kW for the same wind speed profile (Figure 3.45(a)). This reference is relatively well followed, as long as the wind speed is high enough (which is not the case, for example, between 100 s and 140 s). Figure 3.46(b) shows the reactive power, which is always kept at null.

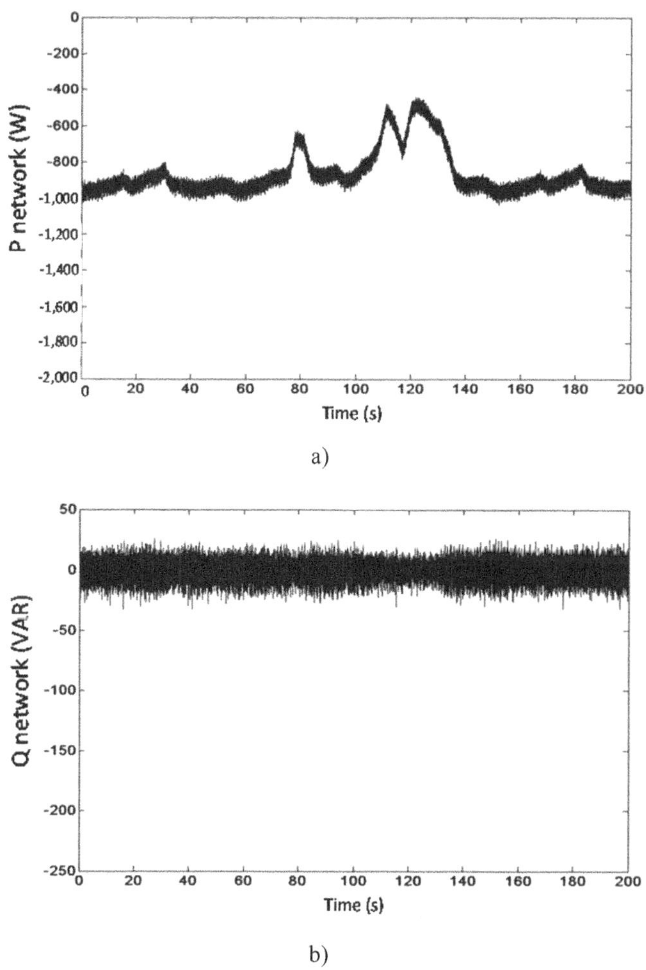

Figure 3.46. *Emulated variable speed wind turbine on a 3 kW test bench (wind profile of Figure 3.44): (a) smooth active power; (b) reactive power regulated at 0*

The test illustrated in Figure 3.47(a) is similar to that of Figure 3.46(a) concerning active power; however, a step of 200 VAR is imposed on the reactive power reference (Figure 3.47(b)). The adjustment of active and reactive powers seems to be uncoupled. However, we have to note that the reactive power which can actually be generated or absorbed by the wind turbine is limited by the value of the generated active power and by the voltage level of the continuous bus [BAR 96].

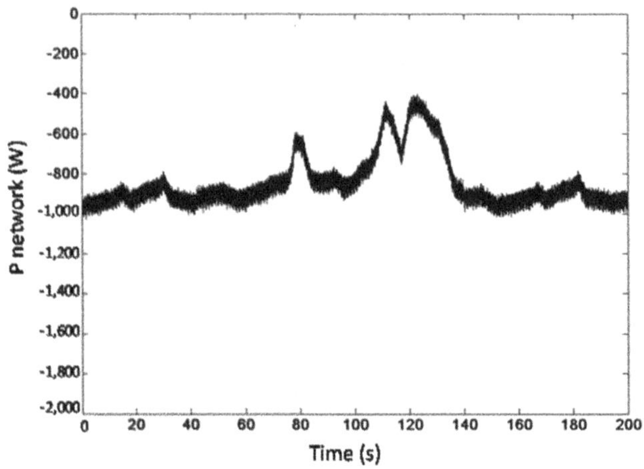

a)

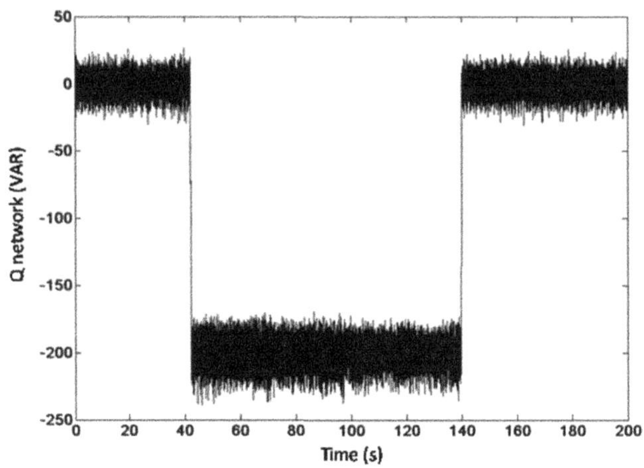

b)

Figure 3.47. *Emulated variable speed wind turbine on a 3 kW test bench (wind profile of Figure 3.45): (a) smooth active power; (b) reactive power regulated at 0, then with a step of 200 VAR*

3.8.5. *Aerogenerators based on a doubly-fed induction machine*

3.8.5.1. *Doubly-fed wound rotor induction machines*

The torque of an induction machine depends on the rotor circuit resistance (equation [3.37]). For a doubly-fed machine, the rotor circuit is accessible thanks to a set of brushes and rings. Thus, for machines with wound rotor circuits, we can vary the mechanical speed of this generator by varying the torque (and thus the speed) by modifying the rotor circuit resistance (r_2 in equation [3.34]). This will lead us to modify the rotor current.

A low-cost solution, but with performances limited by energy dissipation, uses a dissipation resistance via a controlled rectifier (Figure 3.48). This controlled rectifier can be synthetized by a diode bridge associated with a transistor or by a thyristor bridge.

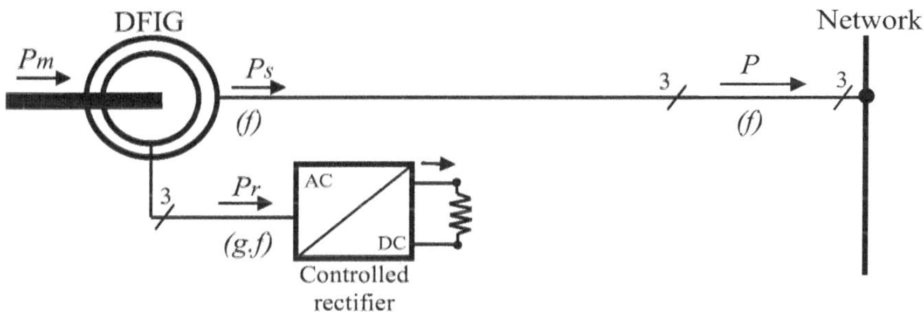

Figure 3.48. *Power flow in a structure with rotor energy dissipation*

Depending on the control of this rectifier, part of the mechanical power, which is proportional to the slip, is dissipated in the external resistance; this deteriorates performances because part of the energy is lost. This simplified structure only allows a very limited adjustment of the speed (in hypersynchronism), which is obtained by an "electrical braking", with the disadvantage of dissipating part of the mechanical power (P_m), but all the while increasing the efficiency of the power generated at the stator, which therefore improves the global efficiency. The power sent on the network (P) thus corresponds to the active power supplied by the stator (P_s), which is equal to the reduced mechanical power of the active power extracted from rotor circuits (P_r).

$$P = |P_s| = P_m - |P_r| \qquad [3.44]$$

The slip expression [3.23] helps to determine the magnitude angular frequency at the rotor:

$$\Omega_r = g\Omega_s = \Omega_s - \Omega \tag{3.45}$$

The power extracted from the rotor circuit is proportional to the slip:

$$P_r = gP_s \tag{3.46}$$

Rather than dissipating this power, it is much more interesting to send it back to the network with the help of two power electronic converters, which are connected by a DC bus (Scherbius system, Figure 3.49). Consequently, the power passing through the rotor circuit is made variable and is bidirectional, if power electronic converters with transistors are used.

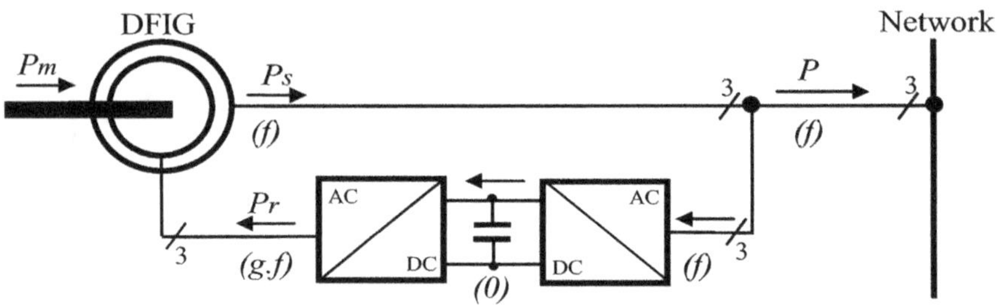

Figure 3.49. *Power flow in a Scherbius system in hyposynchronous mode*

The power bidirectionality of the electronic converters enables the two operating modes: hyposynchronous and hypersynchronous (Figure 3.50).

If the mechanical speed is higher than the synchronous speed ($g < 0$, $\Omega > \Omega_s$), a hypersynchronous operation is obtained for which the rotor power $P_r = gP_s$ is negative (rotor circuit considered in receiver convention), because the stator power is always positive in generator mode (stator current in generator convention). A power is thus extracted from the rotor circuit and is sent to the network through power converters. By suggesting that the losses in the stator and rotor circuits and in the DC bus can be neglected, the total power generated becomes higher than that of the stator (according to Figure 3.49):

$$P = P_s - P_r = P_s - gP_s = P_m \tag{3.47}$$

If ($g > 0$, $\Omega < \Omega_S$), a hyposynchronous operation is obtained and the power then flows from the network to the rotor circuit, because $P_r = gP_s > 0$ (Figure 3.49).

The range of variation of the speed around the synchronous speed is maintained at quite a low value (+/- 25 to 30% around the synchronism) to obtain a good compromise between the improvement of the energy recovery of the variable speed turbine and the cost of the power electronic converter, which is directly related to the maximal slip. The maximum power circulating in the rotor circuit ($P_r = gP_s$) is a part of the total power. Therefore, the electronic conversion chain is indeed sized for a lesser power and thus has a reduced cost.

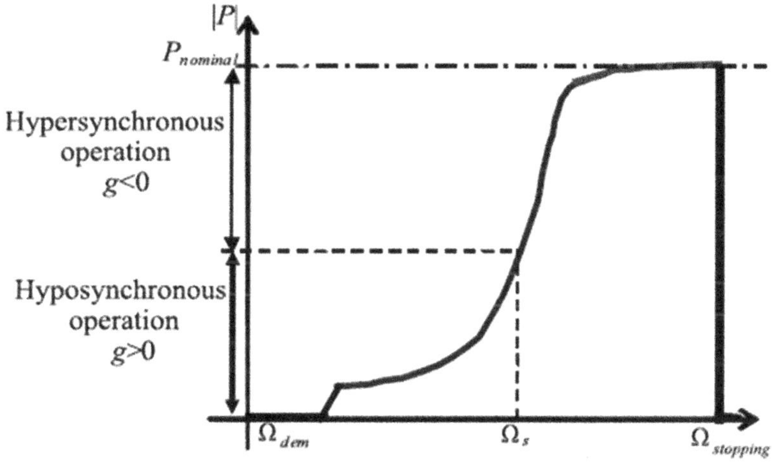

Figure 3.50. *Characteristic of the generated power according to the MADA speed*

The active and reactive power can be controlled independently, thanks to the converter connected on the rotor electric circuit [GHE 11].

However, the disadvantage of this type of generator is that it requires a system of rings and brushes, as well as a multiplier. The maintenance of these pieces of equipment has to be taken into account in the maintenance program, especially for offshore projects located in a saline environment.

3.8.5.2. *Operating characteristics of a wind turbine based on a doubly-fed induction machine*

Figure 3.51 shows the characteristic operating zones, which are measured on a 1.5 MW wind turbine:

– Zone 1 corresponds to the starting of the rotational speed as soon as a significant power is available.

– Zone 2 is where a minimal power extraction is carried out by adapting the generator speed. The mechanical speed is quite variable and corresponds to a large range of variation of the electric power produced. The pitch angle of the blade is kept constant and only the control of the electromagnetic torque of the generator is implemented in this zone.

– Zone 3 corresponds to an almost constant generator mechanical speed. In this zone, the generated power is proportional to the applied torque (of wind power origin). The average speed of the turbine can be adjusted by acting on the blade orientation. The induction machine torque can very quickly become variable, in order to smooth, for example, the power variations or to refine the MPPT in dynamics.

– For Zone 4, the power is limited to its maximal value (1,550 kW) thanks to the orientation system of the blades (pitch).

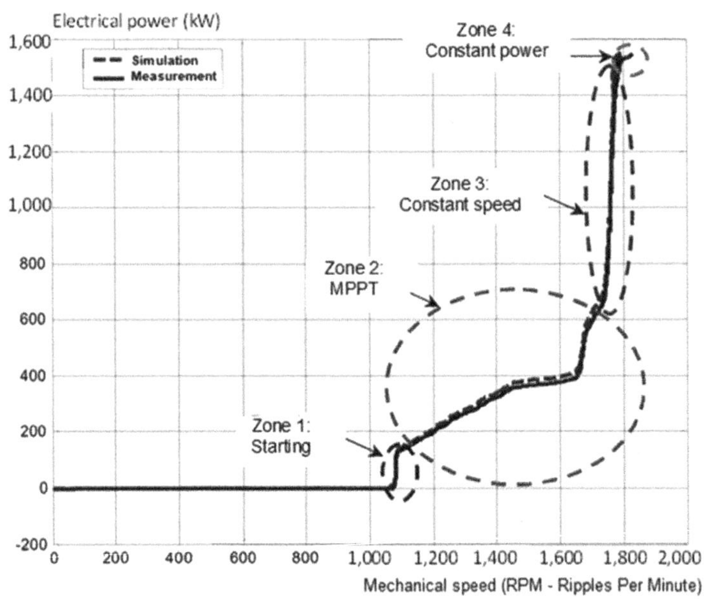

Figure 3.51. *Characteristic operating zones of a variable speed wind turbine, based on an induction machine [EL 04]*

The control of the generated power can thus be carried out by acting on the blade orientation (zone 3 and 4), but also by controlling the torque of the induction generator with the help of a power converter, which is connected to the rotor of the latter (zone 1 to 3). The control of the generated power is then much more precise, as has been illustrated on the power-speed characteristic of the wind measured in Figure 3.52 [EL 04], in comparison to the measurements carried out on a fixed speed wind turbine (Figure 3.37). Figure 3.53 shows a recording of the wind speed, the generated electrical power, the generator speed and the orientation angle of the blades for a 10 h duration, whereas the wind turbine is subjected to a wind speed varying between 2 and 16 m/s. We can note that the maximal power is not exceeded.

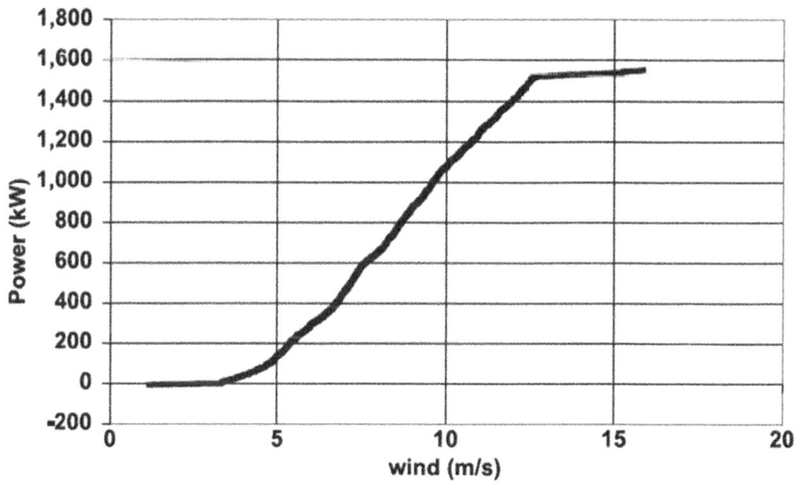

Figure 3.52. *Total power generated measured on a variable speed wind turbine of 1.5 MW according to the wind speed*

The obtained slip can be calculated from Figures 3.51 or 3.53(c) and here varies between +20% and -31%, knowing that the synchronous speed is 1,500 rpm.

Figure 3.53d shows the characteristic of the orientation angle of the blades according to the power. It confirms that the blade orientation system is only implemented from a power higher than 1 MW. For powers lower than this value, the wind turbine is controlled with the help of the generator's electromagnetic torque.

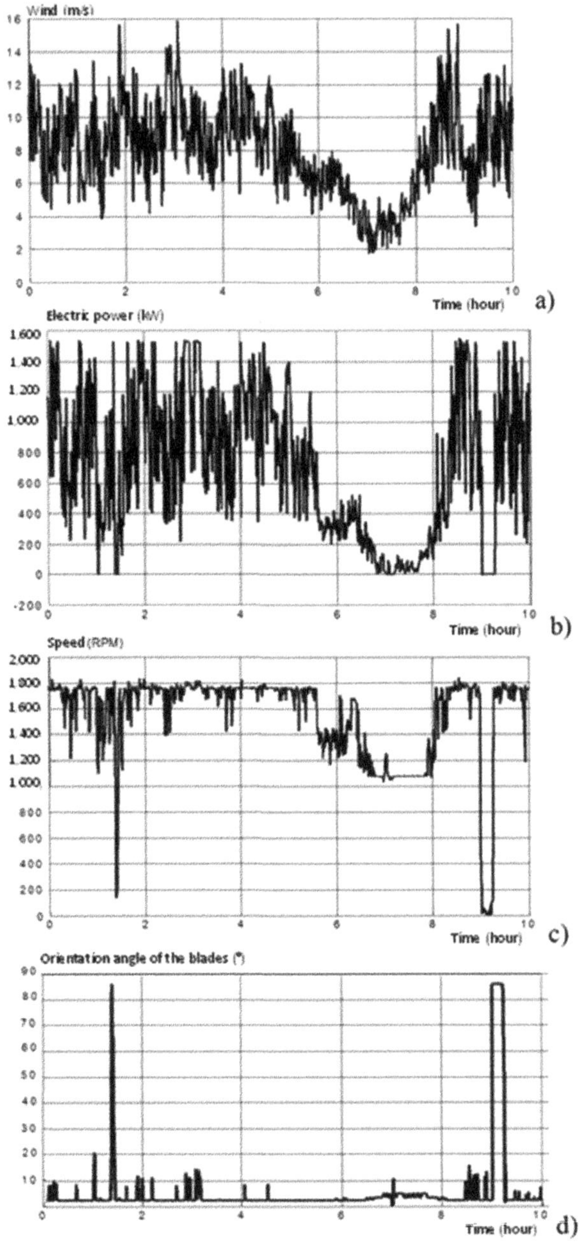

Figure 3.53. *Recording of the wind speed, the electric power, the generator speed and the blade orientation angle for a 1.5 MW wind turbine*

3.8.6. *Aerogenerators based on a synchronous machine*

3.8.6.1. *High power wind turbines*

The disadvantage of wind turbines based on wound rotor induction generators is that they require a system of rings and brushes and a multiplier, which leads to significant maintenance costs, especially for off-shore projects which are located in a saline environment.

In order to limit these disadvantages, manufacturers have developed wind turbines based on synchronous machines with a large number of pole pairs that are directly coupled to the turbine, thus avoiding the multiplier. There are two types of structure for electric machines: radial and axial flow machines (Figure 3.54).

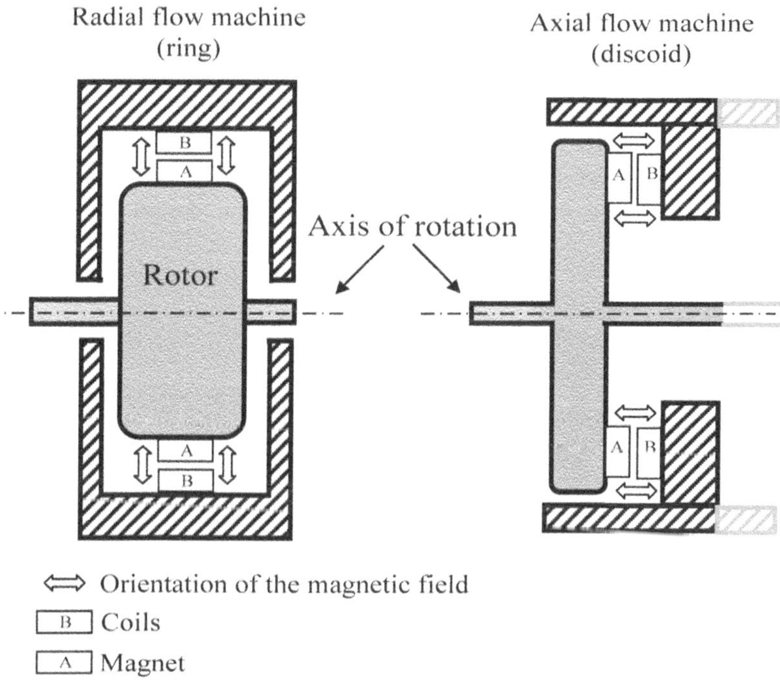

Figure 3.54. *Orientation of the magnetic fields for the two structures*

Direct drive technology was developed for the first time in 1992 by the manufacturer ENERCON by replacing the permanent magnets. They designed a wound inductor with a very large number of poles (ring structure, about 50 poles) on the rotor (Figure 3.55).

Figure 3.55. *Arrangement of the coils on the rotor (photo: Enercon)*

The supply to these coils is carried out by the principle of rotating rectifiers (excitation without a brush). The significant diameter of the machine at radial flow and the absence of a multiplier necessitate a specific design for the nacelle (Figure 3.56).

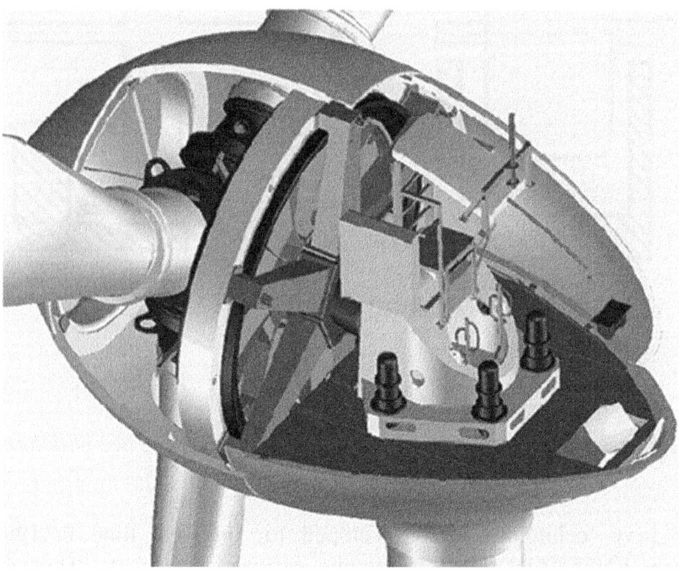

Figure 3.56. *Assembly of the turbine and machine in the nacelle (photo: Enercon)*

The wind farm at Estinnes (Belgium) comprises 11 wind turbines of the E-126 type, each with a power of 6 MW. The turbine has a diameter of 126 m and the nacelle is at a height of 132 m. Therefore the total wind turbine height is 198 m. The new E-126 wind turbines now have a nominal power of 7.5 MW.

Windings are bulky, quite complicated to manufacture and energy consuming, because they are always fed in order to permanently create a magnetic field. This magnetic field can also be created by permanent magnets. The use of Neodymium Iron Boron (NdFeB) magnets provides a powerful magnetic flow, all the while remaining at a small thickness. They increase the power density and thus reduce the machine bulk; it is possible to reduce the mass by 25%. For example, ABB is developing its own Windformer system with a permanent magnet synchronous generator (radial flow). The stator winding is carried out with cables (Powerformer concept) and helps to deliver high voltage (higher than 20 kV) power directly through a diode rectifier for DC energy transportation.

Figure 3.57 shows an exploded view of the nacelle of the future aerogenerator (on the left) and a generator prototype with magnets and a flow concentration with 16 pole pairs.

Figure 3.57. *Synchronous generator with permanent magnets and radial flow [DAH 00]*

For the axial structure (Figure 3.54), the magnetic flow is parallel to the disc's axis of rotation. An example can be found in Figure 3.58, with a rotor with magnets on both sides. This enables the use of two stators and thus an increase of the generated power.

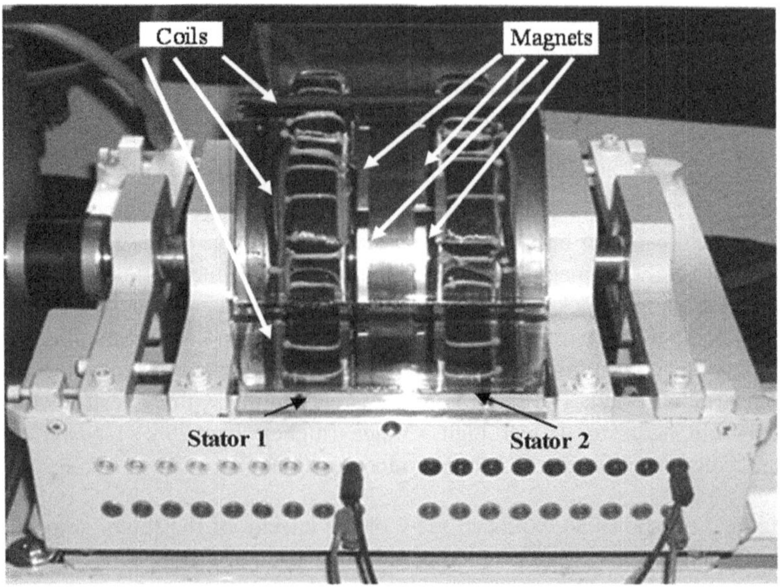

Figure 3.58. *Synchronous generator with permanent magnets and axial flows [VIZ 07]*

Axial structures help to reduce the machine's bulk and mass. Direct Drive Systems have developed the 800 kW DDIS60 wind turbine with a synchronous generator (axial flow), whose rotor is discoid, a large number of permanent magnets and which weighs 15 tons (Figure 3.59). It is also surprising to note that the stator is multi-phased with 9 phases. These phases are organized into three groups of coils, which are connected to the network by three power electronic conversion chains. The current crossing each chain is decreased and therefore electronic components of lower unit power can be used. Moreover, in the case of a coil failing to neutralize a group, the DDIS60 keeps on producing electricity to two-thirds of its capacity. This is a significant technological asset, notably for off-shore wind turbines, which are difficult to access.

The most significant disadvantage of the synchronous machines relies on the fact that they need, for connection to the network, power electronic converters, which are sized for the generator's nominal power. This disadvantage is however an advantage from the point of view of wind turbine control and behavior, when confronted with network disruptions (notably voltage dips). Indeed, interfacing with the network can be entirely controlled via a power electronic converter connected to this network, whereas the converter connected to the generator helps to control the power generated by this generator, by limiting pitch control to a safety function in the case of high winds. The adjustment curve of this type of wind turbine is generally close to the one presented in Figure 3.51.

Figure 3.59. *Assembly of the second disc on the stator (photo: Jeumont Electric)*

3.8.6.2. *Small wind turbines*

Low power wind turbines (from a few hundred Watts to a few kW) have a simplified energy conversion structure in comparison to high power wind turbines. The AC/DC conversion is carried out by a diode rectifier; which is more economical (Figure 3.60) and reliable. However, power transfer is becoming unidirectional and it is no longer possible to directly control the machine currents. We can modify the voltage of the DC bus with the help of the (DC/AC) converter, which would be connected to the network, in order to optimize power recovery. Depending on the value of this voltage, the rectified current can be adjusted, thus allowing power transfer from the machine to the network at an optimized rotational speed. We have to note that thanks to the relatively high inductances of the synchronous machine armature circuit, the direct connection to the DC bus is made possible without any current peaks.

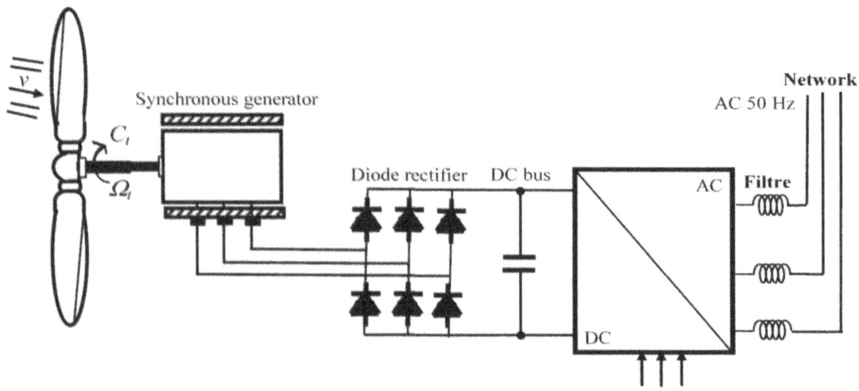

Figure 3.60. *Conversion carried out by a conversion chain with a synchronous magnet machine and a diode rectifier, which is connected directly to a DC bus*

In autonomous situations or to obtain an emergency operation mode, the DC bus can be made up of electrochemical accumulators. Moreover, a transformer can be required to adapt the output voltage of the machine to the DC bus voltage, notably if the battery voltage is low (48 V in Figure 3.61). As long as the phase-to-phase voltage at the transformer output is not higher than the battery voltage, the diode rectifier does not conduct and the batteries cannot be charged; this fixes the minimal wind speed, which can however be relatively low. A chopper can also be interfaced between the output of the diode bridge and the DC bus, in order to better optimize energy recovery.

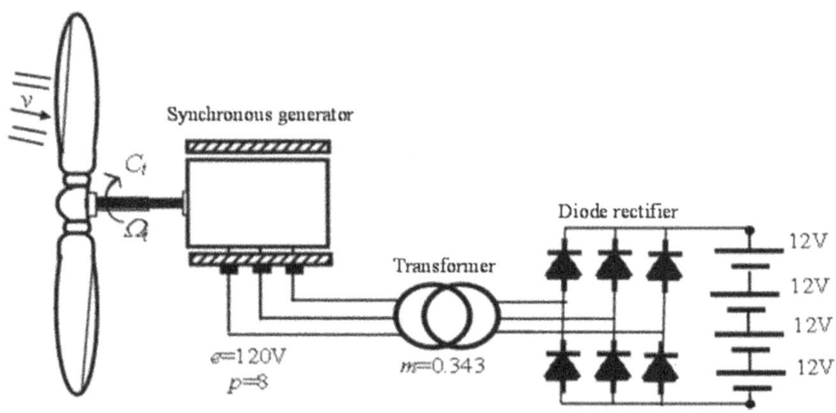

Figure 3.61. *Conversion carried out by a diode rectifier, which is connected to batteries*

As an example, the Aerocraft AC752 wind turbine for 750 W (Figure 3.62) is equipped with a NdFeB magnet generator with 16 poles (maximum speed 600 rpm). It also has a turbine with a 2.4 m diameter, a weight of 43 kg and a tower height of 14 m. It is a stall (dynamic stall) turbine with a power limit obtained by the tail-vane making the wind turbine lose its balance (horizontal deflection) for a wind speed higher than 9 m/s. Figure 3.63 presents the power-speed characteristics of the Aerocraft wind turbines for 750 W, 500 W, 240 W and 120 W.

Figure 3.62. *A 750 W Aerocraft wind turbine*

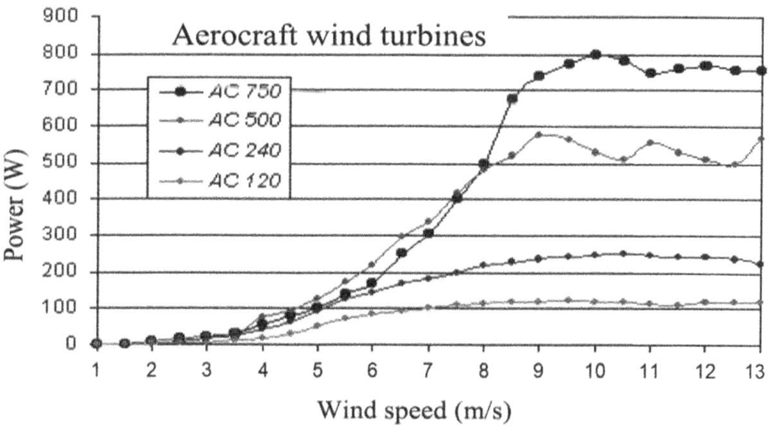

Figure 3.63. *Power-speed characteristics of the Aerocraft wind turbines for 750, 500, 240 and 120 W [GER 01]*

3.9. Offshore wind turbines

3.9.1. *Advantages of offshore wind*

The installation of wind turbines at sea, as illustrated in Figure 3.64, has several advantages: more consistent wind speed; fewer obstacles to disturb the wind; less uncertainty in the wind forecast (on average upwards of 14% whereas onshore it is around 10% [DER 18]); a higher load factor that can be more than double that obtained by onshore wind power; and the remote location of these wind turbines – 10 km or more from the coast and therefore any dwelling.

Figure 3.64. *Offshore wind farm by the Danish manufacturer Vestas [WIN 19]. For a color version of this figure, see www.iste.co.uk/robyns/electricity.zip*

However, because installation on the seabed or in a floating version is more complex, it is more expensive. The same is true for maintenance due to the fact that these wind turbines are more difficult to access. Equally important is the mandate that ensures they do not disturb marine life.

In 2019, the installed offshore wind capacity for Europe was 22 GW from 5,047 wind turbines connected in 12 countries [WIN 20]. The average unit power of installed turbines is 7.8 MW. The UK alone has a capacity of 9.9 GW. The other countries most advanced in the development of this type of wind turbine are Germany with 1.1 GW, Denmark with 374 MW and Belgium with 370 MW. In France, offshore projects are in development; Figure 3.65 shows 6 projects on the Atlantic coast and in the North Sea, the first of which should be operational by 2022.

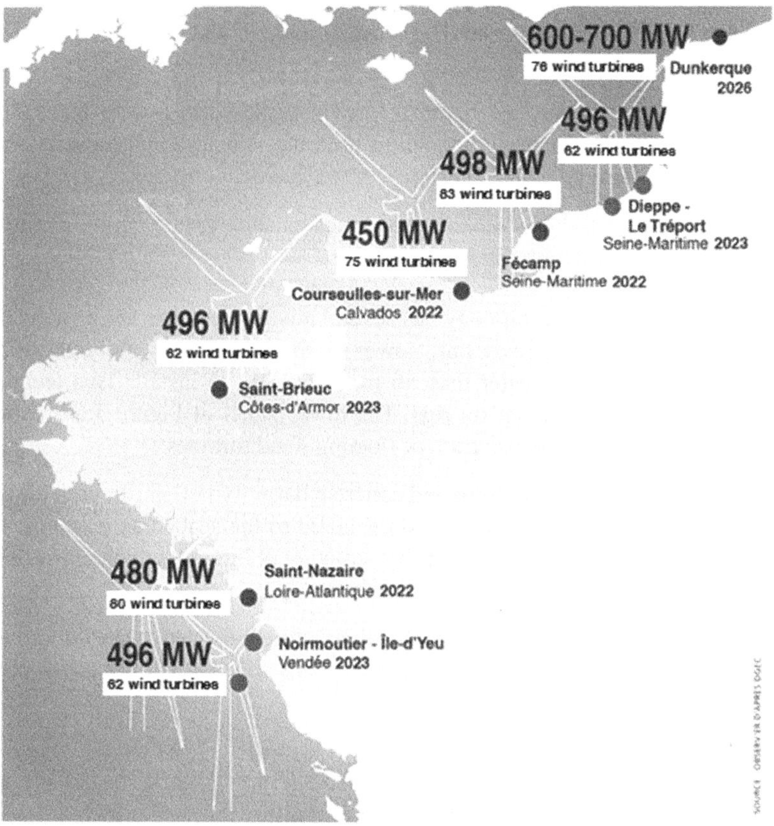

Figure 3.65. *Offshore wind farm projects in France on the Atlantic coast and in the North Sea, with information on the estimated commissioning date [RSS]. For a color version of this figure, see www.iste.co.uk/robyns/electricity.zip*

Most of these wind turbines are placed on the seabed. European capacity projections for offshore wind power by 2030 are 70 GW, of which 40 GW will be floating wind turbines. The global development potential of this latter type of wind power is significant, particularly in Asia and the United States where it could reach 250 GW in 30 years time [LAN 18].

3.9.2. *Types of offshore wind turbines*

The main parameters that determine the type of foundation of an offshore wind turbine are the nature of the soil, the depth of the water and the size of the turbine [DE 18]. The first offshore wind turbines were fixed into the seabed. The most common fixed foundations are as follows (top panel of Figure 3.66):

– gravity bases depend little on the nature of the soil and do not require drilling or hammering;

– monopiles are the most economical, provided the water is less than 30 m in depth and the soil is both strong enough to hold the structure and loose enough to allow for installation at controlled costs;

– jackets or metal trellises replace monopiles as soon as the depth of the water becomes greater and can withstand greater loads.

Floating wind turbine technology has several advantages over fixed foundations: less dependence on soil conditions, lower foundation costs, the only realistic solution for water depths greater than 50 m, and easier to install. This technology benefits from the experience of oil rigs. The lower panel of Figure 3.66 represents some solutions for the implementation of floating wind turbines:

– the semi-submersible platform and concrete barge-type float. In both cases, the floating platform is semi-submerged and anchored to the seabed by a catenary wire or semi-tensioned anchor. The draft of this structure is limited, which allows a more flexible and simpler installation;

– the spar, or pencil, buoy is a cylindrical structure ballasted at the bottom with a catenary anchor to provide stability. In general, this solution is only considered for water depths greater than 100 m.

Figure 3.66. *Different types of offshore wind turbine foundations: fixed or floating [RSS]. For a color version of this figure, see www.iste.co.uk/robyns/electricity.zip*

3.10. Wind farms

3.10.1. *Architecture*

At sites with a significant wind resource, several machines are generally connected by a busbar connection, which is dedicated to a connection station. This enables us to direct the generated power towards the electrical distribution network. This type of connection is thus carried out in alternating mode, as is illustrated in Figure 3.67.

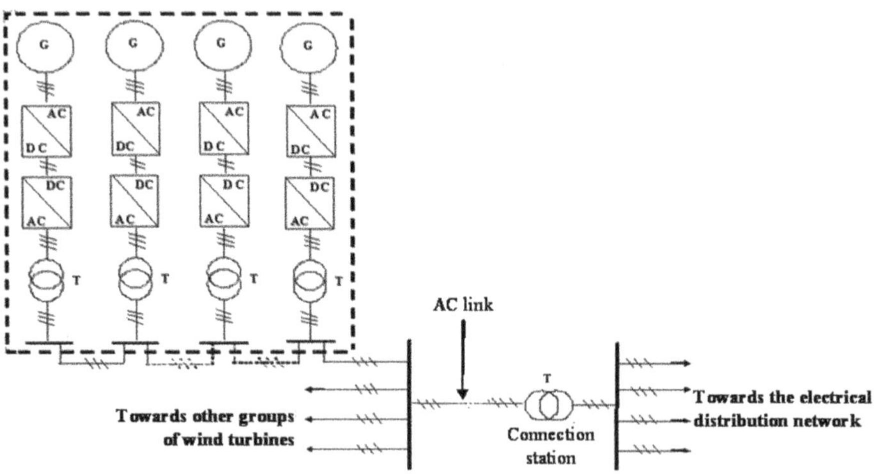

Figure 3.67. *Example of the structure of a wind farm at variable speed with four feeders*

If the wind farm is at sea, wind farm units are generally connected to an offshore platform near the wind turbines as shown in Figure 3.68.

For long distance connections from the offshore farm to the onshore power grid, around 95 km or more [MON 15], it is advantageous to replace the AC link with a DC link. In fact, there is no continuous reactive power which limits the power transmitted through the cable, causing voltage drops, due to the presence of harmonic impedances. For the same power transmitted, the voltage of the cables can be reduced, thus reducing the thickness of the materials insulating the cables. Figure 3.69 illustrates the principle behind this type of connection, already implemented in the Baltic Sea, for example.

Figure 3.68. *Connection platform for an offshore wind farm [WIN 19] (source: Van Oord). For a color version of this figure, see www.iste.co.uk/robyns/electricity.zip*

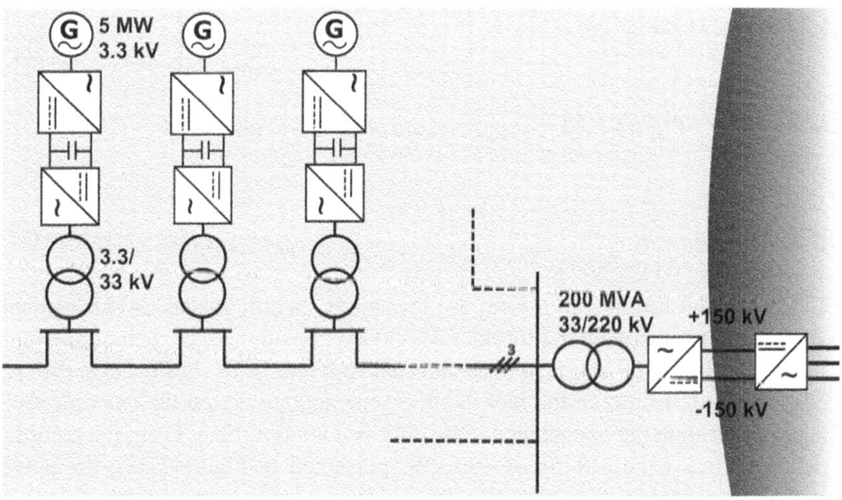

Figure 3.69. *Connection principle for an offshore wind farm via a continuous high voltage link [MON 15]. For a color version of this figure, see www.iste.co.uk/robyns/electricity.zip*

For offshore variable speed wind turbines, where transmission by underwater cable is required, the presence of a DC bus also enables us to consider a series connection, thus creating a connection under high voltage DC (Figure 3.70). This voltage must be sized to enable the total AC power [BOU 09, MON 15]. Let us note that it is a possible structure, but that it is not yet used.

This structure was first imagined for offshore wind farms [VER 05], which are interfaced with the network via only one DC/AC converter. We could imagine, for example, a better contribution of wind farms to voltage adjustment at the connection point. Other possibilities for wind farm structures are presented in [MUL 04].

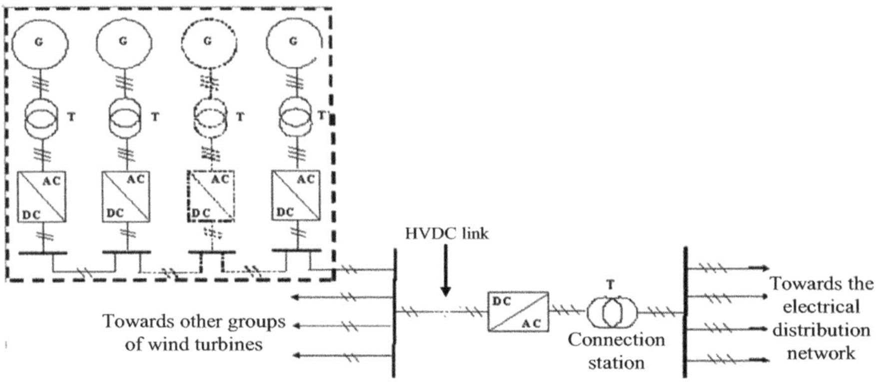

Figure 3.70. *Example of the structure of a variable speed wind turbine farm with DC voltage links*

3.10.2. *Abundance*

Because wind turbines in a farm are located at specific distances (for example, the distance between two wind turbines of 300 kW, positioned one behind the other, must be of at least 168 m), they are subjected to winds with significantly different speeds. This bulk results in the fact that the generated power on the network seems to be smoother than the one generated by only one wind turbine. From the recording carried out for a wind turbine of 300 kW (presented in Figure 3.38), the powers generated by groups of 3 and 10 wind turbines have been reformed and are respectively presented in Figures 3.71 and 3.72. The comparison with Figures 3.38, 3.71 and 3.72 shows a significant smoothing (bulk effect) of the generated power (at least in relative value) when the number of the wind turbines in the farm increases.

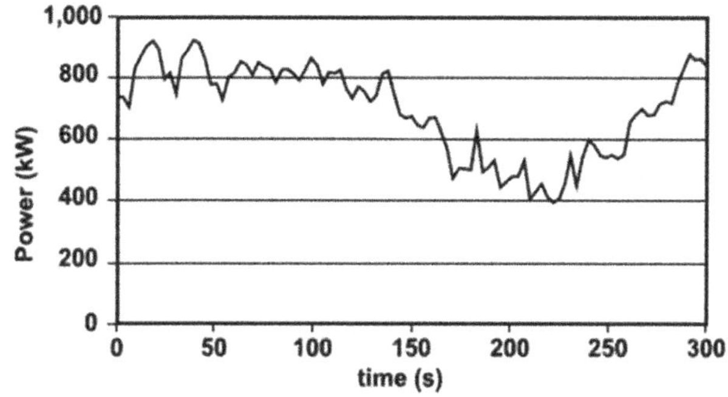

Figure 3.71. *Total power generated by a farm of three 300 kW wind generators*

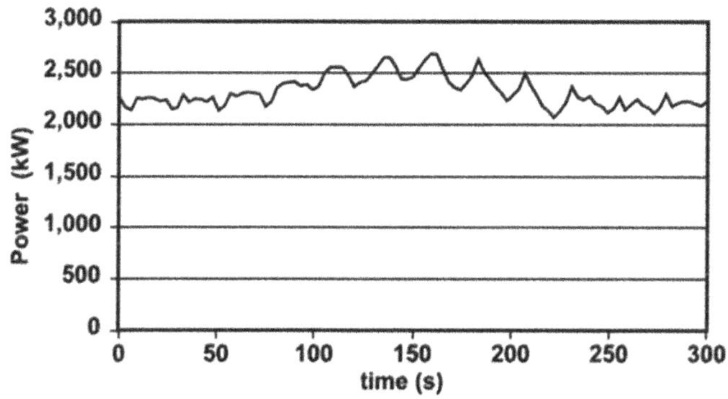

Figure 3.72. *Total power generated by a farm of ten 300 kW wind generators*

3.11. Exercises

3.11.1. *Fixed speed wind turbines*

This exercise draws inspiration from the French *Agrégation interne* exam in electrical engineering from 2001. Figure 3.73 gives the characteristics of the mechanical power, which is supplied by a fixed speed wind turbine on an electrical network at 50 Hz for a wind speed of 15 m/s according to the turbine rotational speed. The nominal operating point is chosen so that the supplied power is at a maximum. An induction machine with two pole pairs is used.

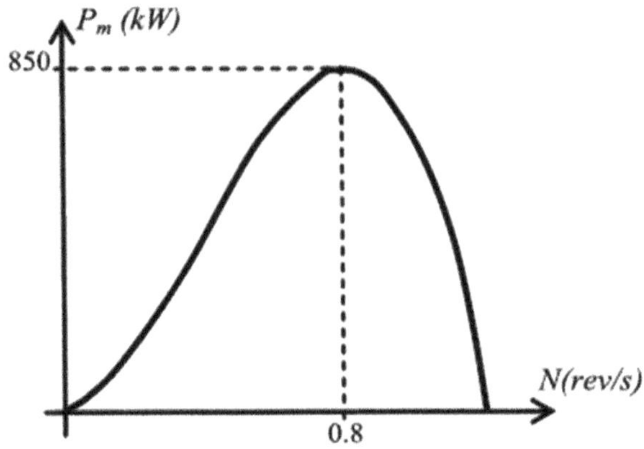

Figure 3.73. *Curve of the power, according to the rotational speed with constant wind speed (15 m/s) supplied by a wind turbine*

A. With a speed multiplier with a 31.7 ratio, determine the optimal generator rotational speed, as well as its slip.

B. Determine the electromagnetic torque of the induction machine, given that the multiplier has an efficiency of 96%. The generator mechanical losses are worth 6 kW at this operating point.

C. At low slip values, in the machine single-phase equivalent diagram, the resistor term $\frac{r_2}{g}$ will be much higher than the inductive term $l_2\omega$. By bringing back the rotor part of the single-phase diagram to the stator side, the expression of the electromagnetic torque is obtained by erasing the coefficient n. The nominal voltage between the phases is worth $U_l = 326$ V; from this the rotor resistance value of the single-phase equivalent diagram can be deduced.

Answers

A. Rotational speed of the induction machine:

$$\Omega = 31{,}7\ \Omega_t = 31{,}7\,.0{,}8.2\pi\ rad/s = 50{,}72\ \pi\ rad/s$$

For a machine with two pole pairs, the synchronous speed is:

$$\Omega_s = 50\ \pi\ rad/s$$

The machine revolves at a higher speed than the synchronous speed of 1,500 rpm for a machine with two pole pairs. This corresponds to a generator's operating mode. The slip is worth:

$$g = \frac{\Omega_s - \Omega}{\Omega_s} = \frac{50 - 50.72}{50} = -1.44\% .$$

B. Mechanical power on the shaft of the induction machine (the negative sign corresponds to a generator's operating mode): $-850 \,.\, 0{,}96 = -816 \text{ kW}$.

Mechanical power transformed into electrical power:

$$P_{em} = P_m - P_{pm} = -816 - 6 \ = -810 \text{ kW}$$

Electromagnetic torque:

$$C_{em} = \frac{P_{em}}{\Omega} = \frac{-810}{50.72\,\pi} 10^3 = -5{,}086 \ Nm$$

C. In the denominator of torque expression [3.46], the inductive term is thus neglected. The expression of the electromagnetic torque can be simplified as follows:

$$C_{em} = 3V_1{}^2 \frac{p\, g}{\omega\, r_2}$$

NOTE.– This expression corresponds to the tangent in $g = 0$ of the torque characteristic depending on the slip represented in Figure 3.37.

It enables us to deduce the resistance value:

$$r_2 = 3V_1{}^2 \frac{p}{\omega} g \frac{1}{C_{em}} = 3\left(\frac{326}{\sqrt{3}}\right)^2 \frac{2}{2\pi 50} \frac{-1.44}{100} \frac{1}{-5{,}086} = 1.9 \ m\Omega$$

3.11.2. *Characterization of a turbine and estimate of the generated power*

This exercise draws inspiration from the eligibility tests for the French *Agrégation externe* electrotechnology and energy exam, option B: electrotechnology and power electronics from 2004. We consider a wind turbine of 1.5 MW comprised of a 77 m diameter turbine with three blades, a speed multiplier of ratio 104.2 and an electric generator. The wind speed at starting is 3 m/s and the rotational speed is from 9.6 to 17.3 rpm. The characteristic provided by the manufacturer of the power supplied by the conversion chain to the network according to the wind speed is given in Figure 3.74. Moreover, the multiplier and generator losses are not taken into account.

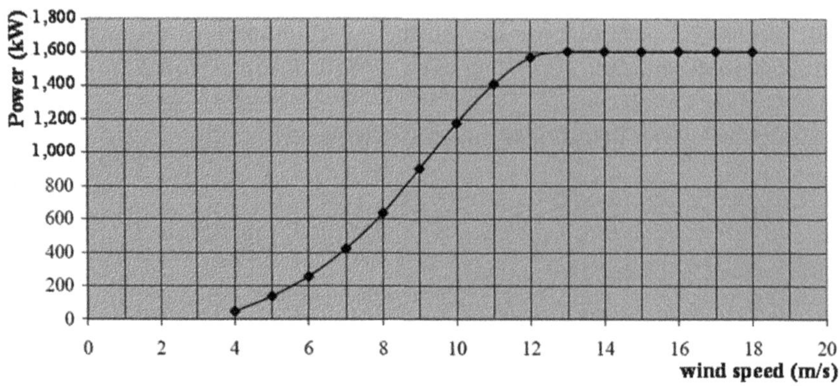

Figure 3.74. *Characteristic of the recovered power according to the wind speed*

A. Operating principle

From the manufacturer's data, calculate the values of the power coefficient C_p and draw its variation according to the wind speed v, by filling in the table below. The air density is: $\rho = 1.225$ kg/m^3.

V (m/s)	4	6	8	10	12	14	16	18
$P_{turbine}$ (kW)	47	257	636	1,177	1,570	1,605	1,605	1,605
Cp								

B. Calculation of the kinetic energy stored in the revolving group

When the wind turbine is at high speed, it is important to know the amount of energy stored in the revolving group for the braking system sizing. To simplify the study, only the moments of inertia of the wind turbine's 3 blades, the blade hub and the generator rotor will be taken into account:

– the moment of inertia of the generator rotor (on the fast side) $J_r = 174.2\ \text{kg.m}^2$;

– the moment of inertia, in relation to the rotational axis, of the hub of the wind turbine $J_m = 9{,}750$ kg.m^2;

– the moment of inertia, in relation to the axis of rotation, of the three blades of the wind turbine $J_p = 4817312$ kg.m^2.

The inertias of the speed multiplier and of the transmission shafts are included in the blade hub and the generator rotor. The wind turbine revolves at a speed of 17.3 rpm.

1) Calculate the kinetic energy stored in the induction machine rotor.

2) Calculate the kinetic energy stored in the set of three blades and the hub.

3) Calculate the total kinetic energy stored in this set.

4) By considering a constant braking effect of 1,600 kW, how long would the braking be in order to evacuate the equivalent of this kinetic energy stored in this revolving set?

C. Calculation of the recoverable energy by the wind turbine

This wind turbine will be established on a site which has previously been characterized by wind measurements carried out over a two year period. The wind speed histogram for this site is presented in Figure 3.75.

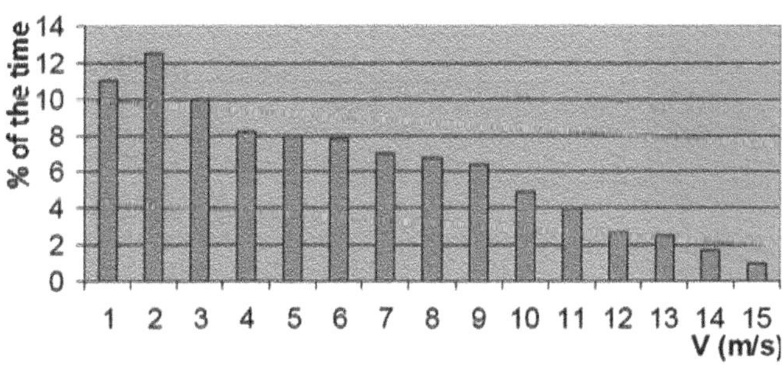

Figure 3.75. *Wind histogram*

1) Complete the table below by filling in the number of hours corresponding to each wind speed (over a year), as well as the recovered energy.

2) Calculate the energy recoverable by this wind turbine on the considered site for a year (365 days). Calculate the maximum energy that could be generated for a year. From this, deduce the loading rate (of use) of this wind turbine.

V (m/s)	P (kW)	% of the time	Time/year in hours	Energy (MWh)
4	47	8.2		
5	138	8		
6	257	7.8		
7	424	7		
8	636	6.8		
9	905	6.6		
10	1,177	5		
11	1,410	4.1		
12	1,570	2.9		
13	1,605	2.5		
14	1,605	1.7		
15	1,605	1		

Answers

A.

$$P_m = C_p \frac{1}{2} \rho \pi R^2 v^3 = C_p 2852 v^3$$

V (m/s)	4	6	8	10	12	14	16	18
$P_{turbine}$ (kW)	47	257	636	1,177	1,570	1,605	1,605	1,605
Cp	0.258	0.417	0.436	0.413	0.319	0.205	0.137	0.096

B.

1) $E_{rg} = \frac{1}{2} J_r \Omega^2 = E_{rg} = \frac{1}{2}$ $174.2 \times (17.3 \times 104.2 \times \pi / 30)^2 = 3.1$ MJ.

2) $E_{rg} = \frac{1}{2} \left(J_m + J_p \right) \Omega^2 = \frac{1}{2} (9{,}750 + 4{,}817{,}312) \times (17.3 \times \pi / 30)^2 = 7.92$ MJ.

3) Total energy: $E_{total} = E_{rg} + E_r = 11$ MJ.

4) $E_{total} = P.t$ hence $t = \frac{E_{total}}{P} = 11/1.6 = 6.9$ s.

C.

1) Total number of hours for a year = 365*24 = 8,760 h.

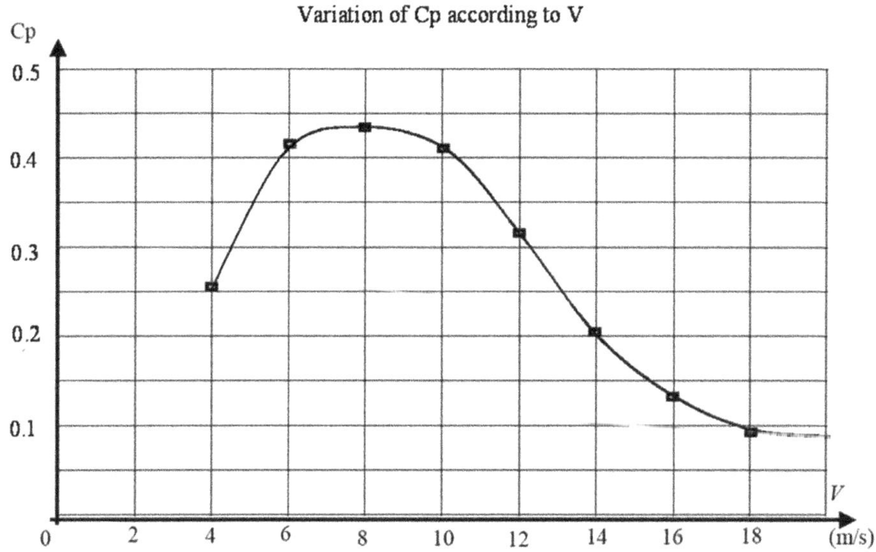

Figure 3.76. *Characteristic of the power coefficient*

<table>
<tr><th>V (m/s)</th><th>Power (kW)</th><th>% of time</th><th>Time/year (hours)</th><th>Energy (MW.h)</th></tr>
<tr><td>4</td><td>47</td><td>8.2</td><td>718.3</td><td>33.8</td></tr>
<tr><td>5</td><td>138</td><td>8</td><td>700.8</td><td>96.7</td></tr>
<tr><td>6</td><td>257</td><td>7.8</td><td>683.3</td><td>175.6</td></tr>
<tr><td>7</td><td>424</td><td>7</td><td>613.3</td><td>260</td></tr>
<tr><td>8</td><td>636</td><td>6.8</td><td>595.68</td><td>378.9</td></tr>
<tr><td>9</td><td>905</td><td>6.6</td><td>578.16</td><td>523</td></tr>
<tr><td>10</td><td>1177</td><td>5</td><td>438</td><td>515.5</td></tr>
<tr><td>11</td><td>1410</td><td>4.1</td><td>359.16</td><td>506.4</td></tr>
<tr><td>12</td><td>1570</td><td>2.9</td><td>254</td><td>398.8</td></tr>
<tr><td>13</td><td>1605</td><td>2.5</td><td>219</td><td rowspan="3">731</td></tr>
<tr><td>14</td><td>1605</td><td>1.7</td><td>148.92</td></tr>
<tr><td>15</td><td>1605</td><td>1</td><td>87.6</td></tr>
</table>

The sum of the energies for a year is worth 3,620 MWh.

2) The maximum energy would be $8760 \times 1.6 = 14$ GWh.

The loading rate is worth 0.258.

3.11.3. *High power variable speed wind turbines*

This exercise draws inspiration from the eligibility tests for the French *Agrégation externe* in electrotechnology and energy, option B: electrotechnology and power electronics. The block diagram of a wind turbine based on a doubly-fed induction generator with two pairs of poles is given in Figure 3.49. The machine is connected to a 400 V network that has a constant frequency equal to 50 Hz.

3.11.3.1. *Nominal operation by neglecting all losses*

The mechanical power supplied by the wind turbine on the level of the generator shaft is worth $P_m = 1{,}572$ kW, for a speed of 1,800 rpm.

1) Calculate the electromagnetic torque of the induction machine.

2) Calculate the power circulating in the rotor circuit. Is it supplied or absorbed by the rotor? In which operating mode is the induction machine found?

3) Calculate the power coming from the stator and the power supplied to the network.

Answers

1) Electromagnetic torque = $C_{em} = \dfrac{P_m}{\Omega} = 1{,}572{,}000/(1{,}800\times\pi/30) = 8{,}340$ Nm.

2) $g = (1{,}500\text{-}1{,}800)/1{,}500 = -\,0.2$.

$$P_{\text{rotor\_gene}} = gP_s = \frac{gP_{\text{m}}}{1-g} = -0.2\times 1{,}572{,}000/1{,}2 = -262 \text{ kW.}$$

The rotor supplies power, hypersynchronous mode.

3) $P_{\text{network}} = P_{\text{m}} = 1{,}572$ kW (assumption of system without losses).

3.11.3.2. *Nominal operation by considering losses*

Stator Joule losses = $P_{Js} = 10.9$ kW

Rotor Joule losses = $P_{Jr} = 11.86$ kW

1) Calculate the mechanical power converted into electric power.

2) Calculate the electric power at the stator.

3) Calculate the power circulating in the rotor circuit.

4) Calculate the power at the transformer input (on the low voltage side), by considering the efficiency of the AC/AC rotor converter equal to 99%.

Answers

1) $P = C_{em}\Omega_s = 8{,}340 \times 50.\pi = 1{,}310$ kW.

2) $P_s = P - P_{Js} = 1{,}299$ kW.

3) $g = (1{,}500\text{-}1{,}800)/1{,}500 = -0.2$.

$P_{rotor\_gene} = g\left(P_s + P_{js}\right) + P_{Jr} = gP + P_{Jr} =$ $-0.2 \times 1{,}310 \times 10^3 + 11.86 \times 10^3 =$ -250.14 kW

4) $P_{network} = P_s - 0.99 P_{rotor\_gene} = 1{,}299\text{k} + 0.99 \times 250.14 \times 10^3 = 1{,}546.6$ kW.

3.11.3.3. *Reduced power operation (low wind) by neglecting all losses*

The mechanical power supplied by the wind turbine on the level of the generator shaft is worth $P_m = 250$ kW, for a speed of 1,200 rpm.

1) Calculate the electromagnetic torque of the induction machine

2) Calculate the power circulating in the rotor circuit. Is it supplied or absorbed by the rotor? In which operating mode is the induction machine found?

3) Calculate the power supplied to the network.

Answers

1) $C_{em} = \dfrac{P_m}{\Omega} = 250{,}000/(1{,}200 \times \pi/30) = 1{,}989$ Nm.

2) $g = (1{,}500\text{-}1{,}200)/1{,}500 = 0.2$.

$$P_{rotor\_gene} = gP_s = \frac{gP_{sm}}{1-g} = 0.2 \times 250{,}000/0.8 = 62.5 \text{ kW}.$$

The rotor absorbs the power, hyposynchronous mode.

3) $P_{network} = P_m = 250$ kW (assumption of a system without losses).

3.11.3.4. *Reduced power operation by considering losses*

Stator Joule losses = P_{Js} = 628 W

Rotor Joule losses = P_{Jr} = 2,025 W

1) Calculate the power transmitted by the revolving field.

2) Calculate the electric power at the stator.

3) Calculate the power involved with the rotor supply.

4) Calculate the power at the transformer input (by considering the efficiency of the AC/AC converter equal to 99%).

Answers

1) $P = C_{em}\Omega_s = 1{,}989 \times 50.\pi = 312.5$ kW.

2) $P_s = P - P_{Js} = 311.9$ kW.

3) $g = (1{,}500\text{-}1{,}200)/1{,}500 = 0.2$

$$P_{\text{rotor\_gene}} = g\left(P_s + P_{js}\right) + P_{Jr} = gP + P_{Jr} = 0.2 \times 312.5 \times 10^3 + 2{,}025 = 64{,}525 \text{ W}$$

4) $P_{\text{network}} = P_s - \dfrac{P_{rotor\_gene}}{0.99} = 311.9 \times 10^3 - 64{,}525/0.99 = 246.7$ kW.

3.12. References

[ADE 09] ADEME, 6329, Guide pratique: l'énergie éolienne, February 2009.

[BAR 96] BARTHOLOMEUS P., LEMOIGNE P., ROMBAUT C., "Etude des limitations en puissance des convertisseurs et apport des techniques multiniveaux", *Actes du colloque Electronique de Puissance du Futur, EPF'96*, Grenoble, pp. 121–126, 1996.

[BOU 09] BOUHALI O., FRANÇOIS B., BERKOUK M. *et al.*, "Power sizing and control of a three-level NPC converter for grid connection of wind generators", *Electromotion Journal*, vol. 16, no. 1, pp. 38–48, January 2009.

[CHA 08] CHAVEROT S., HEQUETTE A., COHEN O., "Changes in storminess and shoreline evolution along the northern coast of France during the second half of the 20th century", *Zeitschrift für Geomorphologie*, Supplementary Issues, vol. 52, no. 3, pp. 1–20, November 2008.

[DAH 00] DAHLGREN M., FRANK H., LEIJON M. *et al.*, "Windformer, Production à grande échelle d'électricité éolienne", *Revue ABB*, no. 3, 2000.

[DE 18] DE ROUX G., GUITTON B., SIMON J. *et al.*, "Eolien en mer: La maturation de la filière au travers de l'expérience d'un développeur énergétitien", *REE*, no. 4, pp. 62–74, 2018.

[EL 03] EL AIMANI S., MINNE F., FRANÇOIS B. *et al.*, "Comparison analysis of control structures for variable speed wind turbine", *Computational Engineering in Systems: CESA'2003*, Lille, France, 9–11 July 2003.

[EL 04] EL AIMANI S., Modélisation de différentes technologies d'éoliennes intégrées dans un réseau de moyenne tension, PhD thesis, Ecole Centrale de Lille, France, 2004.

[FAT 48] FATEEV E.M., *Windmotors and Wind Power Stations*, Mir, Moscow, 1948.

[GER 01] GERGAUD O., MULTON B., BEN AHMED H., "Modélisation d'une chaîne de conversion éolienne de petite puissance", *Electrotechnique du Futur 2001*, pp. 17–22, Nancy, November 2001.

[GHE 11] GHENNAM T., Supervision d'une ferme éolienne pour son intégration dans la gestion d'un réseau électrique, Apports des convertisseurs multi niveaux au réglage des éoliennes à base de machine asynchrone à double alimentation, PhD thesis, Ecole Centrale de Lille and Ecole Militaire Polytechnique d'Alger, no. 162, 2011.

[HAU 06] HAU E., *Wind Turbines: Fundamentals, Technologies, Application, Economics*, 2nd ed., Springer, 2006.

[LAN 18] LANNE M., DELAGE B., COLBOC G., "L'éolien flottant, une chance pour la France", *REE*, no. 4, pp. 75–83, 2018.

[LE 09] LE DOEUFF R., EL-HADI ZAÏM M., *Rotating Electrical Machines*, ISTE, London and John Wiley & Sons, New York, 2010.

[MAR 02] MARTIN J., "Energies éoliennes", *Techniques de l'ingénieur*, B 1 360, 2002.

[MON 15] MONJEAN P., ROBYNS B., "DC connection of offshore wind farms for reduced environmental impact of T&D networks", in BESSEDE J.-L. (ed.), *Eco-Friendly Innovations in Electricity Transmission and Distribution Networks*, Woodhead Publishing, Elsevier, pp. 291–311, 2015.

[MUL 04] MULTON B., ROBOAM X., DAKYO B. *et al.*, "Aérogénérateurs électriques", *Techniques de l'ingénieur*, D 3 960, 2004.

[ROB 06] ROBYNS B., DAVIGNY A., SAUDEMONT C. *et al.*, "Impact de l'éolien sur le réseau de transport et la qualité de l'énergie", *J3eA*, vol. 5, no. 1, 2006.

[ROB 12] ROBYNS B., FRANÇOIS B., DEGOBERT P. *et al.*, *Vector Control of Induction Machines. Desensitization and Optimization through Fuzzy Logic*, Springer Verlag, 2012.

[RSS] REVUE SYSTEMES SOLAIRES, www.energies-renouvelables.org.

[RTE 10] RTE, Le bilan électrique francais 2009, press release, RTE, 12 January 2019.

[TRO 89] TROEN T., PETERSON E., *European Wind Atlas*, RISO National Laboratory, Roskilde, Denmark, 1989.

[VER 05] VERCAUTEREN P. "C-power NV: Le parc éolien offshore en Belgique", *Actes de la journée d'étude SRBE-SEE sur le thème "Eolien et réseaux: enjeux"*, Lille, 22 March 2005.

[VIZ 07] VIZIREANU D., Optimisation de l'architecture des machines synchrones à aimants permanents et attaque directe de l'arbre moteur pour les applications fort couple et basse vitesse, PhD thesis, Ecole Centrale de Lille, 2007.

[WIN 10] WINDFINDER, available at: http://www.windfinder.com/windstats/windstatistic_sevenstones_lightship.htm.

[WIN 19] WIND EUROPE, "Our energy, our future, How offshore wind will help Europe go carbon-neutral", available at: windeurope.org, November 2019.

[WIN 20] WIND EUROPE, "Offshore wind in Europe, Key trends and statistics 2019", available at: windeurope.org, February 2020.

4

Terrestrial and Marine Hydroelectricity

4.1. Run-of-the-river hydraulics

4.1.1. *Hydroelectricity*

4.1.1.1. *Historical background*

The waterwheel was invented for grain grinding (watermills) during the 1st century BCE. From the 11th century, watermills were used as industrial engines. Together with windmills, this was the only type of engine until the invention of the steam engine, in locations such as forges (protective bellows, hammer drive), sawmills, rolling mills, drawing mills, etc.

The foundations of hydrodynamics, fluid flows and the principles of turbomachinery were established in the 18th century by Daniel Bernoulli (1700-1782) and Leonhard Euler (1707-1783).

During the 19th century, the exploitation of hydraulic energy evolved significantly:

– 1827: first hydraulic turbine created by Benoît Fourneyron (1802-1867) in Pont-sur-Orgeon, Haute-Saône, France. This concept increased the efficiency and the captured power by using paddle wheels.

– 1869: first penstock enabling us to use a 200 meter head, all the while producing a mechanical speed of 700 kW.

– 1882: first penstock with a 500 meter head and a production of 1.8 MW.

Chapter written by Benoît ROBYNS and Antoine HENNETON.

The main types of turbines were invented during this century:

– the Francis turbine by James Bicheno Francis (1815-1892);

– the Pelton turbine by Lester Allen Pelton (1829-1908);

– the Kaplan turbine by Viktor Kaplan (1876-1934).

At the same time, the discovery of the laws of electromagnetism and mechanical electrical conversion led to the alternator adjustment. Turbine power significantly increased with the emergence of electricity. Indeed, electricity enabled the transportation of energy (first transmission lines installed from 1883), whereas previously mechanical energy had to be used on the premises.

The development of the first hydroelectric power stations goes back to the end of the 19th century with powers of several kW. In 1950, hydroelectric power generation represented 58% of the total electricity generation in France. In the 1950s, small power stations were closed because of their outdated state and their lack of competitiveness, in lieu of larger and more recent installations. Following the development of the nuclear program in France during the 1970s, the proportion of hydroelectricity in the total electricity production in France went down to 40% in 1970, 20% in 1990 and 11.2% in 2019 [WIK 20].

In 2018, hydroelectricity represented, on its own, more than 16% of the global electricity production and 64% of the production based on renewable energies [HEM 99, DRA 01, MUL 11, WIK 20]. We can distinguish between "large hydraulics" (the most common, i.e. high-power hydraulics (beyond ten megawatts)) and "small hydraulics" (less than 10% of the global hydroelectric production), which comprises lower power stations.

4.1.1.2. *Hydroelectricity [PER 03]*

The energy of a fluid can be broken down into:

– kinetic energy:

$$E_{cin} = \frac{mv^2}{2} \qquad [4.1]$$

– potential energy:

$$E_{pot} = mgh \qquad [4.2]$$

– and pressure energy:

$$E_{pres} = \frac{mp}{\rho} \quad [4.3]$$

with:

– m (kg) = fluid mass;

– v (m/s) = fluid velocity;

– h (m) = fluid water head;

– g (m/s^2) = acceleration of gravity;

– p (Pa) = fluid pressure;

– ρ (kg/m^3) = fluid density.

The total energy per kg of fluid is thus expressed by Bernoulli's equation, by introducing an equivalent height H_b, which is used by hydraulicians [PER 03]:

$$\frac{E}{m} = \frac{v^2}{2} + hg + \frac{p}{\rho} = gH_b \quad [4.4]$$

Hydraulic power is determined from the expression of energy:

$$P = \frac{E}{t} = \frac{gH_b m}{t} = gH_b \frac{m}{Vol}\frac{Vol}{t} = gH_b \rho Q \quad [4.5]$$

with:

– Vol (m^3) = volume;

– Q (m^3/s) = flow rate;

– H_b (m) = gross water head.

Net hydraulic power is obtained by deducting from H_b the head losses ΣH present in penstocks or in the bypass channels. The latter are expressed in meters:

$$H_n = H_b - \Sigma H \quad [4.6]$$

Figure 4.1 shows the structure of a conventional hydroelectric power station, such as a dam.

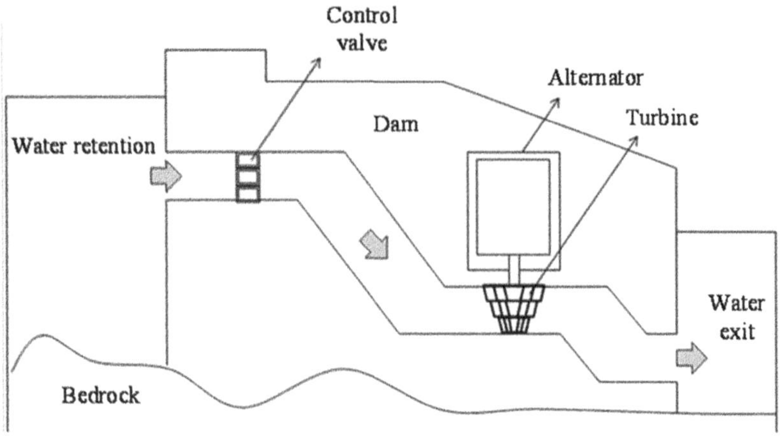

Figure 4.1. *Structure of a large hydroelectric power station (dam)*

4.1.2. *Small hydraulics*

The development of equipment in the framework of large hydraulics is henceforth highly limited because of the few available sites and the consequences of such installations for landscape integrity, water quality, and the subaquatic fauna [KAT 03]. However, a significant potential still remains in Asia, Africa and South America, which should be exploited by "developing countries" with growing energy needs.

Because it has less significant environmental impact, small hydraulics (less than 10 MW) has a strong development potential. We can predict that it will have an important role in future electricity generation by renewable energies; the targeted applications being the supply of isolated non-electrified sites, as well as the fill supply for the interconnected network [ANS 04, PER 03]. Small hydroelectric installations connected to the network combine many advantages, which make them particularly profitable sources of financial income for independent producers. These main advantages are presented below:

– It is easier to find a potential site for the installation of a small plant, because it does not require many facilities. The small size of the installation does not have much of an effect on the rural aesthetics and specific constructions are installed to save fish.

– Many plants were built throughout the 20th century, but they were often abandoned in the 1950s because they were considered to be obsolete and not sufficiently competitive, in comparison to their large-sized counterparts and to some thermal stations. Because of the renewed interest in dispersed electricity production, it could be advantageous (because it is not very expensive) to rehabilitate or renovate them.

– It is possible to add hydraulic turbines to the installations destined to use/have any other applications. This seems sound for example in the case of wastewater treatment plants, whose pressure level must be lowered before treatment. Water turbine action, *in loco* of an expansion valve ("energy absorber"), enables us to build on the energy that was previously lost; and this with relatively low costs and with the hydroelectric power plant using the infrastructures and pipes of the initial equipment [RSS 02a, RSS 02b]. This technical solution is valid for any hydraulic system, whose part of the energy must be dissipated. This solution is applied to several structures, including to the drinking water network and to seawater desalination plants for powers from a few dozen to a few hundred kilowatts.

– New kinds of small plants are appearing on the market. They are floating hydro turbines. They have a unit power of several dozen kilowatts, ready to install, and made up of paddle wheels, which are mounted on floats and drive a synchronous generator. These floating hydro turbines do not require any specific installations, except anchor points which can be mounted on a bank or attached to a bridge pier. Therefore, they are not particularly adapted for the electrification of villages located in disadvantaged countries and endowed with regular watercourses: such installations operate in Congo and Gabon, for example.

Small hydraulics installed all over the world at the beginning of the 21st century represented more than 37 GW. Amongst the European Union, France was the second best equipped country in small hydroelectricity, after Italy which has 3208 MW, with an installed power of 2,065 MW in 2015. The total installed power within the European Community for 2010 was estimated at 13 GW, with an ambition to expand to 17.3 GW by 2020. The amount of small hydroelectricity is evolving relatively slowly because of the administrative procedures and sometimes because of opposition to the building of new power plants. The growth in this area will be partly carried out by renovating old stations. Small hydroelectricity has and will have, however, a significant role in the development of renewable energy sources for the supply of isolated sites, such as remote rural area sources, mountain chalets or villages in countries with a small electrical network, or as a supplementary supply, which could be quickly mobilized, in interconnected networks. France will launch calls for tenders to encourage the development of more than one hundred MW in the period between 2016 and 2024 [RSS].

Small hydroelectric power is often a run-of-the-river layout, i.e. it does not require a storage tank, or just a small tank, with a small height dam most of the time, which helps to deflect part of the flow rate of a river in a conveyance penstock towards the plant. Most of the flow rate flows into the natural course of the river by flowing over this dam. Figure 4.2 presents the structure of a small hydroelectric power plant.

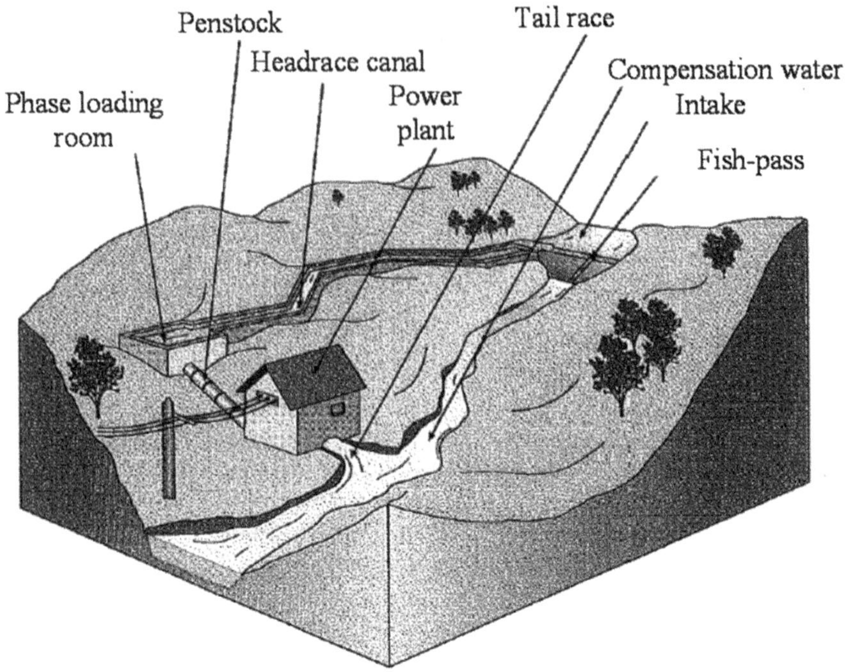

Figure 4.2. *Structure of a small hydroelectric power plant [ADE 06]*

4.1.3. *Hydraulic turbines*

The turbines used for small plants are similar to those encountered in large hydraulics [PAC 95a, ERE 09, PER 03], i.e. Pelton turbines for high heads (up to 500 m), typical of mountainous regions; Francis type machines for average heads (up to 200 m) and Kaplan turbines for small heads (lower than 30 m). There are also Crossflow turbines for head heights lower than 150 m and with relatively low powers. In large hydraulics, head heights can be higher than the previous values.

We can distinguish two types of turbines: action turbines (Pelton, Crossflow), whose fluid pressure at the output of the distributor directing the water on the wheel is equal to the pressure at the output of the wheel, i.e. atmospheric pressure; and reaction turbines (Francis, Kaplan), whose input pressure is higher than the pressure at the output of the wheel.

4.1.3.1. *Pelton turbines*

The Pelton turbine is an action turbine. Water kinetic energy is converted into mechanical energy; the conversion is carried out at atmospheric pressure, with the turbine being dewatered.

The Pelton turbine is made up of a bucket wheel, which is set in motion by one or several water jets coming from one or several injectors. Buckets are faired to obtain the maximum efficiency, all the while allowing water to escape on the wheel sides. A Pelton turbine can be equipped with one to six injectors. The flow rate is generally adjustable with the help of a mobile needle inside the injector, which is moved by a hydraulic or electric servomotor. This needle is controlled by the turbine regulation (Figure 4.3) [PAC 95a].

The maximum efficiency of Pelton turbines is between 84% and 90%. These turbines are used for high heads (from 200 m to 1,800 m), but also in small hydraulics (heads from 10 m to 500 m). The rotational speed of this turbine is high with values sensibly ranging between 500 and 1,500 rpm, depending on the powers and rate flows.

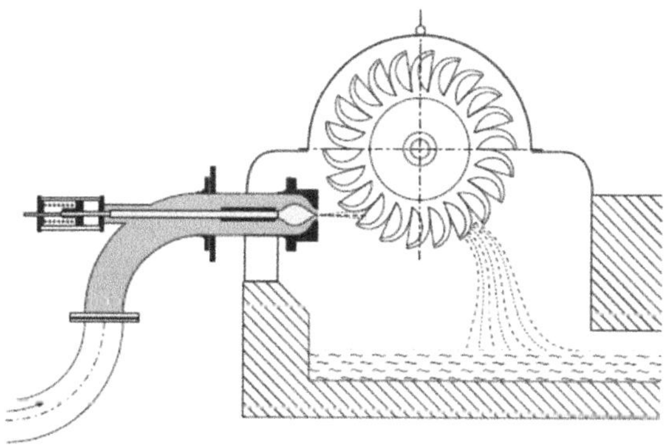

Figure 4.3. *Pelton turbine with an injector [PAC 95a]*

4.1.3.2. *Crossflow turbines*

The Crossflow (Banki-Mitchell) is an action turbine. It is easy to build (Figure 4.4) [PAC 95a]. This is why it was quite popular in developing countries. However, the maximum efficiency of this type of turbine remains average and for a good quality machine, ranges between 78% and 84%. It is typically not used in large hydroelectric power stations. It is used for average heads (up to 200 m), and in small hydraulics for heads of 10 m to 200 m. Its rotational speed is low, at a few hundred ripple/min (rpm).

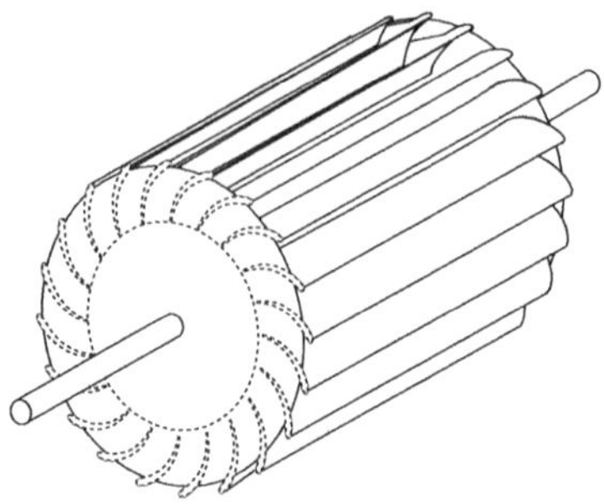

Figure 4.4. *Crossflow turbine [PAC 95a]*

4.1.3.3. *Francis turbines*

The Francis turbine is a reaction turbine. The kinetic energy, as well as the water pressure energy, is converted into mechanical energy, which requires a scroll case, in order to create a whirlpool. Figure 4.5 presents the main components of a Francis turbine.

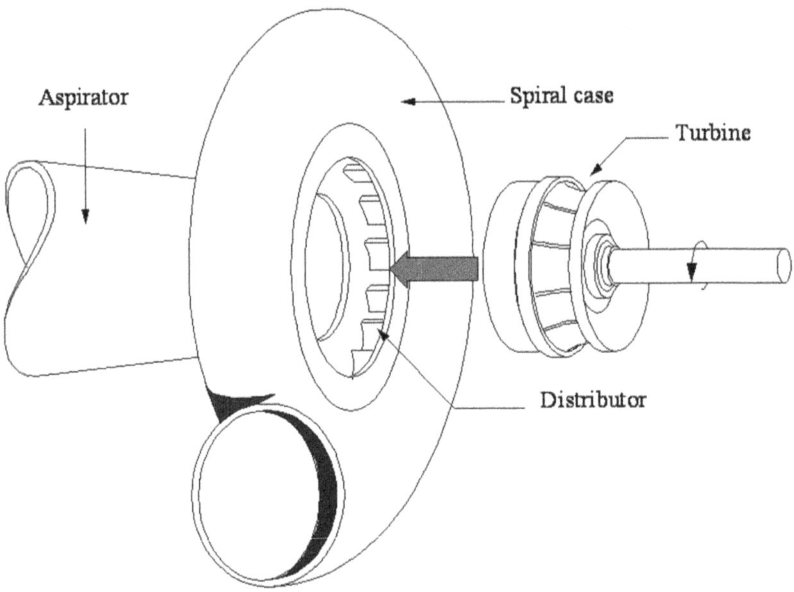

Figure 4.5. *Francis turbine [PAC 95a]*

The maximum efficiency of the Francis turbine ranges between 84% and 90%. It is used for average heads (from 30 m to 700 m), and in small hydraulics for heads from 10 m to 200 m. Its rotational speed can reach 1,000 rpm.

4.1.3.4. *Kaplan turbines*

This reaction turbomachine can be on a horizontal or vertical axis. It is in the form of a turbine with variable step blades (their slope angle in comparison to the rotational plane can vary) [PAC 95a]. The principle of a reaction turbine consists of converting both the pressure energy and the kinetic energy of the water into mechanical energy available on the shaft; the turbine being completely immersed. Rotation is carried out by the whirlpool effect thanks to a scroll case and to fixed or mobile distributor stators (Figure 4.6).

The maximum efficiency of the Kaplan turbine ranges between 84% and 90%. It is used for small heads (from 2 m to 55 m), and in small hydraulics for heads lower than 30m. This turbine is well adapted to run-of-the-river hydraulics. Because of the low speed of this turbine, in low power, it is generally associated with a gearbox to enable the use of a fast generator that is smaller than a slow generator with direct drive (strong couple).

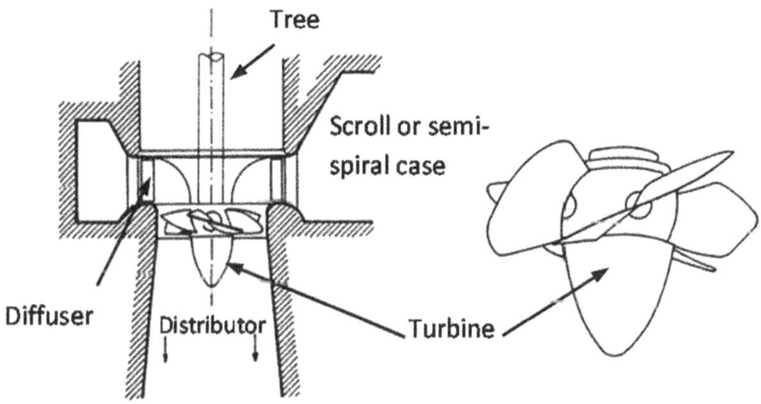

Figure 4.6. *Kaplan turbine [PAC 95a]*

The axis of this turbine can be vertical or horizontal (Figure 4.7). The turbine with a horizontal axis (bulb-type unit) presents a certain number of advantages making it increasingly attractive:

– the device is adapted for very small heads (from 2 m to 15 m) and high flow rates;

– it has improved flow in comparison to a conventional Kaplan turbine from the suppression of the scroll case, the bend at the end of the distributor and the bend of the draft tube;

– there is integration of the alternator into the bulb, which helps to limit the cost of infrastructures (civil engineering);

– the bulbs can be made reversible: they operate in the two directions (they are thus well adapted to tidal power plants), as well as there being the possibility of energy reversibility in centrifugation and pump operation.

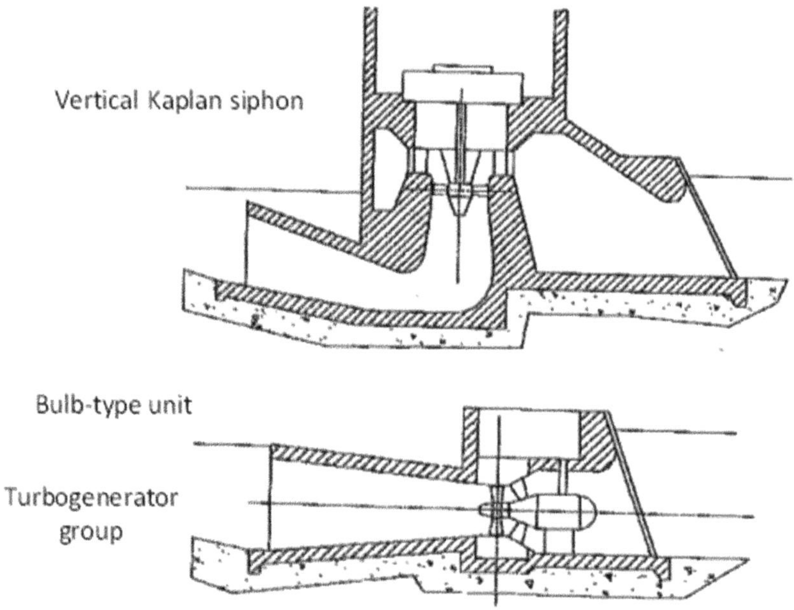

Figure 4.7. *Kaplan turbine with vertical and horizontal axes [ELE 06]*

4.1.3.5. *Efficiency of hydraulic turbines*

Figure 4.8 presents the evolution of the efficiency brought back to the maximum efficiency of the turbine, according to the flow rate brought back to the maximum flow rate. Curve 1 corresponds to a Pelton turbine, curve 2 to a Kaplan turbine, curve 3 to Francis and Crossflow turbines and curve 4 to a reverse pump. Note that the efficiency is highly variable with the flow rate.

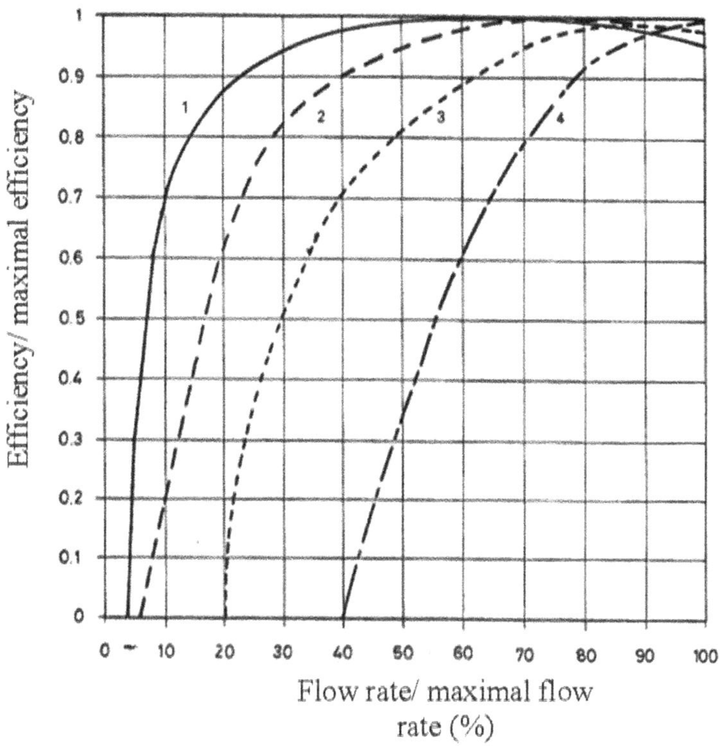

Figure 4.8. *Efficiency of hydraulic turbines: 1) Pelton turbine; 2) Kaplan turbine; 3) Francis and crossflow turbine; 4) reverse pump [PAC 95a]*

4.1.3.6. *Model of a hydraulic turbine*

The considered model of a turbine is a simple static model, which does not take into account some hydraulic parameters, such as inertia and water compressibility, as well as the elasticity from the supply line to the turbine. More specific models are developed in [KUN 94]. Let us assume that the water flow rate, as well as the orientation of the guide vanes and blades, in the case of a Kaplan turbine, are constant. Its torque-speed characteristic is almost linear, as we can see in Figure 4.9, where C_{mec} represents the torque delivered by the turbine and Ω represents the rotational speed [PAC 95a].

The shape of this characteristic, common to all existing categories of hydraulic turbines, enables us to deduce that the supplied mechanical power, noted P_{mec}, has a parabolic shape according to the speed (Figure 4.9). In addition, we distinguish the turbine runaway speed Ω_e, which corresponds to an operation for which the flow rate is non-null, but no load is connected to the generator so that the torque is null. This runaway speed ranges between 1.8 and 3 times the nominal speed. It is often close to 2.

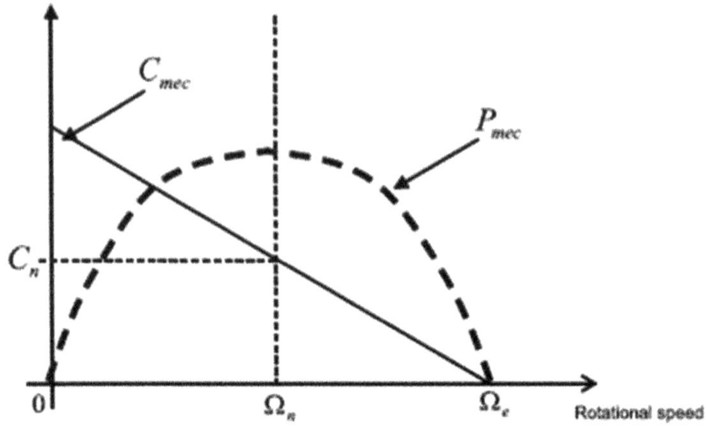

Figure 4.9. *Torque-speed characteristic of a hydraulic wind turbine operating at a constant flow rate*

Generators associated with hydraulic turbines must be sized in order to resist these overspeeds. The equation of the turbine torque-speed characteristic, under nominal flow rate and head, is given below [PAC 95a], all the while considering that a runaway speed of the speed turbine is equal to twice the nominal speed positioned at the maximum power point:

$$C_{mec} = C_n \left(2 - \frac{\Omega}{\Omega_n} \right) \tag{4.7}$$

The index n refers to the nominal magnitudes. In addition, we have:

$$P_{mec} = C_{mec}\, \Omega \tag{4.8}$$

The maximal value of this mechanical power is obtained from the hydraulic power supplied by the river. This mechanical power is proportional to the product of the flow rate by the head height (see equation [4.5]).

The retrievable electric power must still take into account the efficiency of the electric generator.

4.1.4. *Electromechanical conversion for small hydroelectricity*

4.1.4.1. *Operation on isolated loads*

The turbine drives a synchronous alternator. When the plant feeds isolated loads, the generator must ensure the stability of the frequency and the voltage level.

Using a synchronous machine (Figure 4.10) we assume that the turbine operates at fixed speed. This speed is regulated by the adjustment of the combined centrifuged flow rate of the water (if it is a Kaplan turbine) to that of the orientation of its blades. The voltage is controlled by acting on the excitation current of the alternator. Nevertheless, the turbine efficiency depends on both its speed and the hydraulic power [KEL 00]. Therefore, the operating point of this system in this case is not optimized. But it is also possible to regulate the frequency by acting on the electronic adjustment of the ballast resistors, which dissipate the power not consumed by users (Figure 4.11). This solution is very reliable because it avoids having to do any mechanical adjustment on the turbine.

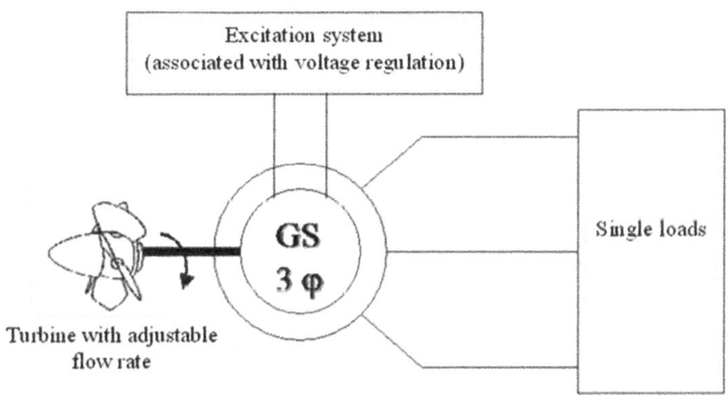

Figure 4.10. *Diagram of a hydroelectric micro-plant based on a synchronous machine*

Permanent magnet synchronous machines could also be used on their own to feed run-of-the-river batteries through diode rectifiers. To feed loads at fixed voltage and frequency, it is necessary to control power electronic converters, as in the case of low power wind turbines.

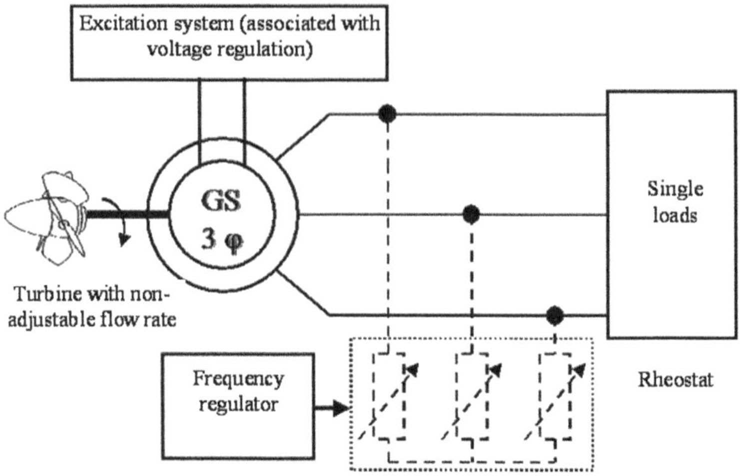

Figure 4.11. *Diagram of a hydroelectric micro-plant based on a synchronous machine*

4.1.4.2. *Operation coupled to the network*

Synchronous generators are generally the conventional solution, when hydraulic power stations are connected to a powerful network (Figure 4.12). They enable us to adjust the production of active power according to the requirements of the network administrator (flow rate adjustment) and to take part in the frequency adjustment, as well as adjusting the reactive power, in order to adjust the voltage. Squirrel-cage induction machines are also very frequently used in small hydroelectricity. The reactive power is compensated by capacitor banks, but it does not participate in voltage regulation.

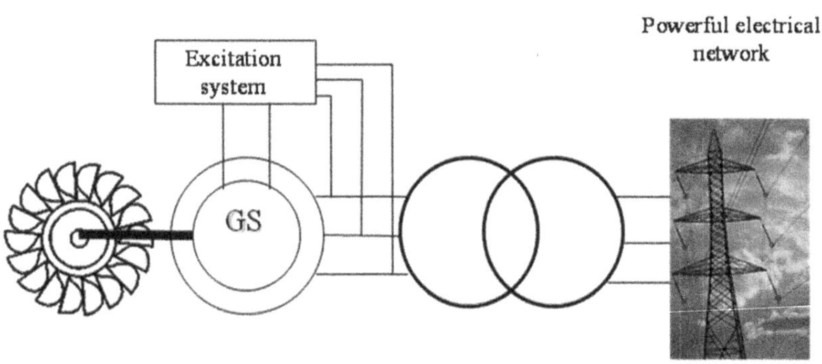

Figure 4.12. *Diagram of a hydroelectric power plant based on a synchronous machine connected to an electrical network*

4.1.4.3. *Variable speed operation*

The structures detailed above are electromechanical groups, i.e. conventional solutions rotating at fixed or almost fixed speed, which do not involve static converters of power electronics. Generally, variable speed systems are increasingly being developed because of their significantly improved energy performance. However, variable speed is not frequently used in small hydraulic power stations, but this situation could change in the future [HEM 99, ANS 04, ANS 06a, ANS 06b, BRE 07].

Figure 4.13 shows that during the change of flow rate Q, the power-speed characteristic of rotation is modified so that the maximum power for each flow rate corresponds to another rotational speed of the turbine [PAC 95a]. Therefore, we can notice that a variable speed operation enables us to maximize the produced hydraulic energy, to generate a reference power when the power station is connected to a powerful network and to adapt to the isolated load, by deleting the mechanical adjustment system of the flow rate.

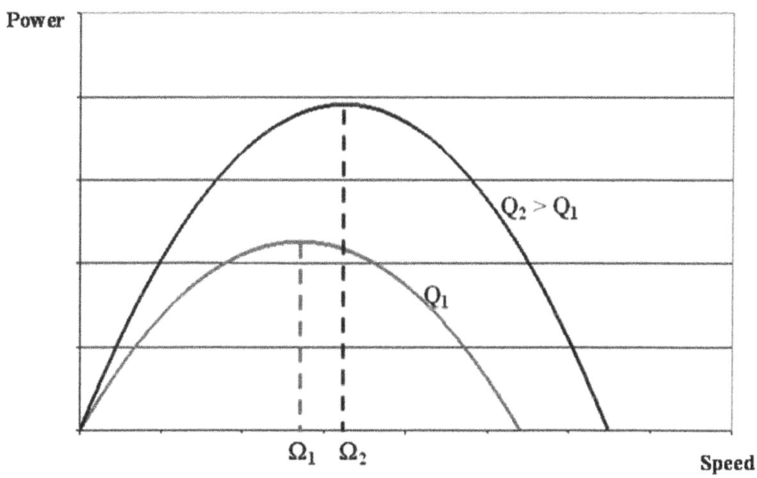

Figure 4.13. *Power-speed characteristic of a hydraulic turbine according to the flow rate. For a color version of this figure, see www.iste.co.uk/robyns/electricity.zip*

4.1.5. *Exercise: small hydroelectric run-of-the-river power station*

4.1.5.1. *Small hydroelectric power station on the Oise river*

Two "run-of-the-river" hydroelectric power stations (without any storage facility for the water) are implanted on the site of the Chigny mill on the river Oise, France

(Figure 4.14). Power station no. 1 was built several dozen years ago to replace an old mill. Power station no. 2 is more recent and will be the subject of this exercise. Floodgates enable us to maintain a constant upstream level (nappe in Figure 4.15) within the limits of their capacity (60 m^3/s) (extract from [RÉS 08]).

In the studied power station, pipes are short and we can therefore initially ignore impedance losses. The minimum and maximum heights of raw water are 1.5 m and 3 m respectively. Hydroelectric power stations cannot operate when there are floods.

The French decree no. 89-804 of October 27, 1989 fixes a compensation water flow rate (oxygen flow rate) to preserve the aquatic environment at:

– 1/10th of the average annual flow rate for any new facility;

– 1/40th for the current facilities.

This minimum flow rate remaining in the natural riverbed between the intake and the restitution of the water downstream of the power station, permanently guarantees the life, circulation and reproduction of the species living in these waters. The compensation water is reached, when the nappe reaches a 7 cm thickness.

Figure 4.14. *Overview of the Chigny site*

The theoretical flow turbine Q_t depends on the reserve flow rate Q_{res} and on the river flow rate Q_{riv}, so that:

$$Q_{riv} = Q_t + Q_{res} \quad [4.9]$$

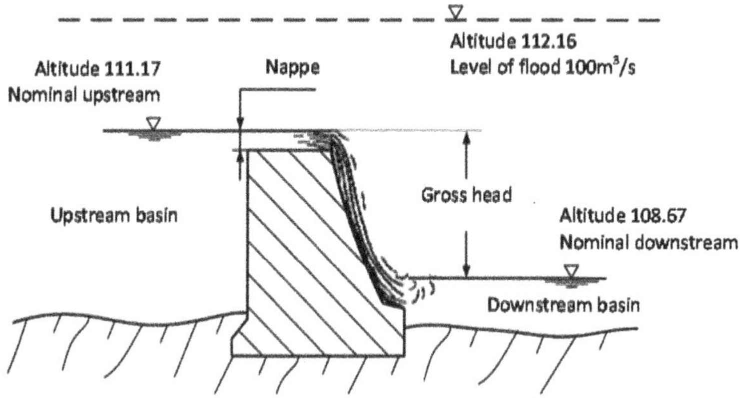

Figure 4.15. *Section of the dam and spillway of power station no. 2*

The effective flow turbine depends on the maximum operation flow rate of the turbine (4 m^3/s for power station no. 1). The maximum flow turbine in power station no. 2 has been set at 8 m^3/s.

We obtain a curve of the so-called flow rates, by sorting all of the instantaneous flow rates measured over a long period of time in decreasing order. The average annual flow rate Q_{ave} observed on the Oise river is 10 m^3/s. Figure 4.16 shows the flow rates of the river over a 10-year period and Figure 4.17 shows the flow rates classified as average over a 10-year period.

The micro power plant is made up of a turbine, a speed multiplier and a generator. Figure 4.18 represents the power transmission chain of this power station.

The technical data of the induction machine are as follows:

– manufacturer: ABB, model: M2FG 355 SA8 B3;

– 160 kW, 759 rpm, 8 poles, 50 Hz, 400/660 V, cos ρ = 0.7;

– resistance of a stator winding: R_1 = 25 mΩ;

– nominal efficiency η_g = 0.93.

1) Determine the nominal gross head noted H_b.

2) Determine the reserved flow rate Q_{res} by taking into account the fact that it is a new installation.

3) With the help of Figure 4.19, determine which type of turbine is best suited to this specific case.

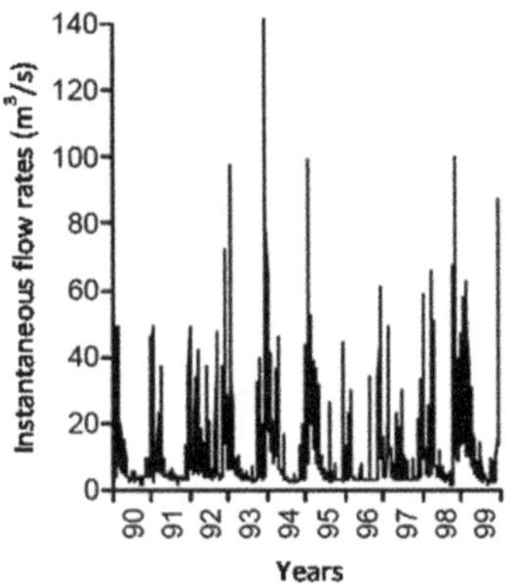

Figure 4.16. *Flow rate of the river over a 10-year period*

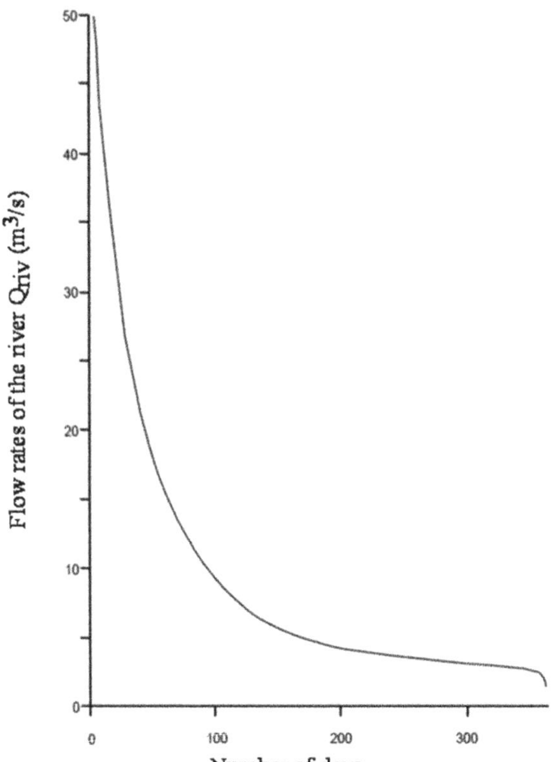

Figure 4.17. *Flow rates classified as average over a 10-year period*

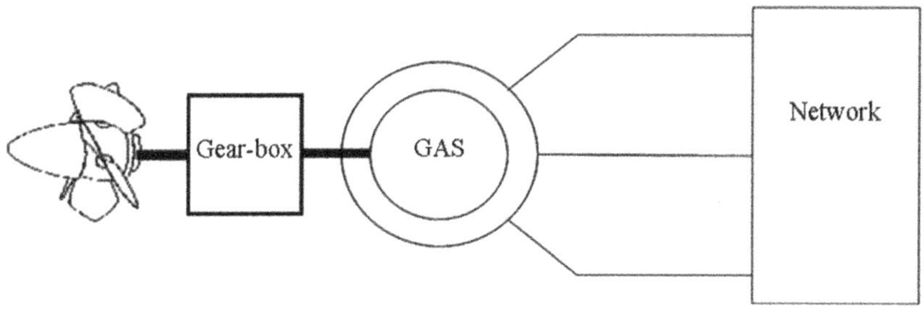

Figure 4.18. *Electromechanical transmission of the hydroelectric power station*

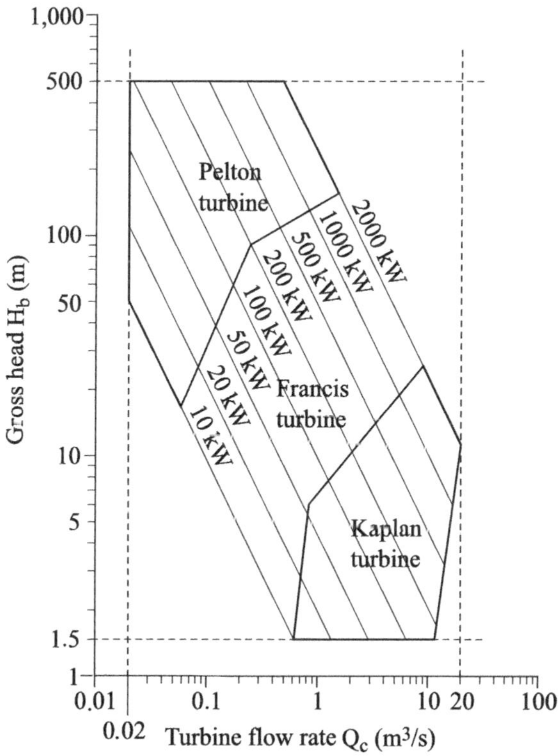

Figure 4.19. *Field of use of the main types of turbine. Estimated available hydraulic power (in kW)*

4) Determine the maximum hydraulic power P_h available for power station no. 2.

5) Determine the power P_t on the transmission shaft of the turbine by considering a maximum efficiency of the turbine $\eta_t = 0.87$.

6) Determine the maximum electric power P_g produced by the micro power plant, with the efficiency of the gearbox being equal to $\eta_p = 0.97$.

7) Determine the height of the gross water head H_b, the produced power P_g, the gross hydraulic power P_h, and the global efficiency η_{hydro} of the hydroelectric micro power plant during the April 8, 2007 reading:

– marker height upstream: 351 cm;

– marker height downstream: 104 cm;

– flow rate: 5.4 m^3/s;

– I generator: 244 A;

– ϕ angle: 54°.

The induction machine is connected to the 3 x 400 V network, whose voltage is supposed to be independent from the production conditions.

8) Determine the efficiency η_{cond} of the pipe bringing the water to the hydraulic turbine for the April 8, 2007 reading. Figure 4.20 gives the impedance losses of the pipe according to the flow rate.

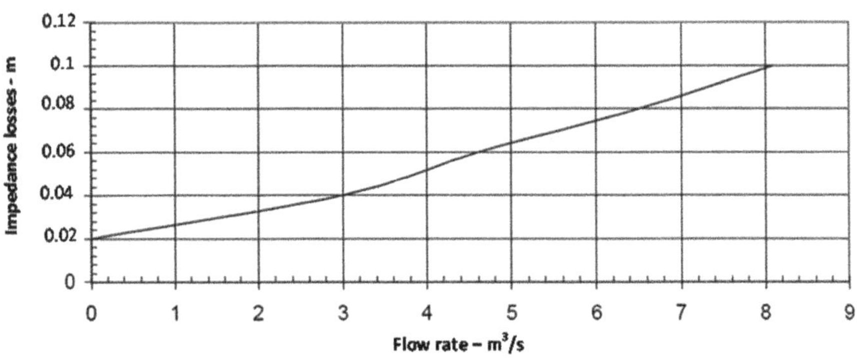

Figure 4.20. *Impedance losses of the pipe according to the flow rate (source: SICAE de l'Aisne)*

9) Calculate the stator Joule losses P_{js} for the operating point from April 8, 2007.

10) Calculate the rotor Joule losses ($P_{jr} = -g \cdot P_{em}$) for the operating point from April 8, 2007 (we will assume stator iron losses equal to 1,070 W and g = - 0.8%).

11) For the operating point from April 8, 2007, calculate the efficiency η_g of the generator by considering mechanical losses of 2,820 W.

12) Under the same conditions, determine the efficiency η_t of the turbine.

13) Calculate the annual energy (in kWh) produced in the year 2006 according to the data in Table 4.1.

Jan.	Feb.	March	April	May	June
66,792	49,666	81,073	80,866	68,406	53,419
July	August	Sep.	Oct.	Nov.	Dec.
23,008	32,008	7,821	1,481	17,387	53,888

Table 4.1. *Electrical energy production of the micro power plant (in kWh) for the year 2006 (source: SICAE de l'Aisne)*

14) Determine the quantity of CO_2 emitted in the atmosphere by the micro power station in 2006 according to the data in Table 4.2, which provides the CO_2 emissions in the atmosphere (in electric g/kWh) according to the sources of energy from the calculation of the life cycle assessment (LCA) of the field (construction of the installation, extraction and transport of the fuel, production of electrical energy, waste processing and storing, deconstruction of the installation).

Nuclear power	Gas	Coal	Oil	Hydraulic power
6	430	800 to 1,050	985	4

Table 4.2. *CO_2 emissions in the atmosphere (in electric g/kWh) according to the sources of energy from the calculation of the LCA of the field (source EDF)*

15) Calculate, with 3 significant numbers, the energy value of oil or fuel oil (in toe – ton oil equivalent), and then the corresponding quantity (in liters), which is saved in 2006 thanks to the hydroelectric power plant by using the data in Table 4.3, and by considering an efficiency of 38.7% for the production of electricity from fuel oil in the thermal power plant. The fuel oil density is considered to be equal to 0.85 kg/l.

1 toe	11,628 kWh thermal	11,000 m^3 of natural gas	7.2 m^3 of wood billet	1.4 ton of coal	7.33 barrels of oil

Table 4.3. *Correspondence of the energy value of several fuels (source: www.energie-rhone.fr)*

Answers

1) According to the data in Figure 4.15:

– Nominal upstream altitude $A_{upstreamN}$ = 111.17 m;

– Nominal downstream altitude $A_{downstreamN}$ = 108.67 m;

– Nominal gross head H_b = 111.17-108.67 = 2.5 m.

2) The reserved flow rate Q_{res} is worth 1/10th of the annual average flow rate Q_{ave}, which is equal to 10 m^3/s.

$$Q_{res} = 0.1\ Q_{ave} = 1\ \text{m}^3/\text{s}$$

3) According to Figure 4.19, for H_b = 2.5 m and Q_t = 8 m^3/s, we will choose a Kaplan turbine.

4) By disregarding impedance losses $H_n = H_b$. The maximum turbine flow rate being 8 m^3/s:

$$P_h = \rho\, g\, Q_t\, H_n = 1000\,.\,9.81.\ 8.\ 2.5 = 196.2\ kW$$

5) Power on the transmission shaft of the turbine P_t:

$$P_t = \eta_{tmax}\ P_h = 0.87.\ 196.2 = 170.7\ \text{kW}$$

6) Maximum electric power (η_p = efficiency of the gearbox, η_g = efficiency of the generator operating at its nominal power, the maximum power being close to this power):

$$P_e = \eta_g\, \eta_p\, P_t = 0.93.\ 0.97.\ 170.7 = 154\ \text{kW}$$

7) Gross head height $H_b = H_{upstream} - H_{downstream}$ = 351 – 104 = 247 cm:

$$P_g = \sqrt{3}\, U\, I \cos\phi = \sqrt{3}.\ 400.\ 244.0.588 = 99.4\ kW$$

$$P_h = \rho\, g\, Q_t\, H_b = 1000\,.\,9.81.\,5.4.\,2.47 = 131\,kW$$

$$\eta_{hydro} = \frac{P_g}{P_h} = \frac{99.4}{131} = 0.76$$

8) Net head height ($\Sigma\, H$ = impedance losses):

$$H_n = H_b - \Sigma H$$

For Q = 5.4 m^3/s, according to Figure 4.20, $\Sigma\, H$ = 0.069 m:

$$\eta_{cond} = \frac{H_n}{H_b} = \frac{H_b - \Sigma H}{H_b} = \frac{2.47 - 0.069}{2.47} \cong 0.97$$

9) The induction machine is connected to the 3 x 400 V network and it is thus connected in a triangle:

R_l = 25 mΩ for a stator winding

Current in a stator winding:

$$I_{1phase} = \frac{I_{1line}}{\sqrt{3}} = \frac{244}{\sqrt{3}} = 141 A$$

Joule losses at the stator:

$$P_{js} = 3\, R_1\, I_{1phase}^2 = 3\,.\,0.025.\,(141)^2 = 1.49\,kW$$

10) Power transferred from the rotor to the stator:

$$P_{em} = P_g + P_{fer} + P_{js} = 99.4 + 1.07 + 1{,}49 = 101.96\,kW$$

Joule losses at the rotor:

$$P_{jr} = -g\, P_{em} = -(-0{,}008).101{,}96 = 0.816\,kW$$

11) Generator efficiency:

$$\eta_g = \frac{P_g}{P_g + P_{js} + P_{fer} + P_{jr} + P_{mech}}$$

$$\eta_g = \frac{99.4}{99.4 + 1.49 + 1.07 + 0.816 + 2.82} = 0.94$$

12) Global efficiency of the hydroelectric micro-power plant:

$$\eta_{hydro} = \eta_t\, \eta_{cond}\, \eta_p\, \eta_g = 0.76$$

Efficiency of the hydraulic turbine:

$$\eta_t = \frac{\eta_{hydro}}{\eta_{cond}\, \eta_p\, \eta_g} = \frac{0.76}{0.97.0.97.0.94} = 0.86$$

13) Energy produced in 2006 = Σ Monthly energy = 535,815 kWh.

14) 1 kWh hydraulic produces 4 grams of CO_2:

Quantity of CO_2 produced in 2006 = 535,815. 4 = 2,143.26 kg

15) To produce the 535,815 kWh hydroelectric of the year 2006 we require a quantity of energy equal (in toe) to:

$$E_{tep} = \frac{535{,}815}{11{,}628 \times 0.387} = 119\ tep$$

As the fuel oil density is equal to 0.85 kg/l, 1 toe corresponds to:

1,000 kg/0.85 = 1,176 *liters*,

we obtain a saved volume of:

$$Vol_{oil.equi} = 119\,.\,1176 = 140{,}000\ \ liters$$

4.1.5.2. *Power supply for a mountain chalet*

A mountain refuge must be energy self-sufficient. The refuge illustrated in Figure 4.21 is supplied by a small hydroelectric power station, photovoltaic panels and a generator (Brevet de technicien supérieur électrotechnique 2012).

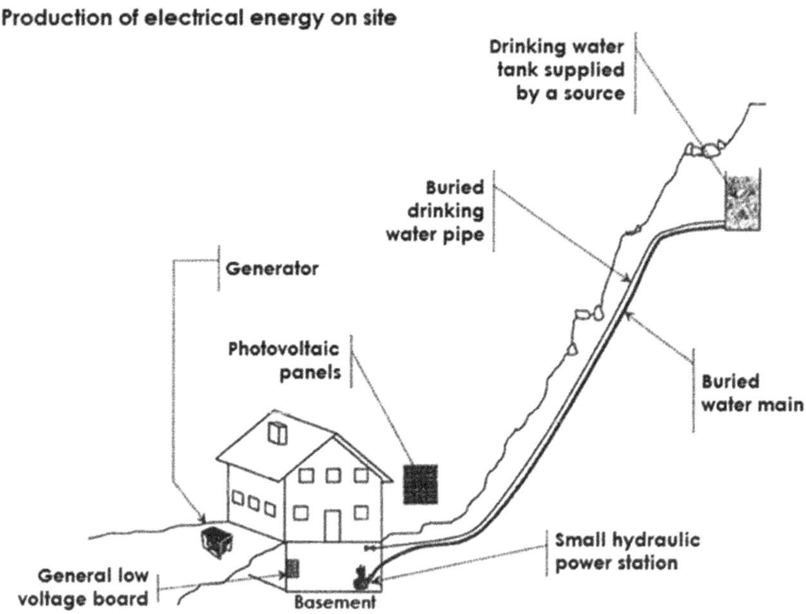

Figure 4.21. *Mountain chalet (Brevet de technicien supérieur électrotechnique 2012)*

The small hydroelectric power station that interests us in this exercise has the following characteristics:

– Pelton turbine with an efficiency of 72%;

– Height of the waterfall: 71 m;

– 230 V single-phase, synchronous, 4-pole, 4500 W alternator.

Household appliances at the refuge:

– Cooling unit: 540 W;

– Washing machine: 1500 W;

– Lighting: 1000 W;

– Extractor hood: 800 W.

1) Determine the hydraulic power required to meet the total demand of electrical loads.

2) Then, what should the flow rate be of the water hitting the turbine?

3) In order to supply the refuge with 50 Hz, what should the speed of the alternator be set to?

4) How does the frequency of the electric voltage change if the power consumed decreases while the flow remains unchanged?

Flow control in the case of the refuge would be too complex and expensive to achieve, thus the frequency of the voltages produced is kept constant by using an electronic device to control the consumption of electrical power. The turbine generator unit has been sized and adjusted so as to take into account the seasonal water flow needed to provide 4500 W, 24 hours a day. The surplus power is consumed through additional electrical loads: an electric cumulus and three heaters. These four devices are fitted with an electronic control card which allows these loads to be controlled from 0 W to 4500 W in increments of 150 W, in order to adapt total consumption to available production. The manufacturer guarantees a frequency variation range of the voltages produced of less than 1 Hz, and a change from 0% to 100% of the electrical load requested in less than 150 ms, following a variation in the power consumed.

The turbine generator unit arrangement responds to the following electromechanical equation, assuming that there is no mechanical loss:

$$J\,(d\Omega/dt) = Ce - Cr$$

with:

1) J, turbine generator unit moment of inertia = 0.5 kgm^2.

2) Ω, angular speed of rotation of the mechanical shaft.

3) Ce, electromagnetic torque of the alternator.

4) Cr, resistant torque.

5) At the nominal operating point, calculate the angular speed Ω.

6) Calculate the value of the resistive torque.

7) At the nominal operating point, calculate the electromagnetic torque of the generator.

When the washing machine is stopped, the loads for regulating the power consumed are not instantly implemented.

8) Determine the resistance torque value just after stopping the washing machine.

9) Determine the rate of change of angular velocity.

10) Estimate under these conditions, how long the frequency of the voltages produced would take to go from 50 Hz to 51 Hz.

11) Will the regulation system be able to respect the anticipated performances for the frequency range of the voltages being produced?

12) Does the choice of a pelton turbine seem appropriate to you? Justify your answer.

Answers

1) $Ph = Pelec/\eta = (540+1500+1000+800)/0.72 = 5.3$ kW

2) $Qh = Ph/(\rho gH) = 5300/(1000.9{,}81.71) = 0.0076$ m3/s or 7.6 l/s

3) $Ns = f.60/p = 50.60/2 = 1500$ rpm

4) The resistive torque of the generator becomes weaker as the electromagnetic torque induces an increase in the speed of the generator, and therefore in its frequency.

5) $\Omega = Ns\ \pi/30 = 157$ rad/s

6) $Cr - Pn/\ \Omega = 4500/157 = 28.7$ Nm

7) $Ce = Cr = 28.7$ Nm

8) $Pelec = 4500 - 1500 = 3000$ W

$Cr2 = Pelec/\ \Omega = 3000/157 = 19.1$ Nm

9) $\Delta\ \Omega/\Delta\ t = (Ce\text{-}Cr2)/J = (28.7 - 19.1)/0.5 = 19.2$ rad/s^2

10) $\Delta\ t = \Delta\ \Omega\ /\ (\Delta\ \Omega/\Delta\ t)\ = (2\ \pi/p)\ \Delta\ f\ /\ (\Delta\ \Omega/\Delta\ t) = (2\pi/2).1/19.2 = 163$ ms

11) 150 ms < 163 ms, the load control has time to react before the frequency varies by more than 1 Hz.

12) The pelton turbine is suitable for drop heights greater than 10 m.

4.1.5.3. *Small hydroelectric power station in Corsica*

The Asinao micro-power plant is located in south-eastern Corsica, near the Ospedale massif near the town of Zonza. It is fed by an outlet on the Asinao river, a tributary of the Rizzanese. This water intake, located altimetrically at an elevation of 1077 m, supplies a pipe with a diameter of 400 mm, intended to support the low flow of the Oso river during the dry period and to fill the Ospedale dam during the wet period [AMO 09].

On the route of the pipe, about 6 km downstream from the outlet, at hill 752, a tapping pipeline allows the water to be diverted to the micro-power plant. The 6 km pipeline induces a pressure drop of 65 m. The nominal water flow rate is 300 l/s.

The turbine is of the Pelton variety with two jets. Its speed is 1000 rpm. Its efficiency as a function of the flow rate is given in Table 4.4.

Flow Rate	300	250	200	150	100	60	50
Efficiency	88	89	89	89	87	83	81

Table 4.4. *Turbine efficiency as a function of flow rate [AMO 09]*

The turbine is coupled directly to an asynchronous machine. This machine, sized for the nominal turbine power and with a nominal voltage of 380 V, is connected to the 50 Hz electrical network via a transformer. The evolution of its efficiency and its power factor as a function of its load is given in Table 4.5.

Load Level	4/4	3/4	1/2	1/4
Efficiency	96	96	95	94
Cos φ	0.89	0.88	0.85	0.7

Table 4.5. *Generator efficiency and power factor as a function of the load level [AMO 09]*

1) Determine the heights of gross and net fall.

2) Determine the nominal hydraulic power.

3) Determine the powers obtained after turbining for each operating point indicated in Table 4.4. Comment on your results.

4) Justify the choice of a Pelton turbine. What is the role of the injectors?

5) Determine the number of poles of the asynchronous machine.

6) Determine the power generated by the asynchronous machine for each load level. Comment on your results.

7) Determine the current of the asynchronous machine for each load level. Comment on your results.

Answers

1) $H_b = 1077 - 752 = 325$ m

$H_n = 325 - 65 = 260$ m

2) $P_h = \rho\, g\, Q\, H_n = 1000.9{,}81.0{,}3.260 = 765.18$ kW

3) $P_t = \rho\, g\, Q\, H_n\, \eta_t$

Flow Rate	300	250	200	150	100	60	50
Efficiency	88	89	89	89	87	83	81
P_t (kW)	673.36	567.51	454	340.51	221.9	127.02	103.3

Table 4.6. *Turbine efficiency and power as a function of flow rate [AMO 09]*

The efficiency of the turbine decreases when the flow decreases.

4) The choice of the turbine is justified on the basis of the diagram in Figure 4.19.

High drop height greater than 25 m.

Low flow rate less than or equal to 0.3 m^3/s.

The injectors allow the flow to be adjusted.

5) 50 Hz electrical network, Nn = 1000 rpm, 6-pole machine

6) 4/4 $P_t = 673.36$ kW

$P_g = P_t\, \eta_g = 673{,}36.0{,}96 = 646.43$ kW

3/4 $Q = 225$ l/s, $\eta_t = 89\%$

$P_t = \rho\, g\, Q\, H_n\, \eta_t = 573.89$ kW

$P_g = P_t\, \eta_g = 573{,}89.0{,}96 = 550.93$ kW

1/2 $P_t = 340.51$ kW

$P_g = P_t\, \eta_g = 340{,}51.0{,}95 = 323.48$ kW

1/4 Q = 75 l/s, η_t = 84.5% obtained by linear extrapolation ((75-60)(0.87-0.83)/(100-60))+0.83

$P_t = \rho\, g\, Q\, H_n\, \eta_t = 161.64$ kW

$P_g = P_t\, \eta_g = 161{,}64.0.94 = 151.94$ kW

The efficiency of the generator remains high over the entire operating range of the hydroelectric plant.

7) 4/4 $I_g = 646.43 / (1{,}732.0{,}89.380) = 1103.5$ A

3/4 $I_g = 550.93 / (1{,}732.0{,}88.380) = 951.2$ A

1/2 $I_g = 323.48 / (1{,}732.0{,}85.380) = 578.2$ A

1/4 $I_g = 151.94 / (1{,}732.0{,}7.380) = 329.78$ A

The cos φ decreases when the power of the generator decreases.

4.2. Hydraulic power of the sea

4.2.1. *Wave power*

4.2.1.1. *Origin and description of waves*

Waves are a form of non-polluting renewable energy. They can be created by:

– the presence of an object in water (example: wake of a boat);

– the crossing of currents;

– seismic and volcanic activity;

– wind blowing at the surface of water.

In this book, we will take a more specific look at waves generated by the wind, because their energy concentration is the highest in comparison to other waves. They represent a natural conversion of the wind power coming from wind blowing on the ocean surfaces, which itself comes from the conversion of part of the solar power. With these two successive conversions, energy is very concentrated. The average flow of the wave power is generally five times denser in surface waters than that of the wind power located 20 m above the sea surface, and 10 to 30 times denser than the solar power flow.

When the wind blows on a smooth surface of water, air particles rub water molecules. This frictional force between air and water, associated with the surface tension of water and with gravity, spreads over the surface, thereby forming small wrinkles or oscillations. The latter are called capillary waves. They will become increasingly well formed as the wind keeps on blowing. The size of the waves depends on the wind speed, on the distance over which it blows – called the fetch – on the time during which the wind blows, as well as on the depth and the topography of seabeds. When waves get further away from their place of origin and still keep on propagating freely without wind, they are called a swell. In deep waters, a swell can cover long distances, almost without any energy loss.

On the contrary to what we might think, energy (and not water) is propagating on the surface of the oceans. According to the theory by the English astronomer and mathematician George Biddell Airy, at each point of the ocean, water particles are moving according to a circular trajectory, when the wave passes by. Its diameter is at its maximum at the wave surface and decreases exponentially with the depth (Figure 4.22).

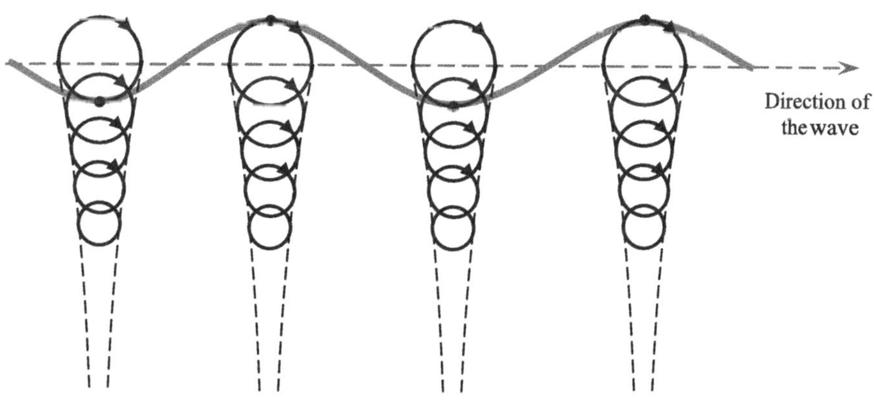

Figure 4.22. *Trajectory of the water particles at the surface of the sea and according to the depth*

Waves are scientifically considered to be surface gravity waves. A wave corresponds to the propagation phenomenon of a disturbance – the disturbing element being the wind – from one point of a material environment (water) up to another point, which does not induce any global motion of the environment itself. Waves are thus interface oscillations, which are maintained by an exchange between the kinetic energy of water particles with a circular trajectory and the gravitational potential energy of those located at the top of the wave (Figure 4.23). On average, the kinetic energy of a wave is equal to its potential energy. The waves are called gravity waves, because their potential energy is due to the Earth's gravitational force.

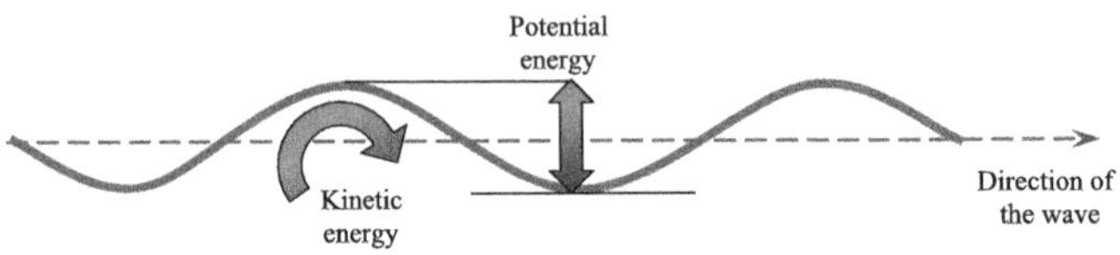

Figure 4.23. *Overview of the wave energy*

According to a simple model, established by Airy, a regular wave can be characterized, for a depth of local water h, by the following parameters (Figure 4.24):

– its wavelength L, representing the distance between two crests or two successive hollows;

– its crest-to-crest period of time T, representing the time taken by a wave to cover the wavelength;

– its frequency f, which gives the number of waves passing through a specific point;

– its amplitude A, representing the distance between the crest and the average sea level;

– its height H, representing the distance between the hollow and the crest;

– its speed c.

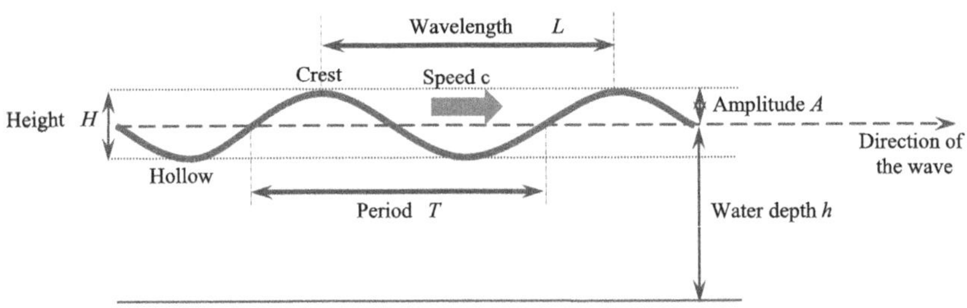

Figure 4.24. *Characterization of a wave*

Similar to a sine wave moving in a homogeneous environment, we can associate two wave speeds to the waves: phase and group speed. If we observe any specific point of the wave (for example, the crest), it will give the impression of moving on the surface of the water with a certain propagation speed: this is the phase speed, which corresponds to the sine curve translation. As for any physical wave, waves are actually made up of a packet of monochromatic waves propagating at various speeds. This packet moves at a different speed from the plane waves constituting it: this is the group speed. The energy transported in the waves moves at this group speed. In deep waters, when the depth is two times more significant than the wavelength, the phase speed c is two times greater than the group speed c_g:

$$c = 2c_g = \frac{g}{2\pi}T \qquad [4.10]$$

Each wave is different (in amplitude, period and direction) than the preceding and following one. Therefore, the sea state, made up of several waves, can be defined by two statistical parameters based on the power spectrum of the waves:

– the significant height H_s which is the average of the crest-to-hollow heights of the third of the highest waves. This is sometimes also noted as $H_{1/3}$;

– the peak period T_p is the peak period of the swell power spectrum.

4.2.1.2. *Potential*

4.2.1.2.1. Energy and power

For a sine wave with a crest-to-hollow height H, the energy density corresponding to the average energy E contained in a meter square at the surface of the water is:

$$E = \frac{\rho g H^2}{8} = \frac{\rho g A^2}{2} \qquad [4.11]$$

with ρ as the water density at the sea surface, ranging between 1,020 and 1,029 kg/m^3 according to the temperature and the salinity, and g as the gravity acceleration worth 9.81 m/s^2.

The potential energy, related to the water height ranging between the hollow and the crest, represents half of this average energy. The other half corresponds to the kinetic energy due to the water motion.

The power corresponds to the energy per period and is determined by dividing the energy density by the period of the wave:

$$P = \frac{E}{T} = \frac{\rho g H^2}{8T} = \frac{\rho g A^2}{2T} \quad [4.12]$$

4.2.1.2.2. Vertical distribution of energy

We saw previously that water particles move according to a circular trajectory, whose diameter decreases exponentially with the depth. Consequently, the energy density also decreases with the depth. To correctly size a conversion system of the waves, which would be immersed, it is thus necessary to know the energy density available at the depth of implantation and the operation of the converter. For waters whose local depth h is higher than half of the wavelength L, the relationship between the energy at the surface and the energy at a depth d is:

$$E(d) = E_{surface} \cdot e^{\frac{-2\pi d}{L}} \quad [4.13]$$

In deep waters, we estimate that 95% of the transported energy is located between the surface and a depth equal to a quarter of the wavelength.

4.2.1.2.3. Transported power per meter of a wave front

The primary energy resource contained in the swell is typically quantified by calculating the power transported per meter of front wave, i.e. per meter in a plane perpendicular to the wave propagation direction. It can thus be obtained by multiplying the energy density by the group velocity:

$$P_{wave\_front} = c_g E = \frac{\rho g^2 H^2 T}{32\pi} = \frac{\rho g^2 A^2 T}{8\pi} \quad [4.14]$$

A 2 m high wave with a 10 second period will thus have an energy density of 5 J/m^2 and a wave front power per meter of 40 kW/m.

4.2.1.3. *Global resource*

The global distribution of the available power has been established by measuring the wave parameters over several dozen years, mainly with the help of measurement buoys installed at various points of the globe. They are also called wave recorders. They periodically record the data over a given period and transmit them by radio wave or satellite. The height and period of the waves are then deduced from the spectral or temporal analysis of the temporal sequences of the surface evolution of the sea. The main propagation direction is directly measured with the help of directional sensors.

Depending on the site topology, other characterization methods of the state of the sea are also used, such as the use of pressure sensors or of immersed probes, laser remote sensing or radar altimeter.

These various readings enable us to establish a world atlas of the average power of ocean wave fronts (Figure 4.25).

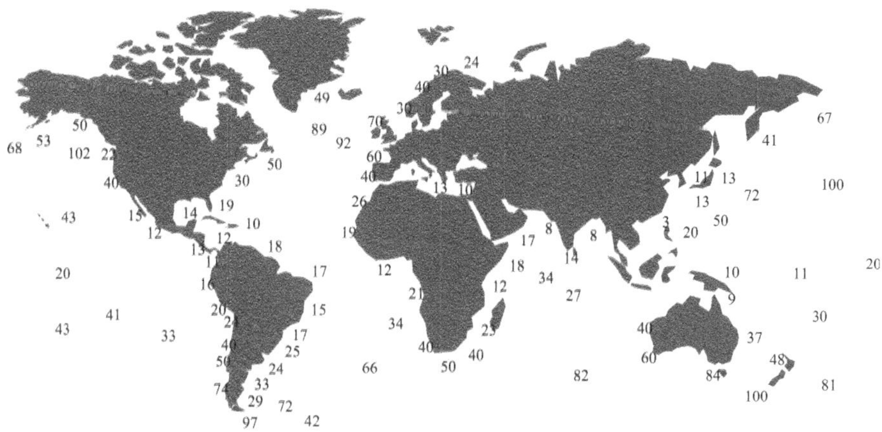

Figure 4.25. *Global distribution of the average wave power in kW/m [FAL 99]*

The global resource is evaluated at 22,000 TWh/year. In France, exploitable sites, mainly located on the Atlantic Coast, where the wave power is estimated at 45 kW/m, represent a potential of 40 TWh/year for an installed power of 10 to 15 GW [ADE 10].

4.2.2. *Energy of the continuous ocean currents*

4.2.2.1. *Description of the current phenomenon*

These ocean currents correspond to the motion of sea waters and transport a large quantity of energy. The solar radiation unequally distributed at the surface of the Earth, the Coriolis force related to the rotation of the Earth and the gravitational force are the source of this phenomenon. The combination of these various elements leads, between the Equator and the poles, to:

– temperature differences and thus air density differences, thereby creating winds leading to surface currents, affecting the waters up to 400 m depth, i.e. about 10% of the waters (Figure 4.26);

– temperature and salinity differences and thus water density differences, causing the thermohaline circulation at the origin of the depth currents, which affect the waters at depths exceeding 400 m, and which represent about 90% of ocean currents (Figure 4.27).

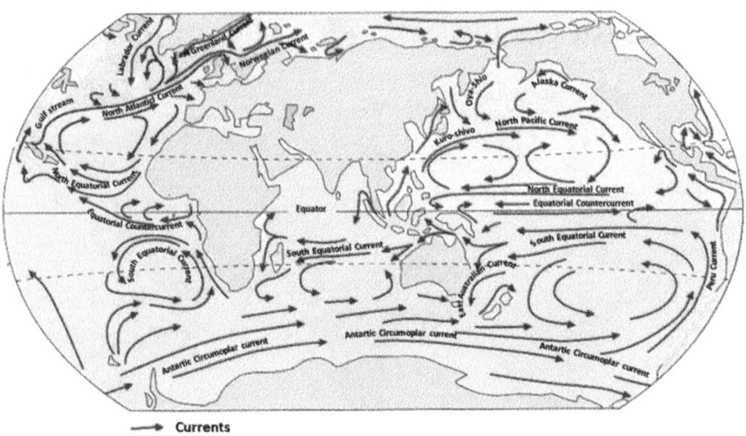

Figure 4.26. *Surface currents [EDU 20]. For a color version of this figure, see www.iste.co.uk/robyns/electricity.zip*

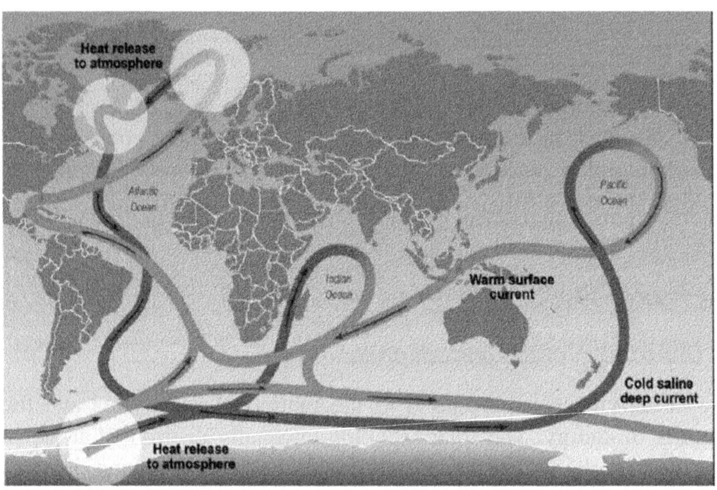

Figure 4.27. *Thermohaline ciruclation [SWI 06]. For a color version of this figure, see www.iste.co.uk/robyns/electricity.zip*

4.2.2.2. *Potential*

The energy of the currents comes from the kinetic energy of the water in motion, which can be expressed in the following form:

$$E_c = \frac{1}{2} m V^2 \qquad [4.15]$$

with m as the mass of water and V as the current speed.

The instantaneous power recoverable by the helix of an underwater turbine of surface S placed perpendicularly to the current is thus:

$$P_r = \frac{1}{2} \rho S V^3 \qquad [4.16]$$

with ρ as the density of sea water.

As for a wind turbine, the power extracted by an underwater turbine must take into account a power coefficient C_p, corresponding to the global efficiency of the underwater turbine:

$$P = \frac{1}{2} C_p \rho S V^3 = \frac{1}{2} C_p \rho \pi R^2 V^3 \qquad [4.17]$$

with R as the radius of the underwater turbine and $C_p<0.59$ (Betz limit).

The most well-known example of a warm ocean current is the Gulf Stream, which is located in the Atlantic and goes from the east of Florida to the high latitudes of Europe. Propagating at a speed, which can exceed the 2 m/s, it transports a thermal energy of about 1 million billion watts, i.e. a thousand times the global production of energy [SWI 06].

4.2.3. *Tidal energy*

4.2.3.1. *Tide phenomenon*

Tides are oscillatory motions of the water level, which are comparable in surface to that of the swell, but which have a very significant wavelength, which is always greater than the ocean depth. The variation of the sea level is due to:

– the respective gravitational actions of the Moon and the Sun on the liquid particles of the Earth;

– the sea and coast configuration.

The amplitude of the tide at a given point varies according to the relative position of these three stars:

– the highest tides then occur during the alignment (or syzygy) of the Moon and the Sun in conjunction (new moon) or in opposition (full moon), when their attractions are adding up: these are spring tides;

– the lowest tides occur when the Moon and the Sun are in a quadrature position, i.e. when they form a right angle with the Earth (first or last moon quarter), when their attractions have separate orientations: these are neap tides.

The effect of the Moon on the Earth's bodies of water is significantly higher than that of the Sun because of the Moon-Earth distance, which is much shorter than the Sun-Earth distance. The periodicity of the tide thus follows that of the "moon day", whose value is 24 h 50 min.

Depending on the place, there are four different types of tide according to the Moon's phases:

– Semi-diurnal tide: two high tides and two low tides per day of identical heights.

– Semi-diurnal tide with diurnal inequality: two high tides and two tides per day, but with important differences of height and duration between the increasing and decreasing tide.

– Mixed tide: a period of semi-diurnal tide and a period of diurnal tide follow one another during a lunation.

– Diurnal tide: only one high tide and only one low tide per moon day, with a tidal range that is generally lower than the semi-diurnal tide.

A tide, at a given date and in a given place, is defined by:

– its tidal coefficient. Introduced by the physician Laplace, this coefficient is a number without dimension ranging between 20 and 120. It is associated with the amplitude of the semi-diurnal tide, which is predominant on the Channel and Atlantic coasts. Calculated for a high sea, it helps to quantify the significance of the tide. For neap tides, the coefficient is lower than 50, and for spring tides, it is higher than 90;

– its tidal range, which corresponds to the difference of the water height between the high tide (HT) and the low tide (LT);

– time of the high tide;

– time of the low tide.

The tide itself corresponds to the vertical component of this oscillatory motion. Tidal currents are formed by the horizontal component and constitute another aspect of the gravitational effect, in relation to the motion of the three stars on the oceans.

4.2.3.2. *Tidal amplitude*

The evolution of the sea level is not linear and can be represented by a variation curve of the tidal amplitude, according to the time, called the tidal curve (Figure 4.28). Its reference level is the chart datum, which corresponds to the lowest level reached by the sea, i.e. the water head of the low tide of the highest tide of coefficient 120. The approximate duration of a tide is 6 hours for a rising or lowering.

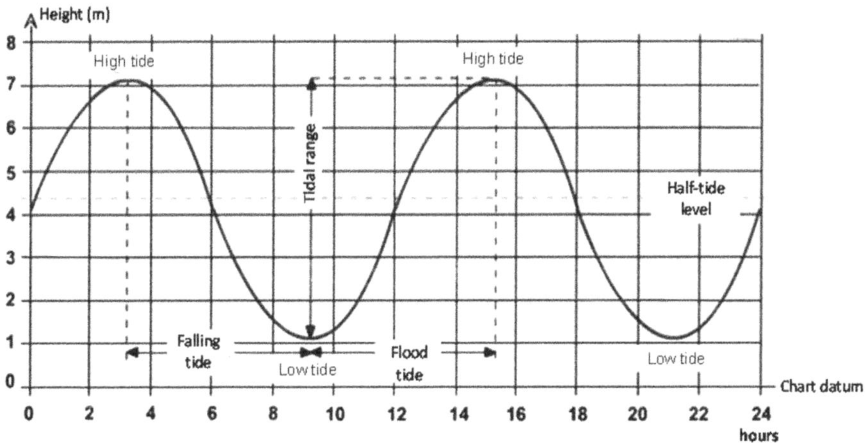

Figure 4.28. *Curve of semi-diurnal tide [ENM 11]*

The spectral representation, which results from a calculation, can also be used to characterize the tidal phenomenon. It has the advantage of giving energy according to the frequency and showing that its amplitude is more significant at lower frequencies.

The tidal rise at any time t can be broken down into an unlimited sum of strictly periodical elementary oscillations, which are called harmonic components. It is expressed by the formula called the "harmonic tidal formula":

$$h(t) = Z_0 + \sum_j \sum_i A_{ij} \cos\left(V_{ij} - G_{ij}\right) \qquad [4.18]$$

with:

– Z_0 as the average level of periodic oscillation;

– A_{ij} and G_{ij} being called harmonic constants: the amplitude and the phase of an elementary wave depending on the considered point, respectively;

– V_{ij} as the astronomical argument related to time t;

– the index i characterizing the nature of the wave;

– the index j related to the period of the wave ($j = 0$ for annual waves, $j = 1$ for "diurnal" waves that have a period close to 24 h, $j = 2$ for "semi-diurnal" waves with a period close to 12 h, etc.)

The sinusoidal variation of the height can be approximated by a broken line by applying the "rule of twelfths":

– 1st hour of the tide: rising or lowering of 1/12 of the tidal range;

– 2nd hour of the tide: rising or lowering of 2/12 of the tidal range;

– 3rd hour of the tide: rising or lowering of 3/12 of the tidal range;

– 4th hour of the tide: rising or lowering of 3/12 of the tidal range;

– 5th hour of the tide: rising or lowering of 2/12 of the tidal range;

– 6th hour of the tide: rising or lowering of 1/12 of the tidal range.

4.2.3.3. *Tidal currents*

The tidal phenomenon produces level fluctuations and currents corresponding to the motions of bodies of water subjected to the mentioned gravitational effects. The recoverable kinetic energy in tidal energy is quantified in the same way as for high sea currents (section 4.2.2.2). The power available for an underwater wind turbine is similar to that of a wind turbine (see equation [4.17]).

4.2.3.4. *Potential*

The tidal range of a tide has the advantage of being specifically predictable, because the motion of the Moon and the Sun can be determined over several centuries. We consider a site that is exploitable from a tidal range over 5 m. The global exploitable resource is estimated at 380 TWh/year for a peak power of 160 GW. The potential in the United Kingdom is estimated at 6 GW and at 2 GW in France [ADE 10].

The observed tidal ranges around the world are very variable according to the typology of the seas. In deep waters, the tidal range is generally low or even non-existent, of about a few dozen centimeters at a maximum. On the coasts, the values are much higher, up to several meters (Figure 4.29). The most significant tidal range is found in the Bay of Fundy, on the Atlantic Coast of Canada: it reaches 16 m.

Figure 4.29. *Amplitudes in meters of the average spring tides around the world [GIB 55]*

The French resource is mainly located on the Brittany coast, where tidal ranges can reach 14 m, as in the Mont Saint-Michel bay (Figure 4.30). The French tidal power production is located in the La Rance estuary and is about 500 TWh/year.

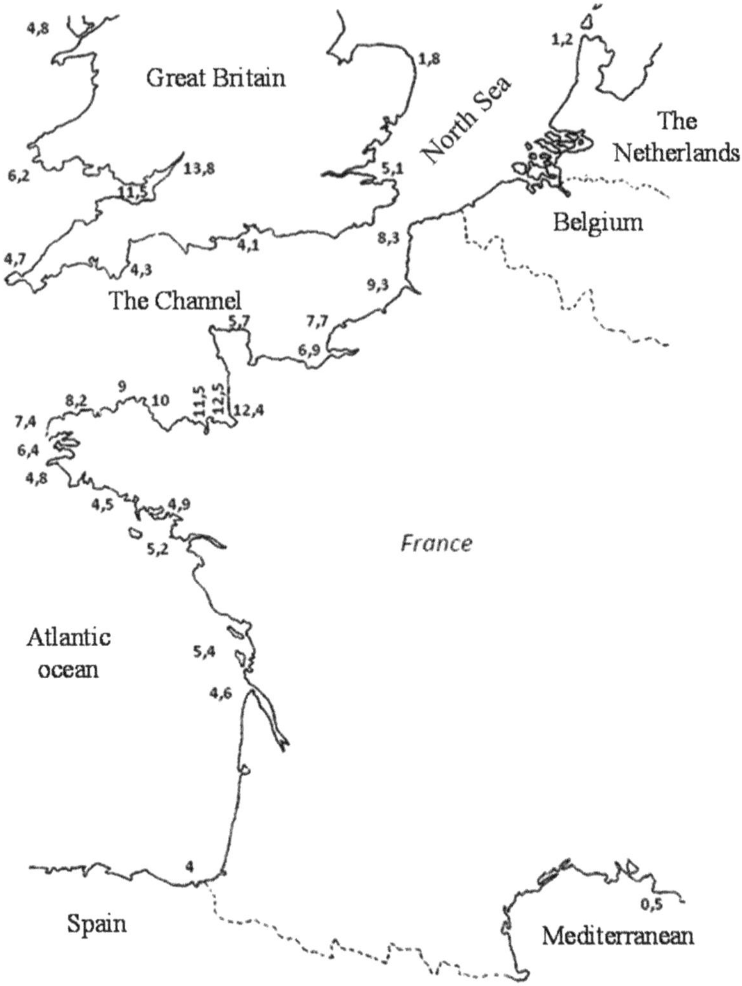

Figure 4.30. *Amplitudes in meters of the average spring tides on the coasts of France and Great Britain [GIB 55]*

The resource related to the tidal current is studied similarly to that of the tidal range and can also be represented as an atlas with propagation speeds expressed in m/s. The French coast potential is greater than 6 GW, with favorable zones on the coasts of Brittany and Normandy (Figure 4.31). The technically exploitable sites, where the current speed is higher than 1 m/s and the water depth is at least 20 m, have a potential of 14 TWh/year for 2.5 to 3.5 GW of installed power [ADE 10].

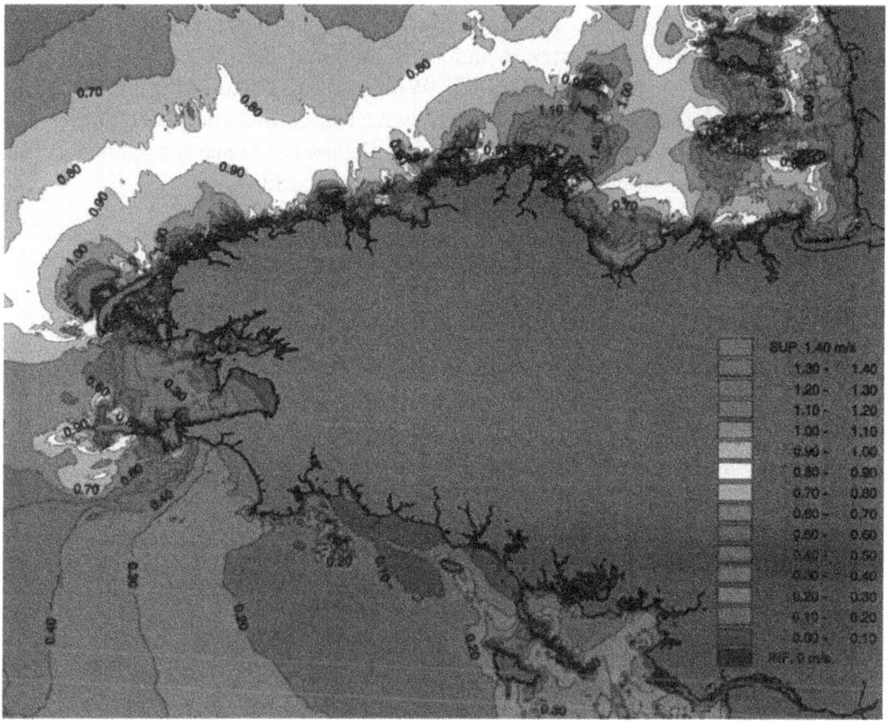

Figure 4.31. *Intensity of the tidal currents in Brittany in m/s [BRE 09]. For a color version of this figure, see www.iste.co.uk/robyns/electricity.zip*

4.2.4. *Wave production, wave-power generator*

4.2.4.1. *Recovery of the primary energy*

4.2.4.1.1. Historical background

The first system devoted to the recovery of energy from waves appeared in the 18th century (Figure 4.32). Girard, father and sons, filed a patent for it on July 12, 1799 for 15 years. It was called "Pour divers moyens d'employer les vagues de la mer, comme moteurs" (Various ways to use waves as motors) [PÈR 23].

In 1885, Don José Barrufet, a Spanish inventor, patented a device called a "Marmotor", a series of floating buoys in the ocean, whose undulatory motion from top to bottom generated by the wave driving force is transformed via mechanical structures and a generator into electricity [BAR 85].

In 1965, Yoshio Masuda, a Japanese naval officer, invented the first oscillating water column (OWC) converter, which was designed for the electric supply of buoys or offshore installations [MAS 65]. With studies carried out in the 1940s, he is considered to be the father of the modern wave energy recovery technologies. His converter is based on an air turbine. This principle has been used for the installation of more than a thousand navigation buoys around the world.

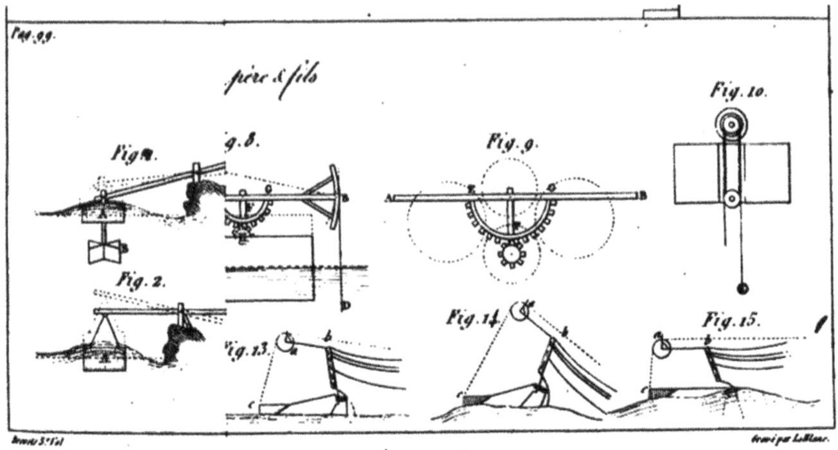

Figure 4.32. *Plate 5 of Girard father and sons' patent [PÈR 23]*

In response to the oil crisis, research in the field of wave energy extraction really started in the 1970s in several countries. In Japan in 1976, after the success of Masuda's air turbine buoys, JAMSTEC (Japan Marine Science and Technology Center created in 1971), built the first floating large scale converter prototype to be deployed offshore. KAIMEI is a large barge of 80 m × 12 m, which was used in 1978 and 1979 as a test platform with up to 9 OWC and equipped with various air turbines onboard.

In 1970, the "Ocean Engineering Program" of the US Naval Academy was implemented. Michael McCormick, the first program manager and who was also considered as one of the pioneers of OWC systems, developed self-rectifying air-turbines for OWC. Following the offer from Japan to the United Kingdom, Canada, France and the United States to participate in a test program under the auspices of the International Energy Agency, he designed a bidirectional turbine for the KAIMEI project. He is also the author of some of the first journal articles on the subject [MCC 74] and of the first book about wave energy conversion [MCC 81].

In 1974, within the "Wave Power Group" of Edinburgh University, Stephen Salter, from the department of mechanical engineering, invented and tested a nodding floater, an anchor float with a hydraulic conversion system, known under the names of "Salter's Duck", "Nodding Duck" or "Edinburgh Duck". This system was the subject of a first patent in 1975 [SAL 75], but remained a laboratory test prototype on a reduced scale and was never installed at sea on a full-scale.

From 1975 to 1982, the British government authorized a major research and development program on wave energy, under the management of Clive Grove-Palmer. The Department of Energy (DOE) was financially supporting several projects at this time:

– Stephen Salter's Duck;

– the NEL OWC, a water column system developed by the National Engineering Laboratory (NEL) in East Kilbride, Scotland;

– the wave-contouring raft, generally called the Cockerell raft, an articulated float (or raft) developed by Sir Christopher Cockerell, a British engineer;

– the HRS Rectifier, a rectifier in the shape of a large hollow rectangular box, with a series of narrow compartments developed by R.C.H. Russell in 1975, within the Hydraulics Research Station.

This first generation of systems is mainly shoreline based, the major drawbacks being their visual impact and the obligation to adapt to coastal topology. Since the beginning of the 21st century, a second generation of systems has appeared and is mainly located in nearshore and offshore zones, where the potential of available primary resources is much more important because of the depth of the seas. This second generation of systems will be presented in the following sections.

4.2.4.1.2. Classification and technologies of converters

Throughout the last century, a wide variety of concepts and devices have been developed and/or tested. The various energy conversion technologies are based on the use of one or more wave characteristics, which are mainly:

– wave curvature, which can be used to create mechanical torque around an axis;

– the surface sine-wave oscillation, which generates vertical motion;

– breaking waves, which create horizontal motion.

To recover the force supplied by the wave, we have to implement a force reaction system. To achieve this, wave power plants are generally made up of the following elements:

– an absorber, in charge of directly "absorbing" the wave energy. It can be the surface of the sea itself, a body at the free surface of the sea or immersed (buoys, floats, bodies put in motion by the swell, flapping wing, etc.) or a structure manufactured as rigid or flexible (capture chamber, counter-reaction mass, etc.);

– a reaction point, which can be a fixed structure (concrete structure with foundations or earth), anchored to the bottom (by gravity anchor or with piles) or with inertia;

– a "working fluid", used to convert the energy extracted from the waves and, which can be sea water, air or hydraulic oil;

– the mechanical conversion equipment, which can typically be in the current devices, a (air or water) turbine, a pump or a hydraulic motor;

– the electromechanical converter, which can be a rotating, linear synchronous or induction generator.

The force reaction resulting from the relative motion between the absorber and the reaction point will enable us to mechanically activate the mechanical conversion equipment with the help of the working fluid. Then, the electric generator, mechanically coupled to the mechanical conversion equipment will ensure the electromechanical conversion. The various stages of wave energy recovery thus lead to three successive energy conversions. Losses, mainly under the form of heat occur at each conversion stage. Figure 4.33 proposes a non-exhaustive classification of the wave power recovery systems using these various conversions.

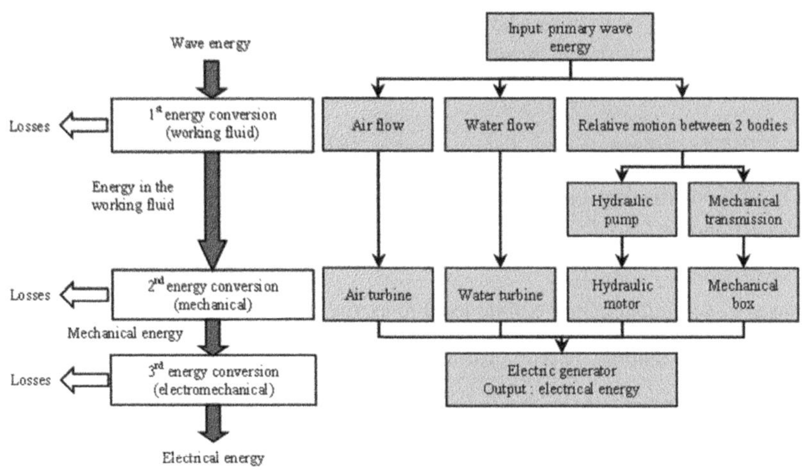

Figure 4.33. *Stages of wave energy recovery. For a color version of this figure, see www.iste.co.uk/robyns/electricity.zip*

Wave energy converters (WECs) can be classified according to the following criteria:

– their size and orientation: if their size is relatively small in comparison to the wavelength, we can speak of an absorber, otherwise it is an attenuator when the system is parallel to the wave propagation direction or a "terminator", if it is perpendicular;

– their location: onshore or shoreline, nearshore or offshore.

In this book we are interested in the following systems:

– the oscillating water column (OWC) exploiting the sea level variations on the compression of a volume of air. It is an onshore, nearshore or offshore technology;

– the breaking wave ramp system (or OTD for overtopping device) exploiting the breaking waves, which is positioned on the coasts or which can sometimes be floating offshore;

– the body systems driven by the swell or direct wave generators (or WABs, Wave Activated Bodies, or Direct Wave Energy Converters), whose oscillating bodies are at the surface or immersed according to the system and are generally installed at sea.

By taking an electrotechnician point of view, these three different system concepts can be divided into two sub-categories depending on whether the electromechanical conversion is ensured by a continuously rotating electric generator or a limited motion electric generator (linear or angular).

4.2.4.2. *Electromechanical conversion with limited linear motion*

The main devices implementing an electromechanical conversion system are typically based on Archimedes' principle and exploit the vertical motion of the waves, which is called a heave motion. The most widespread system is that of the heaving buoy. These systems exploiting the potential energy of the waves are basically made up of three elements:

– a mobile part at the surface or immersed (float), designed to follow the vertical motion of the surface water;

– a fixed part with an anchor point to the seabed;

– a linear permanent magnet generator.

One of the components is moored or fixed at the bottom. Therefore, this type of converter falls into the category of systems with bodies moved by the swell with an external reference. It also qualifies as an absorption point system or "localized absorber", because it is able to convert energy at one point of the ocean in all directions. Depending on whether it is at the surface or immersed, the float will move in various directions according to the wave propagation direction.

4.2.4.2.1. Immersed float system

The float immersed under the water surface at a given depth rises and lowers itself according to the force exerted by the quantity of water located above it. When the crest of a wave gets above the float, the mass of water increases and pushes the float towards the bottom. Conversely, in the presence of a wave hollow, the mass decreases and the float goes back up (Figure 4.34). The float, constituting the mobile part of the converter, thus acts as a piston and is directly coupled to the mobile part of the linear generator, which thus constitutes the damper of the vertical translation motion. This prevents us from using an intermediate mechanical conversion system, with gears or of the hydraulic type. The drive is described as direct.

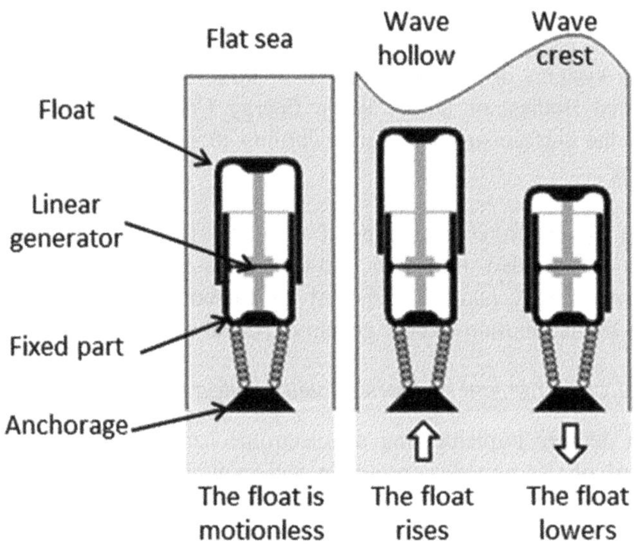

Figure 4.34. *Absorber with immersed float [POL 05]*

The force exerted by water can be expressed using Newton's second law:

$$F = \rho HSg \quad [4.19]$$

with ρ as the density of water, H as the wave height, S as the surface of the float in m^2 and g as the acceleration of gravity.

The power transmitted to the float is obtained by multiplying the force by the float speed:

$$P = Fc = \rho HSg.\frac{2L_f}{T} \qquad [4.20]$$

with L_f as the length of the float and T as the period of the wave.

The Archimedes Wave Swing (AWS), commercialized by the Scottish company AWS Ocean Energy Ltd, is an example of a localized absorber, operating according to this concept of an immersed float. The fixed part anchored to the bottom of the sea is a cylindrical chamber full of air, on which a second mobile cylinder (used as a float) slides. It oscillates as a piston under the variations of the water height and compresses air in the chamber, when it lowers in the presence of a wave crest. When the water pressure decreases in the wave hollow, the air compressed in the chamber acts as a spring and lifts the float. Electromechanical conversion is ensured by a synchronous linear generator with permanent magnets set on the float and whose stator windings are located on the fixed part. In 2004 a full-scale prototype of a 1 MW nominal power was deployed off the Portuguese coast at a depth of 40m and connected to the electricity power network.

Developed from 1999, by the Australian company Carnegie Wave Energy Limited, CETO (named after the Greek Sea Goddess and meaning marine monster in Ancient Greek) is also an absorber that is completely immersed and anchored at the bottom of the sea. But the vertical heave motion of the spherical buoy, instead of directly driving an electric generator, enables us to operate a pump sending sea water under pressure via a pipeline towards an onshore station. The latter uses it to operate a desalination system by reverse osmosis for the production of soft water and a Pelton turbine for the production of electricity (Figure 4.35).

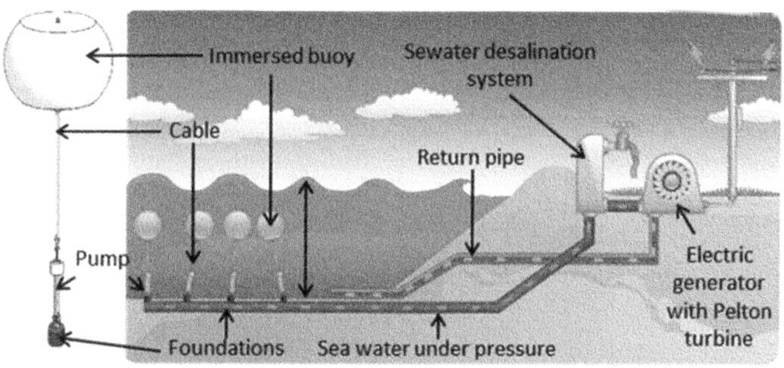

Figure 4.35. *Block diagram of the CETO technology [BAB 11]. For a color version of this figure, see www.iste.co.uk/robyns/electricity.zip*

Between 2003 and 2008, three prototypes were built and tested *in situ* in the Australian waters of Fremantle that have enabled us to validate the concept, which has now been marketed. Three 240 kW 5th generation CETO 5 units operated throughout 2015 at the Garden Island site in Western Australia for the simultaneous production of electricity and desalinated water, as part of the Perth Wave Energy Project. In 2019, the Albany Wave Energy Project (AWEP) planned to install and operate the latest generation, CETO 6, over a 12-month period with a nominal capacity of 1.5 MW. However, it was halted following the withdrawal of funding from the Western Australian government [CAR 20]. Carnegie continues to develop its technology, notably in collaboration with Wave Hub, the test site connected to the network, created in 2010, at a depth of 50 m, and located 16 kilometers off the north coast of Cornwall in the southwest of England [WAV 20].

4.2.4.2.2. Surface float system

The float placed at the surface rises and lowers with wave action. This motion is transmitted via a cable to the mobile part of a linear waterproof generator fixed at the bottom of the sea, which converts the translation motion into electrical energy. The mobile part is equipped with permanent magnets, which induce currents in stator windings. A retraction spring located at the bottom of the generator enables us to pull the piston down, as the buoy goes down in the wave hollow. A compression spring located at the top of the generator enables us to dampen the piston, as a limit during high amplitude waves (Figure 4.36).

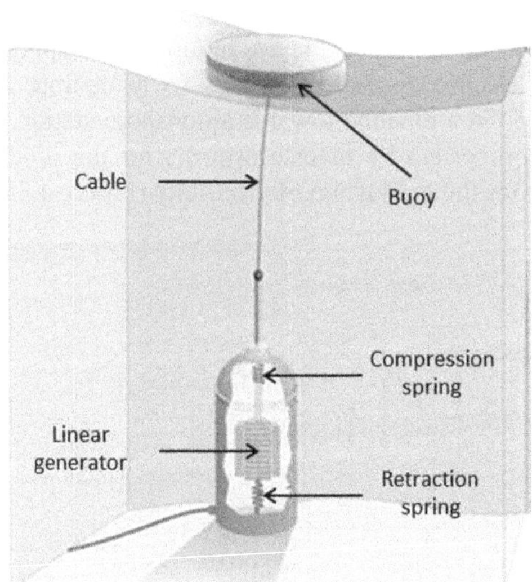

Figure 4.36. *Absorber with a surface float [DAN 11]*

In the framework of the Lysekil project, Uppsala University developed a point absorber based on a vertical synchronous linear generator, which is fixed at the bottom of the sea by gravity foundations and driven via a chord and piston by a buoy at the surface (Figure 4.37). The buoy has six degrees of freedom and therefore, the chord's mechanical guiding system enables us to absorb the horizontal forces induced by the movement of the buoy at the surface. A rotation system ensures the guiding between the mover and the stator.

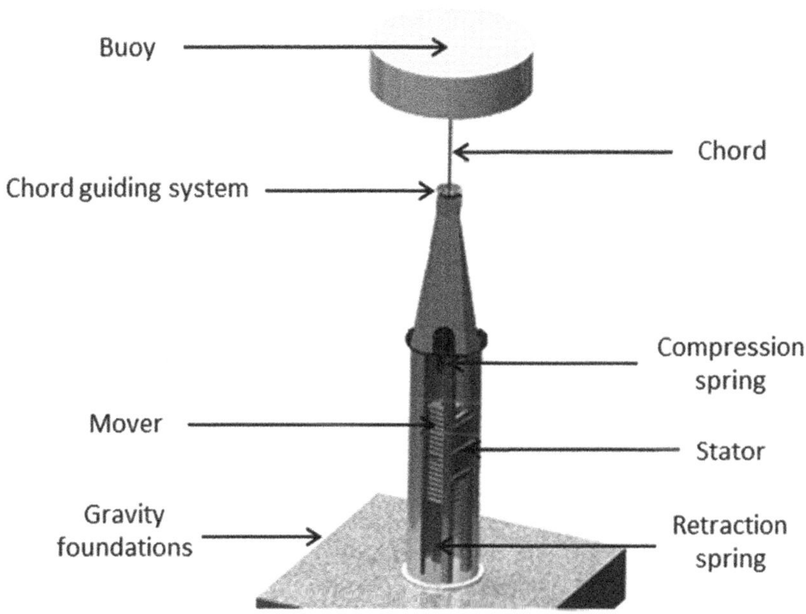

Figure 4.37. *Uppsala concept [RAH 10]. For a color version of this figure, see www.iste.co.uk/robyns/electricity.zip*

Between 2006 and 2010, four experimental prototypes of 10 kW, out of the 10 planned in total until 2014, have been deployed at a depth of 25 m on the research site Lysekil, located 100 km from the West Swedish coast. WEC are connected in parallel by an individual cable to an underwater sub-station, which is itself connected by a single underwater cable to a coastal station.

Powerbuoy and Wavebob systems are respectively developed by the American company Ocean Power Technologies (OPT) and the Irish company Wavebob. They are two examples of point absorbers based on surface buoys.

4.2.4.2.3. Principle of the linear generator

We can describe the principle of a linear electromagnetic structure by relying on that of the rotary structures, because magnetic interactions are exactly the same. The main differences between the linear and rotating structures are as follows [SAI 76, KAN 04]:

– The structure is linear and developed along the air gap, whose thickness is generally more significant than that of a rotary motor, because of mechanical constraints, which are more difficult to control than in rotating concentric architectures.

– The mover can be made up of a simple solid conductive plate, in which the induced currents circulate in the case of induction machines.

– The stator can be broken down in two parts, positioned face-to-face on both sides of the mover, to balance the normal efforts of attraction, to increase the induction in the air gap and to ensure flow closing in the active zones contributing to energy conversion.

– When the rotor is shorter than the stator, the latter can be fixed and the rotor can be mobile. Conversely, when the stator is shorter than the rotor, the stator can be mobile and the rotor fixed. These two linear motor categories are respectively called "short rotor" (or "long induction unit") motor and "long rotor" (or "short induction unit") motor.

– The motion of the linear generator is a bi-directional translational motion and is generally relatively short (between 2 and 3 meters).

– The stator induction is called "sliding", when it is "rotating" in rotary structures.

– There are end-effects, under the form of field discontinuity, because of the "open" structure of the magnetic circuit.

The structure can also have a tubular shape, due to the mover and stator being cylindrical and concentric.

As for conventional rotary machines, the operation is reversible: a linear structure is able to operate as a motor or generator. In the framework of this book, we will pay particular attention to the generators used to convert the mechanical energy contained in the linear motion (resulting from the swell) into electric energy.

In the context of wave-powered generators exploiting linear generators, the motion of the mobile part is bi-directional and necessarily at variable speed depending on whether this mobile part accelerates or decelerates at the beginning or end of the stroke, corresponding to a wave crest or hollow. The induced voltage is thus at variable amplitude and frequency (Figure 4.38) and the machine must be supplied by an electronic converter, which regulates the currents to control damping in the two directions of motion. Considering the low motion frequencies, the direct coupling to the electrical network of a linear generator cannot be considered.

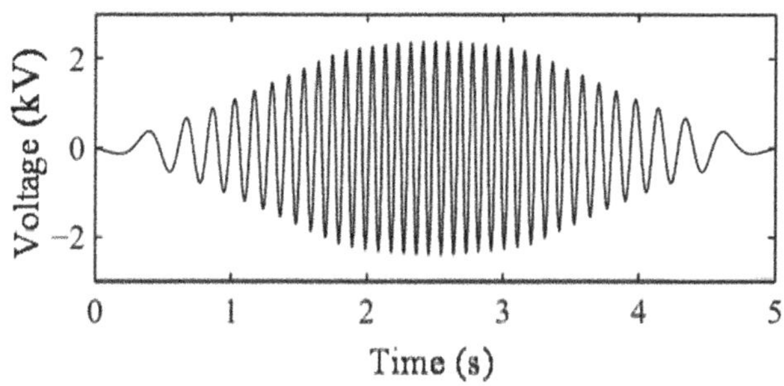

Figure 4.38. *No-load voltage of the Archimedes Wave Swing according to the time on a half-period of the wave [POL 02]*

Due to the direct drive, the mover speed, activated under the buoyancy effect, is directly related to the vertical component of the variation speed of the sea surface. Because it is about 1 m/s, the generator speed is 30 to 50 times slower than that of a conventional rotary generator. Therefore, the effort levels, at a given power, are very high and the dimensions of the linear generator should be larger, in order to obtain an electric power identical to that of a conventional generator operating at a higher speed.

There are various types of linear generators, which can be used for wave energy conversion [VIN 07]:

– induction (asynchronous);

– permanent magnets (synchronous);

– variable reluctance (category of synchronous).

Concerning electromagnetic architectures, we can distinguish generators:

– at longitudinal or axial flux (the travelling field is directed along the axis) (Figure 4.39(a));

– at transversal or radial flux (the flux circulates in a plane perpendicular to the motion direction) (Figure 4.39(b)).

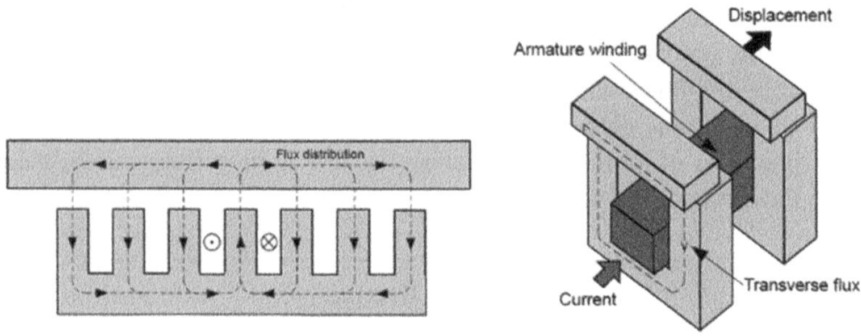

Figure 4.39. *Linear generator with longitudinal and transverse flux [CHE 06]*

Permanent magnet synchronous machines are considered to be the best suited for the direct conversion of wave energy [POL 05].

4.2.4.3. *Electromechanical rotary conversion*

These conversion systems generally implement an air, water or oil turbine, which drives a conventional electric generator in noticeably continuous rotation.

4.2.4.3.1. Oscillating water columns

Oscillating water columns (OWC) are onshore or can be floating. They are made up of (Figure 4.40):

– a chamber open at its lower end under the level of the sea and open or closed depending on the systems on its upper end;

– a partially submerged structure, called a chamber or collector, open on the sea;

– an air turbine coupled with a generator and located in the upper opening adjoined with the chamber.

When the waves enter the chamber (Figure 4.40(a)), the water level increases, leading, by pressure, to an increase in the air present in the column. The air under pressure is propelled towards the top of the column, where the turbine is located. This turbine is then put into rotation by the flow. A generator coupled with the turbine produces electricity. When a wave recedes (Figure 4.40(b)), the air flow in the column is reversed and once again crosses the turbine. The latter is designed to rotate in the same direction, regardless of the airflow direction. The water surface thus behaves as a piston, which alternately drives and draws air in a cylinder.

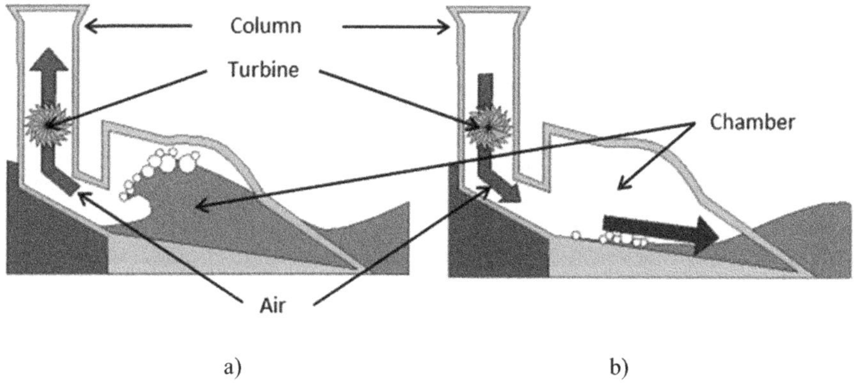

Figure 4.40. *Oscillating water column system [ADE 04]. For a color version of this figure, see www.iste.co.uk/robyns/electricity.zip*

The calculation of the kinetic energy contained in the air flow present in the chamber is similar to the one used for wind turbines. The power removable from the air flow present in the chamber of an OWC can be expressed by the following relationship:

$$P = pvS + \frac{\rho v^3 S}{2} = \left(p + \frac{\rho v^2}{2}\right) vS \quad [4.21]$$

with P as the power available on the turbine in the chamber, v as the air speed, S as the surface swept by the turbine, p as the air pressure and ρ as the air density.

PICO is the first *onshore* experimental structure with a nominal power of 400 kW to use the concept of the oscillating water column. It was commissioned in 1999 as a European demonstration pilot project and is located on Pico Island in the Azores, where the primary resource is evaluated at 13.4 kW/m (Figure 4.41).

The power generation unit is located immediately on the back of the collector and is made up of a single Wells turbine with 8 blades with a horizontal axis that is 2.3 m diameter, which is directly coupled to a dual-fed wound-rotor induction 400 kW generator operating under 400 V between 750 and 1,500 rpm maximum. It is connected to the local network of EDA (Electricidade dos Açores) at the connection point of Cachorro on the north coast of the island.

Figure 4.41. *Front and back views of the PICO [TCT 05]. For a color version of this figure, see www.iste.co.uk/robyns/electricity.zip*

The wave power plant experienced several successive periods of malfunction between 2004 and 2006, associated with design problems, and again in 2013 due to electrical problems that required repair works. It begins to produce autonomously at nominal power as of 2008 (Figure 4.42). Following the structural wear and tear of the submerged parts that led to a partial collapse of the plant in April 2018, the Portuguese non-profit association WavEC, created in 2003, renamed WavEC Offshore Renewables in 2013, and in charge of operations since 2004, took the decision to disconnect the project from the network and close the site down [PIC 20].

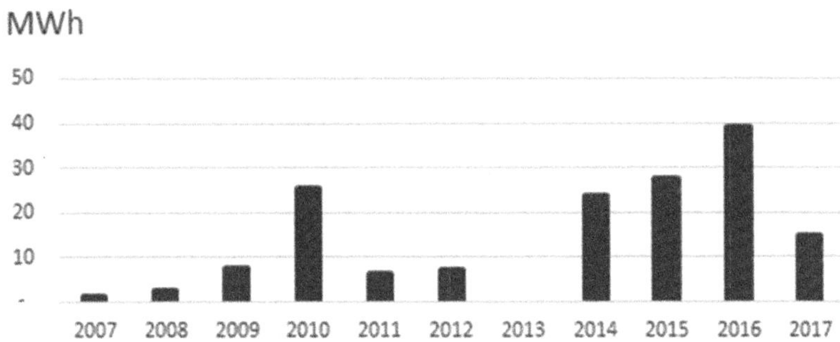

Figure 4.42. *Annual electricity production from the Pico power station [TET 20a]*

Based on the prototype of 75 kW built in 1991 by Queen's University Belfast (QUB), the LIMPET 500 (Land Installed Marine Powered Energy Transformer – 500 kW) by the British company Wavegen is the second example of an onshore 500 kW system installed in 2000 on the coast of the Isle of Islay, to the east of Scotland, where the potential is of 10 to 20 kW/m (Figure 4.43).

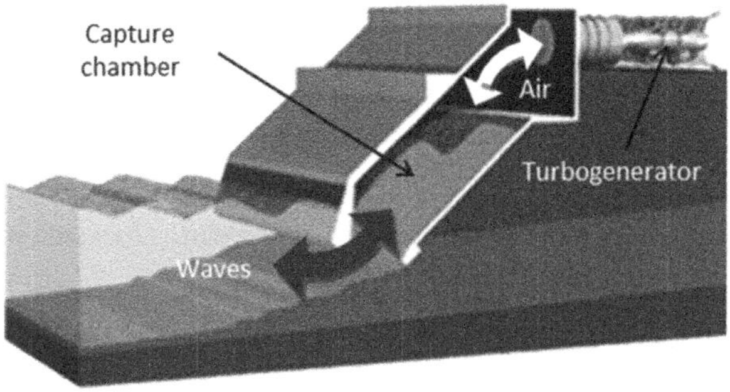

Figure 4.43. *3D view of the LIMPET 500 [TCT 05]. For a color version of this figure, see www.iste.co.uk/robyns/electricity.zip*

Its operation relies on the principle of three oscillating water columns, whose section of 6 m by 6 m of each collector with a 40° slope enables a total absorbing area of 169 m^2. Electromechanical conversion is ensured by two Wells turbines with 7 blades of 2.6m diameter each, which are directly coupled to a wound-rotor induction generator of 250 kW. This generator rotation speed varies between 700 and 1,500 rpm for a nominal voltage of 400 V. The system is connected to the British three-phase 11 kV electrical network.

The concept of floating wave power generator Hydro Air Concept Energy (HACE), developed by the French company of the same name created in 2013, is the first offshore system made up of several submerged vertical oscillating water columns (Figure 4.44(a)) with the ability to harness the energy from small and medium amplitude waves. A 50kW star-shaped prototype (Figure 4.44(b)) was anchored 500 m from the coast in the Port La Rochelle in August 2018. After suffering damage in October 2018, it was repaired and put back into service in May 2019 [HAC 20].

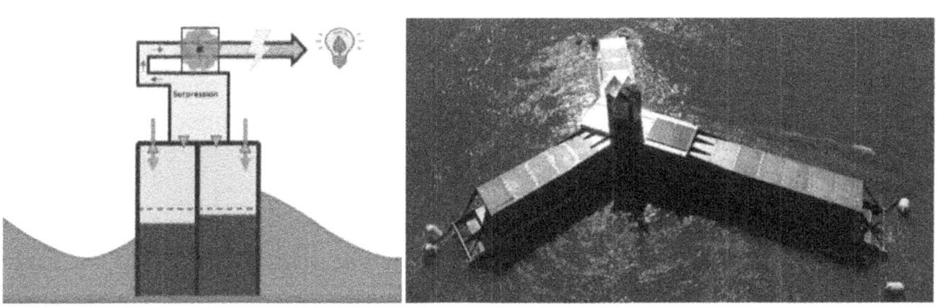

a) b)

Figure 4.44. *HACE multi-chamber offshore system [HAC 20]. For a color version of this figure, see www.iste.co.uk/robyns/electricity.zip*

4.2.4.3.2. Breaking wave system

Breaking wave systems, also called funnel canal systems or TAPCHAN (TAPered CHANnel) systems, have a similar operation mode to that of a low-head hydroelectric power station. Installed on the coast, they are generally made up of (Figure 4.45):

– an open dam with a funnel canal, whose largest part is located on the sea side;

– a high reservoir along a cliff or floating with an adjustable flotation level depending on the sea state;

– a water turbine coupled with an electric generator.

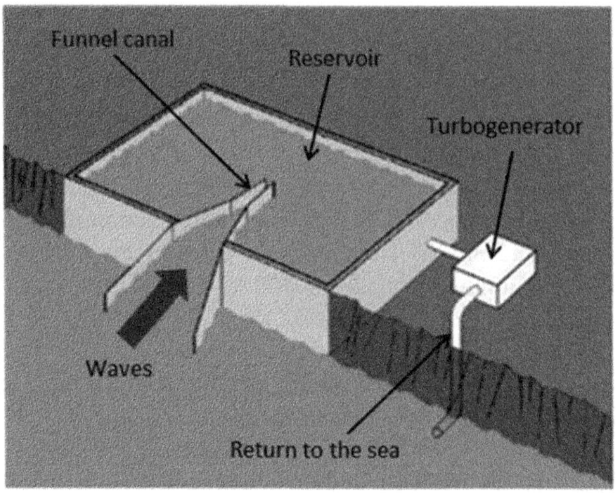

Figure 4.45. *Wave-breaking system [BOY 96]. For a color version of this figure, see www.iste.co.uk/robyns/electricity.zip*

Waves break on the funnel canal, making the level of the reservoir rise. Then, the water flows into the reservoir located at the back of the canal. The wave energy is thus stored in the form of potential energy, which is then converted, when the reservoir water returns to the sea by activating the turbine connected to an electric generator.

Installed in 1985 on the coast of Toftestallen in Norway, the Tapchan Channel is an example of an onshore system with a breaking wave ramp; able to produce 350 kW, it was destroyed during a storm in 1991.

Breaking wave ramp systems, floating and moored at sea, operate following the same principle. The ramp is equipped with deflectors and constitutes the dam, which helps to transport and store water in the reservoir (Figure 4.46). The potential energy of water is converted by a low-head set of turbines, which are located at the bottom of the reservoir.

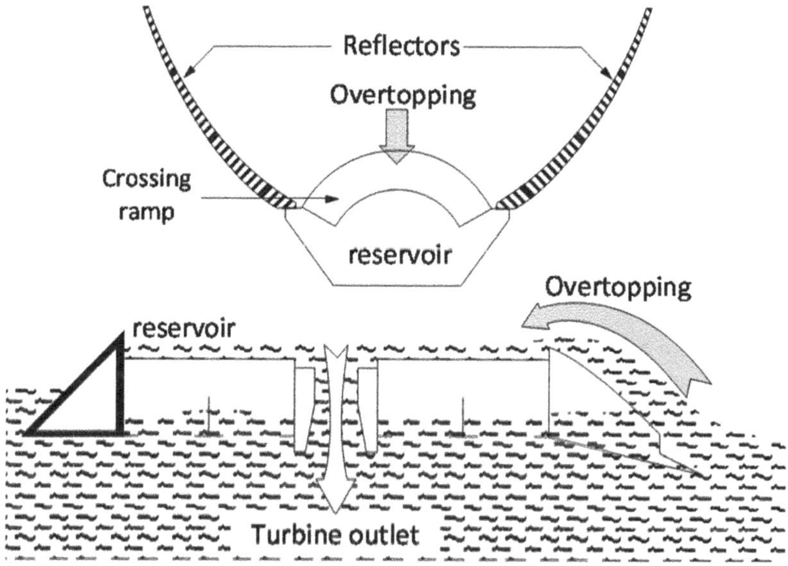

Figure 4.46. *Top and side view of a crossing ramp [TED 07]*

The Wave Dragon is an offshore system with a wave-breaking ramp that was developed by Danish company Wave Dragon ApS. It was first tested from 2003 to 2005 at the Nissum Bredning Fjord test site in the northern part of Denmark in the form of a pilot 2/9th prototype of 20 kW, and was connected to the local electrical network. The reservoir has a capacity of 8,000 m^3 and its flotation level is adjustable from 3 to 6 meters above the sea level, depending on the swell characteristics. The potential energy of the water stored in this reservoir is converted by centrifugation by a set of 16 low-head Kaplan turbines, which are directly coupled to permanent magnet synchronous generators. The annual production is estimated at 20 GWh for a site with a potential of 36 kW/m [WAV 09]. In 2009, a full-scale pre-commercial demonstrator with a 7 MW power unit was installed at Milford Haven Bay in Wales for a test period of three to five years [AQU 08]. Since 2013, studies have been carried out on the Milla Fjord site for the development of a commercial farm with an expected capacity of up to 112 MW, installed in groups of 4 units of 7 MW [TET 20b].

4.2.4.3.3. Systems with bodies moved by swell

There are numerous and varied systems with bodies moved by the swell. Equipped with surface or immersed floats, they exploit the relative motion between several bodies animated by the swell or the motion of a single body, according to an external (Figure 4.47(a)) or internal (Figure 4.47(b)) reference.

a) b)

Figure 4.47. *Systems with bodies put in motion by the swell with (a) external and (b) internal reference [AUB 09]*

SEAREV (*Système Électrique Autonome de Récupération d'Énergie des Vagues* – French Autonomous electric system for wave energy recovery) is a system with bodies put in motion by the swell with internal reference (gravity). It was designed by the LMF (*Laboratoire de Mécanique des Fluides)*, renamed in 2012 to LHEEA (*Laboratoire de recherche en Hydrodynamique, Énergétique et Environnement Atmosphérique* – French research laboratory specializing in hydrodynamics, energy and the atmospheric environment) from the Ecole Centrale de Nantes, as early as 2002 [COR 15]. It is mainly made up of:

– a float with a sealed-hull placed on the surface of the water, which is faired and simply moored to a line, in order to be able to naturally direct it in relation to the main propagation direction of the waves;

– a pendulum wheel of about 400 tons oscillating around an axis of rotation, which is positioned in the float;

– a hydropneumatic regenerative system in the first version and an all-electric system in the version under study. It can be controlled in "latching mode" to benefit from parametric resonances, when the excitation frequency is different from its own frequency.

The float and pendulum wheel, which are indirectly put in motion by the action of the waves on the float, oscillate in relation to one another because of the waves. The wheel directly drives an electric generator, which carries out an optimized damping according to the state of the sea. A "block-latch" system enables us to block or latch the pendulum wheel, in order to amplify its motion to increase the energy efficiency of the converter.

In the case of indirect drive (solution studied by the LMF between 2002 and 2006 for the first generation G1, which was tested in a tank at a scale of 1/12 in October 2006), the pendular wheel activates hydraulic actuators, which pump oil from a low-pressure reservoir in the pneumatic accumulators and send it back under pressure to a hydraulic motor, which drives an induction generator (Figure 4.48).

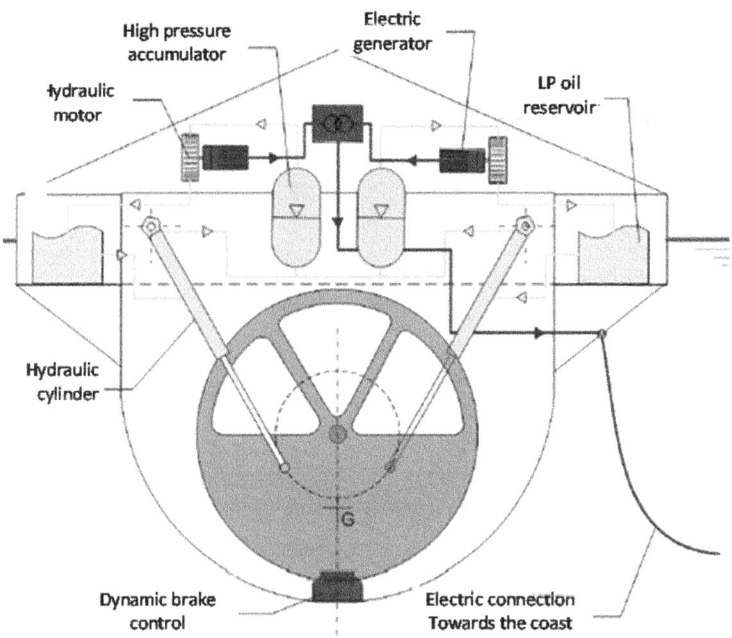

Figure 4.48. *Cross-sectional diagram of the SEAREV G1 with Indirect drivo [COR 15]. For a color version of this figure, see www.iste.co.uk/robyns/electricity.zip*

Between 2006 and 2007, to overcome the problems of maintaining hydraulic components and their inferior performance in comparison to electrical components, a direct drive solution for the second generation G21 was studied in partnership with the Laboratoire SATIE of the ENS of Cachan. It was tested in a tank at a scale of 1/25th from July to October, 2007. This "all electric" version, whose shape and dimensions are also changing, incorporates a three-phase permanent magnet synchronous generator directly coupled with the wheel, thereby at the same time playing the role of an electromechanical converter and a damper of the relative motion and an energy storage system made up of supercapacitors to smooth the large fluctuations in instantaneous electrical power due to direct drive (Figure 4.49).

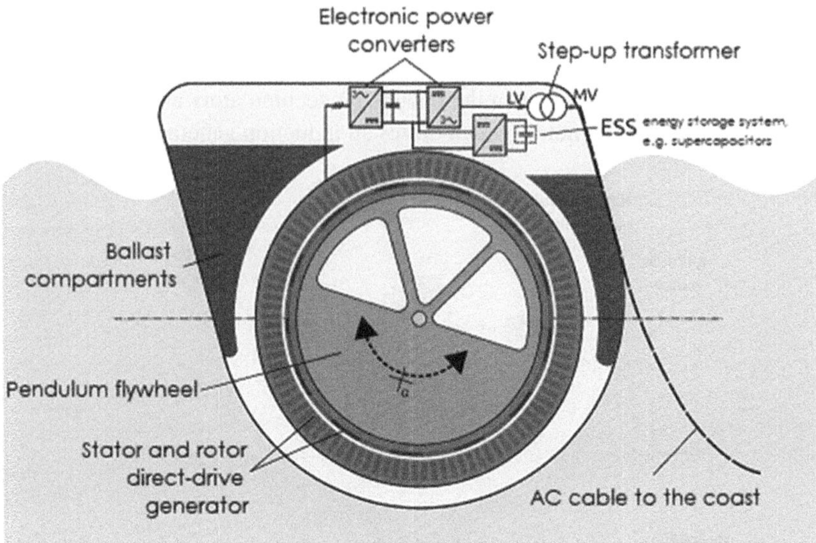

Figure 4.49. *Block diagram of the SEAREV with direct drive [AUB 11]. For a color version of this figure, see www.iste.co.uk/robyns/electricity.zip*

Between 2007 and 2009, the third generation G3 was developed with optimizations to the shape and mechanical parameters for the proven principles of the G21 model.

The maximal average power of the SEAREV prototype at scale 1 should be approximately 600 kW, thereby authorizing an annual production of 1,450 MWh/year for a site with a 20 kW/m potential.

The Oceanswing company was created in 2013 to assess the economic viability for the industrialization of the technology and to conduct the sea tests of the proposed prototype at the SEM-REV sea test site of Centrale Nantes. This test site was established in Le Croisic and connected to the network, launched in 2007, and operational since 2015 [SEM 20].

The PELAMIS, also called the "sea snake", is a modular system with direct drive with bodies moved by the swell developed by the Scottish Company Pelamis Wave

Power (formerly Ocean Power Delivery, created in 1998 and renamed in 2007). Anchored to the seabed with un-stretched moorings, so as to be oriented in the direction of the main waves, its half-submerged and articulated structure is made up of (Figure 4.50):

– 4 cylindrical sections, called floats with a total length of 150 m and a weight of 700 tons (the version P2 includes 5 sections and has a length of 180 m, for a weight of 650 tons or 1,300 tons with ballasts);

– hinged joints placed between the adjacent floats and each integrating an independent module of hydroelectric conversion with a unit power of 250 kW, made up of cylinders, intermediate storage under pressure and hydraulic motors each driving a three-phase induction machine;

– a three-phase 10 kV transformer located in the nose of the front float and, which sends electric energy towards a coastal station via an underwater electric cable.

Figure 4.50. *Description of the PELAMIS P-750 [MUL 09]. For a color version of this figure, see www.iste.co.uk/robyns/electricity.zip*

Under the swell effect, cylinders are put in motion between themselves according to two degrees of freedom. The power resulting from the relative motion between the cylinders is converted on the level of each hinged joint by four hydraulic cylinders: two are damping the vertical heaving motions and two are damping the transverse sway motions (Figure 4.51). The oil pumped under pressure between 100 and 350 bars by cylinders is stored in the high pressure intermediate accumulators and activates two hydraulic motors, which drive two fixed speed induction generators (1,500 rpm for about one slip) of 125 kW.

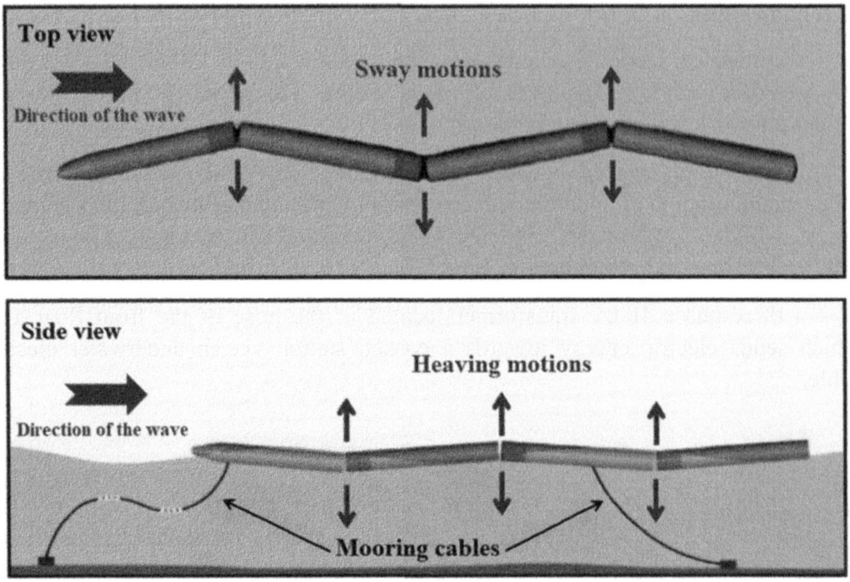

Figure 4.51. *Top and side views of the Pelamis [PEL 11] motions. For a color version of this figure, see www.iste.co.uk/robyns/electricity.zip*

The annual production of the P-750 is estimated at 2.7 GWh for an installation on a site with a potential of 55 kW/m. Pelamis Wave Power tested its first full-scale P1 prototype at EMEC (European Marine Energy Center) between 2004 and 2007. The first commercial wave farm, with a total power of 2.25 MW and using the technology of 3 Pelamis P-750, was implemented in July 2008 in Aguacadoura, a few kilometers north of Portugal and only worked for a few months. Following technical problems, the machines were brought back to shore for repair in November 2008 and remained out of service due to financial difficulties faced by the company running the project. Two second-generation P2 machines, which incorporated an additional articulation and float for identical nominal power, were tested at EMEC from 2010 to 2014, the year in which PWP closed its doors, owing to a lack of funding needed to continue the development of its technology [EME 20]. Nevertheless, the Pelamis is the first wave generator to have reached the industrial stage.

4.2.4.4. *Characteristics of electric production*

The electric power supplied by the systems with breaking waves or with a crossing ramp is smoothed and regular, because it does not directly depend on the wave fluctuation, because of the water storage in a reservoir.

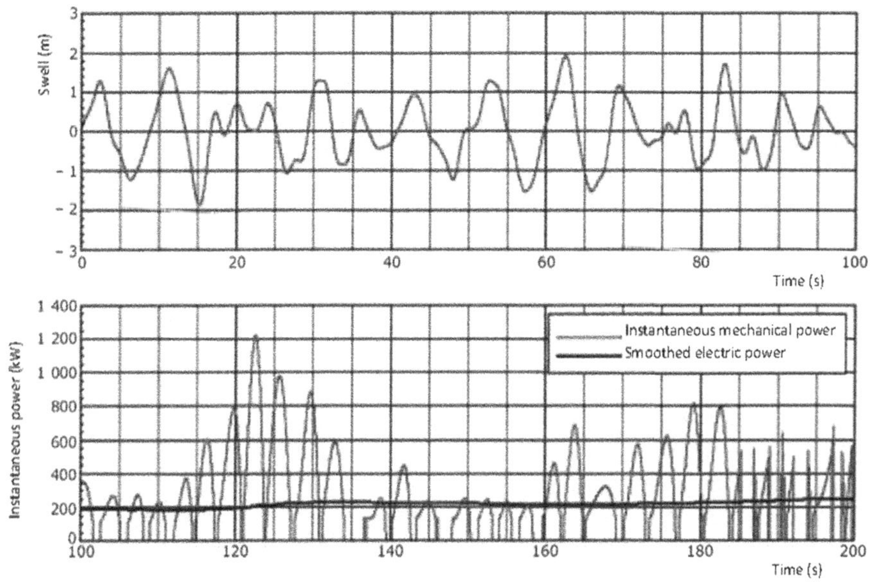

Figure 4.52. *Instantaneous mechanical power and smoothed electric power by a SEAREV according to the swell [MOU 08]. For a color version of this figure, see www.iste.co.uk/robyns/electricity.zip*

Conversely, the instantaneous mechanical power produced by the systems with bodies moved by the swell is similar to the swell, i.e. fluctuating and oscillating (Figure 4.52). To be able to send it back under the form of electric power on the network, we thus have to add smoothing systems into the energy conversion chain. Indirect drives can take on part of this smoothing function on the level of the mechanical conversion of the wave energy. The disadvantage of indirect conversion is increased losses due to the multiplication of the conversion stages; this multiplication also affects the reliability of the system.

Direct conversion is currently not frequently used and would allow us to free ourselves from this problem and would require less maintenance operations. However, to date there is no standard electric generator (where the motions are linear or rotary pendular) adapted to the recovery of the waves' pulsating energy.

4.2.5. *Production by sea currents*

4.2.5.1. *System of mechanical conversion: underwater turbines*

The concept of the underwater turbine can be compared to that of an underwater wind turbine, which converts part of the kinetic energy of the seas' currents. Underwater turbines can be classified into various system categories:

– horizontal axis turbines, which are the equivalent of conventional horizontal axis wind turbines installed onshore or offshore. They can be equipped with fairing to exploit the acceleration of a fluid in a tube, whose diameter is narrowing, called the "Venturi effect";

– vertical axis turbines, with an operation mode similar to that of the onshore Darrieus vertical axis wind turbines;

– flapping or oscillating wing systems, also called "hydrofoil" or "hydroplane" systems;

– paddle wheel systems.

Except for the paddle wheel underwater turbine, which is installed at the surface on a floating structure, underwater turbines are generally immersed and depending on their installation place and depth, they can also be classified according to their fixation or anchoring system (Figure 4.53) amongst the following categories:

– underwater turbines mounted on a pile/piles anchored to the seabed, for which the upper end emerges above the surface;

– floating underwater turbines anchored to the seabed;

– gravity underwater turbines, which are completely immersed and placed at the bottom of the sea.

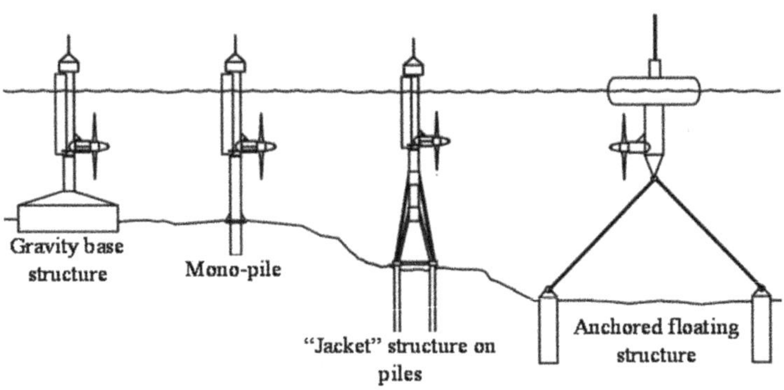

Figure 4.53. *Fixation or anchoring system of the underwater turbines [MUL 09]*

4.2.5.2. *Electromechanical conversion*

4.2.5.2.1. Horizontal axis rotor turbine in the current direction

Horizontal axis underwater turbines are made up of an axial flow turbine, whose axis of rotation is positioned in the direction of the current (Figure 4.54).

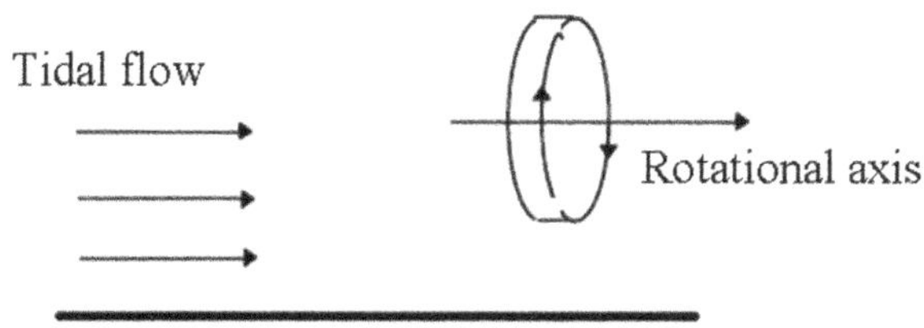

Figure 4.54. *Horizontal axis underwater turbine [BRY 05]*

The British company Marine Current Turbines, founded in 2000, has developed two prototypes of underwater turbines based on the concept of the horizontal axis turbine.

Installed in 2003 in the Bristol Channel off the coast of Lynmouth, in the south-west of England, SeaFlow is the first underwater turbine to be commissioned in the world (Figure 4.55). It is made up of a single turbine with two adjustable blades with an 11 m diameter, assembled on a single pile of 42.5 m driven into the seabed and emerging above the sea surface. It is sized to produce 300 kW, with an induction generator that has a nominal speed of 1,000 rpm, for sites of a depth ranging between 15 and 25 meters and currents of about 2 to 3 m/s maximum. The system can slide vertically to allow access to the rotor, generator and gearbox for maintenance and inspection, thereby avoiding the need to resort to a team of divers. The control of the blade step at 180° authorizes the operation of the turbine in the two directions, according to the change of direction of the current between the flow and reflow. As for a wind turbine, the adjustment can also be carried out according to the current speed, in order to optimize the performances, but also to limit the power during currents that are too strong.

Inspired by the SeaFlow project, SeaGen is considered to be the world's first commercial turbine prototype, developed by Marine Current Turbines (Figure 4.56). It is made up of two twin turbines that have a diameter of 16 m for a total power of 1.2 MW. First designed to be installed on a single pile of 3 m diameter, it is supposed to operate on sites with a current speed of about 3.5 m/s. SeaGen has produced, since its commissioning in July 2008 in Strangford Lough North Ireland, 11.6 MWh during its eight years of operation. The facility was dismantled in stages between

2016 and 2019 by Simec Atlantis Energy (formerly Atlantis Resources), which bought Marine Current Turbines in 2015 from Siemens, who acquired it in 2012.

Figure 4.55. *SeaFlow underwater turbine immersed during operation and lifted for maintenance [SEA 05]. For a color version of this figure, see www.iste.co.uk/robyns/electricity.zip*

With the E-Tide or Blue Concept project, the Norwegian company Hammerfest Ström is proposing a three-blade horizontal axis underwater turbine positioned at the bottom of the sea by gravity base foundations (Figure 4.57). The control of the blade orientation enables the operation of the underwater wind turbine in the two rotational directions and enables the generator located in the nacelle in fixed position to supply optimized power.

Figure 4.56. *Single-pile concept of the SeaGen underwater turbine [SEA 10]. For a color version of this figure, see www.iste.co.uk/robyns/electricity.zip*

Figure 4.57. *Hammerfest Ström underwater turbine [AND 11]. For a color version of this figure, see www.iste.co.uk/robyns/electricity.zip*

The HS300, a 300 kW prototype, was installed in 2003 and was tested for four years in Kvalsund, in the north of Norway. Inspired by this prototype, the first pre-commercial demonstration version of a 1 MW power, the HS1000, was installed in 2011 at the EMEC (European Marine Energy Centre) in the north of Scotland [EME 20]. In 2012, *Hammerfest Ström* became *Andritz Hydro Hammerfest*, a subsidiary of the Austrian industrial group *Andritz*.

As part of Phase 1A, with an authorized capacity of 86 MW for the MeyGen experimental offshore tidal turbine farm project, planned to reach a total of 398 MW over three phases, through four 1.5 MW tidal turbines on a gravity base – including one supplied by Simec Atlantis Energy (AR1500) and three supplied by Andritz Hydro Hammerfest – were submerged to a depth of 50 m in 2016 and 2017 in the Pentland Firth strait, to the north of Scotland; and in March 2018, commissioned for an expected operating period of 25 years. Production reached 17 GWh in June 2019 [MEY 20].

On the French side, the first prototype of a 1/3 scale marine tidal turbine was the Sabella D03. It was tested in 2018 over the course of a year, at a depth of 25 meters, in the Odet estuary in Brittany. It is 3m in diameter, with a power of 30 kW (Figure 4.58(a)). Developed by the Quimper Company, it is based on the turbine concept patented by *Hydrohélix Energies* in 1999. The Sabella D10, the pre-industrial version which was 10 m in diameter with 1 MW in power, (Figure 4.58(b)), is the focus of two separate proofs of operation campaigns in the passage du Fromveur, off the coast of the island of Ouessant; one ran for the year 2015-2016 and the other for a period of three years as of the end of 2018, as part of the ICE project (Intelligent Community Energy) [SAB 20].

a) b)

Figure 4.58. *Prototype Sabella D03 (a) and tidal turbine Sabella D10 (b). For a color version of this figure, see www.iste.co.uk/robyns/electricity.zip*

4.2.5.2.2. Turbines with a rotor axis perpendicular to the current

Vertical axis underwater turbines are also called cross-flow turbines, because their rotation axis is positioned perpendicularly to the direction of the current (Figure 4.59). The advantage of vertical axis turbines is to be able to capture the kinetic energy of the currents in any direction and to thus be impervious to current orientation. This is an advantage in the case of tidal currents.

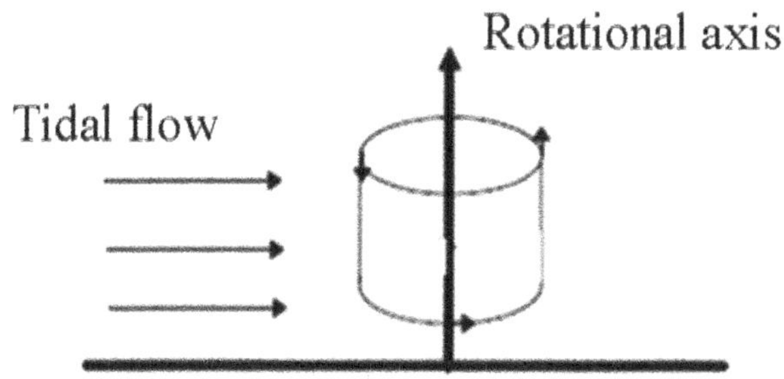

Figure 4.59. *Vertical axis underwater turbine and transverse flow [BRY 05]*

There are several systems operating as Darrieus wind turbines, and are based on a Darrieus turbine (Figure 4.60(a)) or its by-products, such as Gorlov turbines (Figure 4.60(b)), Kobold turbines (Figure 4.60(c)) and Achard turbines (Figure 4.60(d)).

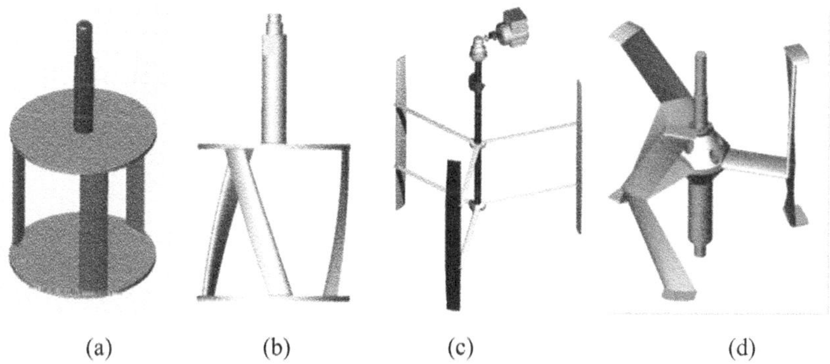

Figure 4.60. *Cross-flow turbines: (a) Darrieus; (b) Gorlov; (c) Kobold; (d) Achard. For a color version of this figure, see www.iste.co.uk/robyns/electricity.zip*

Assembled under the form of a squirrel-cage, the blades of a Darrieus turbine are straight, while those of a Gorlov are helical. Achard turbines have blades with free extremities with an aircraft wing profile and which are connected from their center to the turbine axis by arms in the aircraft wing profile. The advantage of this turbine is its ability to progressively get into rotation.

In the framework of the Enemar project, the Italian company Ponte di Archimede International SpA has developed a prototype of a vertical axis underwater turbine on the basis of a three-blade Kobold turbine with 25 kW power. Installed in

2001 in the Mediterranean in the Messine straits located between Italy and Sicily, it is made up of a floating platform under which the turbine is fixed, which is coupled to a three-phase synchronous generator (Figure 4.61). The power produced for a current of 3.5 m/s is 120 kW, i.e. an efficiency of about 23%. The advantages of the system are its capacity to get into rotation autonomously, because of the significant starting torque and its ability to operate independently from the current direction.

Figure 4.61. *Enemar – Kobold system [CAL 07]. For a color version of this figure, see www.iste.co.uk/robyns/electricity.zip*

Initiated by the Laboratoire des Écoulements Géophysiques et Industriels (LEGI) in 2001, in the framework of a research program on the recovery of kinetic energy from the ocean and river currents, the HARVEST project (Hydroliennes à Axe de Rotation VErtical STabilisé, i.e. Stabilized vertical rotational axis underwater turbines) proposes a French concept of vertical structure under the form of a tower containing a stack of Achard vertical axis turbines connected on a single rotational axis to a single generator at the upper end [ANT 07].

Created in 2010 to industrialize the concept of vertical architecture with multiple turbines, the Grenoble-based company HydroQuest markets a range of river tidal turbines on floating barges, with a lifting system that is moored to the riverbed in order to recover the kinetic energy of watercourses (HydroQuest River). And then a range of marine tidal turbines, an extrapolation of the fluvial concept and assembled onto a tripod that is placed on the ocean floor, so as to recover the energy generated by tidal currents (*OceanQuest*). They have the distinctiveness of a double counter-rotating axis with transverse flow, on an adjustable height of one or two stages that allow it to adapt to the water height, to provide different electrical powers, and to also be equipped with asymmetrical hydrodynamic side fairings that accelerate the speed of the water (Figure 4.62).

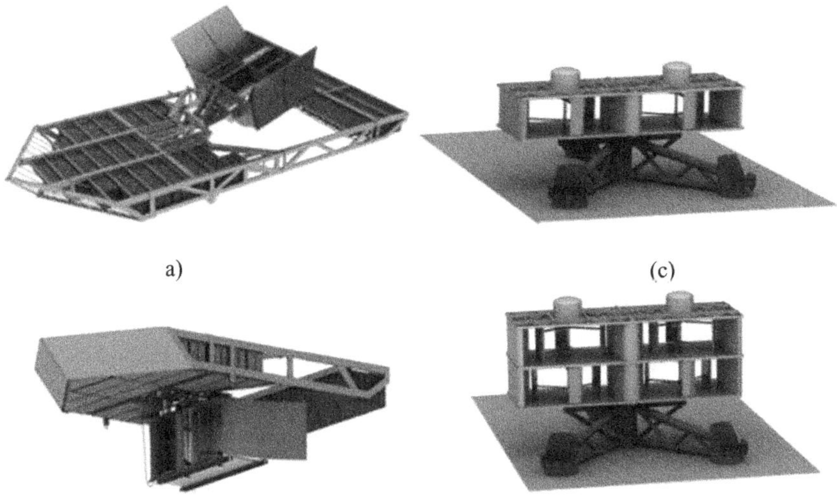

a) (c)

b) (d)

Figure 4.62. *(a) HydroQuest River tidal turbine in raised position (b) or operating (c) and OceanQuest marine tidal turbine with one (d) or two stages [HYQ 20]. For a color version of this figure, see www.iste.co.uk/robyns/electricity.zip*

Successful HydroQuest deployments:

– a 30 kW protoype in the supply channel of the EDF Pont-de-Claix hydroelectric power station near Grenoble in 2010;

– a 12 kW river tidal turbine, acquired by EDF, on the Oyapok river in the village of Camopi in Guyana in April 2013;

– a 40 kW river tidal turbine on the Loire river in Orléans, commissioned in October 2014 and connected to the Enedis network in September 2015, for a period of experimentation which lasted four years, thereby ending the R&D phase and allowing for the transition to phase commercial;

– an 80 kW estuarine tidal turbine which inaugurates the Site Expérimental Estuarien National pour l'Essai et l'Optimisation d'Hydroliennes (*National Estuary Experimental Site for the Testing and Optimization of Tidal Turbines*) or

SEENEOH, a tidal turbine park located in the Gironde estuary in Bordeaux, at the beginning of 2018 for a test period of one year;

– the first river farm in the world, made up of four tidal turbines with a combined total production of 320 kW, in the riverbed of the Rhône at Caluire and Cuire, in the metropolis of Lyon, in October 2018. *Hydrowatt*, the Lyon hydroelectric division of the renewable energy producer group *UNITe*, will operate the facility over an 18 year period;

– a marine tidal turbine with a power of 1 MW on the EDF demonstrator fleet at Paimpol-Bréhat in 2019 (section 4.2.5.2.4.).

4.2.5.2.3. Oscillating systems

Flapping wing or "hydroplane" systems are different from conventional underwater turbines because they have the particularity of exploiting an oscillating motion instead of a rotating motion, in order to capture the kinetic energy from the currents. They are made up of a supporting structure fixed on the seabed by gravity, on which an arm equipped with a wing is fixed (Figure 4.63).

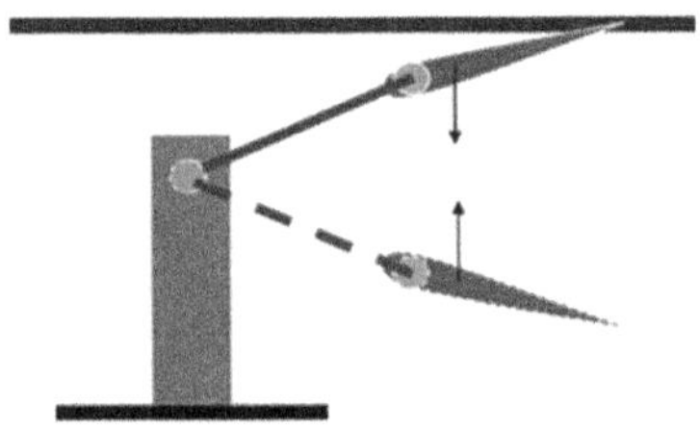

Figure 4.63. *Flapping wing or "hydroplane" system [BRY 05]*

The water bearing capacity on the wing, whose leading angle is adjustable via a hydraulic mechanism, causes vertical oscillations of the arm, which leads to the pump operation of a fluid under pressure supplying a hydraulic motor coupled with an electric generator. The main advantage of this type of system is the almost total absence of cavitation risk, because of the shape of the wing.

The Stingray system developed from 2002 onwards, by the English company Engineering Business Ltd, continues with this concept of an immersed wing placed in front of the sea current (Figure 4.64).

The electric generator of the prototype tested in 2003 is sized to provide an average nominal power of 150 kW for currents of about 2 m/s.

Figure 4.64. *Oscillating Stingray system [TEB 03]*

4.2.5.2.4. Fairing systems

The operation principle is based on the concentration of the flow going through the underwater turbine with the help of a Venturi shaped duct (Figure 4.65).

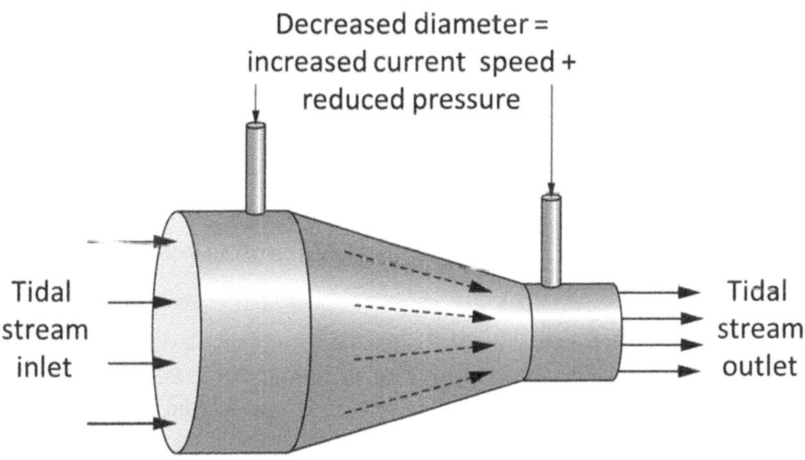

Figure 4.65. *Underwater turbine with Venturi duct [BRY 05]*

Created in 2008 by EDF, the demonstrator tidal park project on the Paimpol-Bréhat site in Brittany welcomed its first prototype for this technology for an underwater turbine with a Venturi effect of 500 kW, called Arcouest. This system was previously developed by the Irish company OpenHydro for the exploitation of tidal currents of about 3 m/s at a depth of 35m. A first immersion took place at the end of 2011/early 2012 and a second between December 2013 and April 2014, giving rise to two conclusive test campaigns.

Based on the prototypes of 250 kW that have been tested at the EMEC in Scotland since 2006 and in the Bay of Fundy, Canada, since the end of 2009 – diameters of 6 and 10 meters respectively – the OpenHydro underwater turbine's particularity is to have an open center. Having an external diameter of 16 m, it is made up of a rim-type (the rotor is directly installed at the periphery of the wheel) permanent magnet synchronous generator, which relies on a tripod supporting structure on the seabed (Figure 4.66). With a total power of 2 MW, the annual production is estimated at 3,000 MWh [ABO 09, EDF 10].

Figure 4.66. *Drawing of the Paimpol-Bréhat underwater turbine [EDF 15]. For a color version of this figure, see www.iste.co.uk/robyns/electricity.zip*

Between early 2016 and late 2017, two tidal turbines designed by DCNS (a French company which bought OpenHydro in March 2013, later renamed as the Naval Group in June 2017), with a diameter of 16 meters and a power of 1MW, were tested at a depth of 35 to 40 meters off Ploubazlanec on the Paimpol-Bréhat experimental site and connected to the network (Figure 4.67). The experiments will be stopped and the tidal turbines will be taken out of the water due to a corrosion problem on the system that fixes the coils in the stator.

Figure 4.67. *Immersion of the second DCNS / OpenHydro tidal turbine of the EDF tidal turbine demonstrator fleet on May 29, 2016 [NAV 16]. For a color version of this figure, see www.iste.co.uk/robyns/electricity.zip*

Converted by EDF into a test site that is available to other operators, the Paimpol Bréhat site hosted the Hydroquest Ocean tidal turbine in 2019 with two vertical axes and hydrodynamic fairings measuring 20 meters wide and 11 meters high (Figure 4.68). With 1 MW of power and manufactured by French partners HydroQuest and CMN (*Constructions Mécaniques de Normandie*), it was commissioned in July 2019 and its electricity production fed into the French national grid [HYQ 19].

Figure 4.68. *Immersion of the Hydroquest Ocean tidal turbine [HYQ 19]. For a color version of this figure, see www.iste.co.uk/robyns/electricity.zip*

4.2.5.2.5. Paddle wheel systems

Developed in 2004 by the French company Aquaphile, the particularity of the Hydro-Gen 1 machine turbine is to be a floating system at the surface, in the shape of a twin-hull nozzle anchored to the seabed and on which a large paddle wheel is fixed (Figure 4.69(a)). There are two versions of the system: one for tidal currents, which can alternately rotate in the two directions and one with asymmetric paddles for sea and river unidirectional currents. An electric generator coupled with the wheel converts the water kinetic energy into electric current. This paddle wheel system is more adapted to high currents and waters that are not too deep. On the contrary Hydro-Gen 2, which is the second version with a retractable helix, can go in and out of the water (Figure 4.69(b)).

a) b)

Figure 4.69. *(a) Hydro-Gen 1 and (b) Hydro-Gen 2 [HYD 20]. For a color version of this figure, see www.iste.co.uk/robyns/electricity.zip*

A first 7 kW prototype with a 6m diameter and a 10m length was tested in 2006 in the waters off Brittany. This simple and low-cost system is, today, marketed for power ranges between 10 kW and 300 kW with an evolution perspective of up to 1 MW.

4.2.5.3. *Comparison between wind power and tidal power production*

With identical sizes and fluid speeds, the productivity of an underwater turbine is higher than that of a wind turbine, because the water density is about 800 times higher than the air density.

However, the average current speed in exploitable sites is three to five times lower than the optimal velocity of wind turbines. Moreover, the rotation velocity of an underwater helix is limited to 10 m/s to avoid the cavitation phenomenon at the tip of the blade (formation of small vapor bubbles within the liquid). This phenomenon is well-known in the marine propulsion field, which generates shock waves, leading to metal corrosion and the degradation of the efficiency of hydraulic systems.

Because of the high water density, for an identical power, the dimensions of an underwater turbine will be much lower than those of a wind turbine (Figure 4.70). But with a stronger fluid pressure on the rotor and with lower current speeds, the torque exerted on the motor shaft will be higher.

Moreover, the production from ocean currents is predictable and therefore much less random.

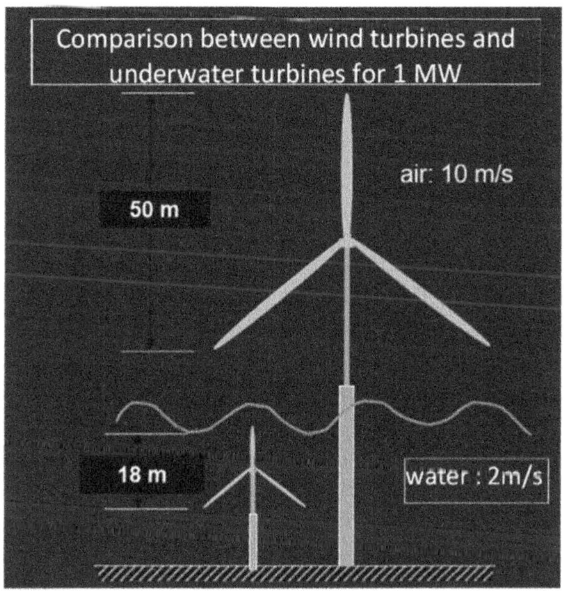

Figure 4.70. *Comparison of the dimensions between a wind turbine and an underwater turbine for a 1 MW power [MAÎ 08]*

4.2.6. *Tidal production*

There are two methods for extracting tidal power:

– the capture of potential energy related to the variation of the sea level with the help of tidal barrages;

– the capture of kinetic energy of tidal currents by underwater turbines (just studied in the previous sections), which is a lighter recovery technology than the tidal factory.

In this section, we will only focus on the capture of potential energy by tidal barrages.

4.2.6.1. *Historical background*

The first devices of potential energy capture from tides were tide mills. They appeared in France in the 12th century, mainly on the Brittany coast because of the high tidal amplitude. With the development of mills during the 16th and 17th centuries, up to 100 mills were built to recover the energy generated by the flow and reflow. Flour mills, built on four or five floors, replaced them during the 19th century. Today some of these tide mills still remain in France, the United Kingdom and Portugal.

Tide mills are built in estuaries and bays, at the foreshore, which corresponds to the tidal range zone. This zone is defined by the maximum levels of high and low tide.

Built in dry-dock between 1961 and 1966 after 20 years of studies and feasibility studies, the La Rance Tidal Power Plant, between Dinard and Saint-Malo, is a concept of an industrial size tide mill with a total power of 240 MW. The La Rance site has the advantage of having a significant current, whose natural flow rate is 18,000 m^3/s, and an average tidal range of 8.5 m, which can reach 13.5 m during the highest equinoxial tides, for a low seabed depth of about 35 m. This was the first factory in the world to produce electricity from the tidal force, as well as from rising and falling tides (double cycle effect). The annual production, provided by the 24 bulb-type units with a unit power of 10 MW, is about 520 GWh/year, i.e. 30% of Brittany's electricity production in 2010 and 90% of the global marine electricity production [SSO 09]. In the 2000s, nine bulb groups were restored. In 2013, a major renovation program of €100 million over 10 years began for the 15 remaining bulb units, six pumping valves and automation systems, and for the replacement of high-voltage electrical connections, without interrupting production.

The potential energy of the tides is relatively underexploited at the global level, in particular because of the limited number of sites with sufficient tidal range, combined with the high investment costs and environmental impact of the structures. Only six dams are operational, including one in Europe, three in Asia, one in North America and one in Russia (Table 4.7).

Year Established	Name or Location	Country	Total Power (MW)	Number of turbines x unit power (MW)	Average Tidal Range /max. (m)	Basin / Cycle
1966	Estuaire de La Rance	France	240	24 x 10	**8.5 / 13.5**	Single / Double effect
1968	Kislaya Guba Fjord	Russia	1.7	1 x 0.4 (1968-2004) 1 x 0.2 (2004) + 1 x 1.5 (2007)	2.3 / 3.9	Single / Single effect
1972	Haishan	China	0.25	1 x 0.125 (1972) + 1 x 0.125 (1975)	4.9	Dual / Single effect
1980	Jiangxia	China	4.1	1 x 0.5 (1980) upgraded to 0.7 (2015) + 1 x 0.6 + 3 x 0.7 (1985) + 1 x 0.7 (2007)	5.1 / 8.4	Single / Double effect
1984	Annapolis Royal	Canada	20	1 x 20	6.4 / 8.7	Single / Single effect at ebb tide
2011	Sihwa Lake	South Korea	254	10 x 25.4	5.6 / 7.8	Single / Single effect at rising tide

Table 4.7. *Tidal power plants in operation [WAN 19]*

Several projects have been or are being studied in the areas with the greatest tidal amplitudes (Table 4.8).

The operating principle of the coastal tidal power plants is taken up in projects of artificial lagoons, which take the form of a circular offshore structure, which is smaller than a dam. This type of project is still under study on several sites [ADE 10]:

– in Swansea Bay, Wales, with a lagoon of 5 km^2 supplying 24 turbines of 2.5 MW, and in Rhyl with 432 MW;

– in Yalu, China, 1 km off the coast, with a power of 300 MW.

Country	Location	Total Power (GW)	Number of turbines x unit power (MW)	Average Tidal Range (m)	Cycle
South Korea	Garolim	0.52	20 x 26	4.7	Single effect at ebb tide
	Ganghwa	0.42	14 x 30	5.5	Single effect at ebb tide
Russia	Mezen	24	(200x10x3) x 4	9.1	
	Tugur	7.8		6.8	Single effect
	Penzhin	87		6	
Great Britain	Mersey	0.7	28 X 25	6.5	Single effect at ebb tide
	Severn Estuary	8.64	216 x 40	7.8	Single effect at ebb tide

Table 4.8. *Tidal power plants planned [WAN 19]*

4.2.6.2. *Conversion of the potential energy*

A tide mill is made up of the following elements:

– a cove or basin to store the water transferred at rising tide;

– a dike to retain the basin water and to isolate it from the sea;

– a gate for opening and closing the basin;

– a paddle wheel.

The operating principle of the tide mill is similar to that of a conventional water mill and is broken down into the following three stages:

Stage 1: Filling – at rising tide, the gate is open and the basin fills up with the flow. The wheel does not rotate.

Stage 2: Waiting – at low tide, the gate is closed and there is a wait for the sea level to lower until the wheel has dried out.

Stage 3: Emptying or turbine operation – after opening the gate, the paddle wheel driven by the ebb tide thus produces a driving force enabling us to convert the energy of the water fall, which is stored at rising tide.

Mills only produce once a tide, two times a day with low tide, when the basin empties towards the sea. This is a simple effect emptying cycle. The Korean power plant in Sihwa operates by following this principle.

Tidal power plants are the modern adaptation of the tidal mill. Different types of plants are distinguished according to their architecture, single or dual basin, and their mode of production, unidirectional or bidirectional:

– Single basin with a single-effect cycle at ebb tide (Figure 4.71(a)): this operation is the oldest and is identical to that of the tidal mill presented previously. The Annapolis plant in Canada falls into this category.

– Single basin with a single-cycle at rising tide (Figure 4.71(b)): unlike the tidal mill, the basin is isolated from the sea during the rising tide and the valves are only opened at high tide to allow for the flow of water into the basin via the turbines. The disadvantages of this method are lower electricity production in addition to an unfavorable impact on the environment and navigation, due to the average water level of the basin being kept lower over longer periods. The South Korean factory in Sihwa operates on this principle.

– Single basin with a double-cycle (Figure 4.71(c)): the production of electricity is generated at rising and falling tides and is therefore more important thanks to the reversible operation of bidirectional turbines, which also offer the possibility of incorporating additional pumping phases, in order to raise or lower the level of the basin. The La Rance plant uses these different operating modes in both directions of the tide, which we will detail in the rest of this section.

– Connected dual basin (Figure 4.71(d)): a so-called upper basin and a so-called lower basin are connected to each other by a dam in which there is a unidirectional electricity production unit. The upper basin is filled at high tide and its valve is closed as soon as the tide lowers, until the next high tide. The lower basin is emptied at low tide and its valve is closed as soon as the tide rises until the next low tide. The level difference is thus permanently maintained between the two basins and energy production is ensured by emptying the upper basin into the lower basin. The operation of a dual-basin system is similar to that of a single-cycle single basin. The production capacity is lower but obtained over a longer production range, or even continuously, as opposed to a single basin system. Pumping from the lower basin to the upper basin can also be carried out as in Pumping Energy Transfer Stations (STEP). The Haishan plant in China is the only existing dual-basin plant.

– Paired dual basin with single effect cycle (Figure 4.71(e)): Two physically separated basins are located so as to be out of phase with each other with respect to the tidal cycle. Electrically interconnected, they operate in a paired fashion, one producing at rising tide and the other at ebbing tide.

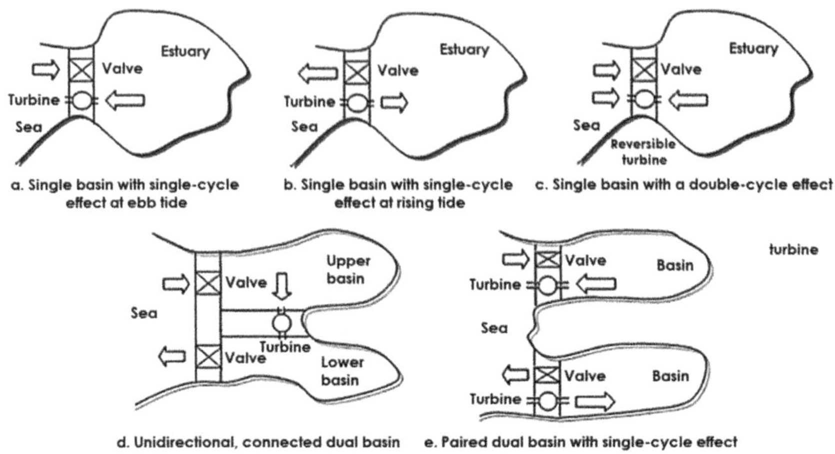

Figure 4.71. *Types of tidal power plant according to infrastructure and production mode [KHA 17]*

The La Rance Tidal Power Plant operates according to the principle of the double effect cycle, thanks to 24 bulb-type units, which can rotate in two directions during the filling and emptying of the basin. It is made up of (Figure 4.64):

– a basin with an area of 22 km^2 and a height ranging between 0 and 13.5 m for a live storage of 184 million m^3;

– a hollow dike, comprising 28 spans and a road with four lanes connecting Dinard to Saint-Malo, in which the electric production factory is found, which hosts 24 bulb-type units, three transformers and a control room;

– an inactive dike in rip-rap with a 163m length between the hollow dike and the Chalibert rock;

– a mobile dam 115 m in length with six wagon-type gates, each maneuvered by an oil servomotor, for the quick emptying or filling of the basin with a total rate flow of 9,600 m^3/s under a static head of 5 m;

– a navigation lock for the circulation of boats between the basin and the sea.

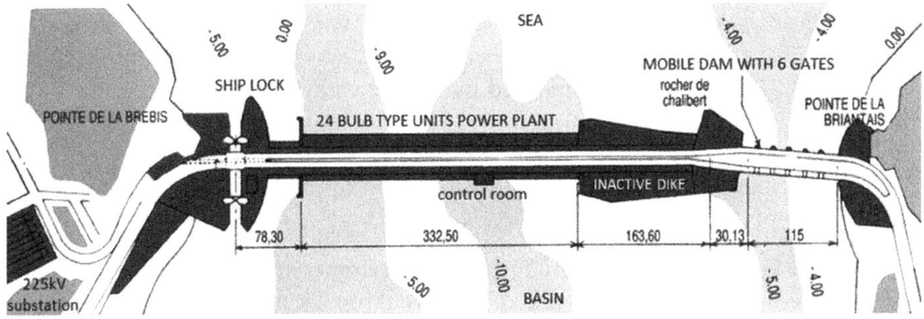

Figure 4.72. *Installation of the Rance Tidal Power Plant [DE 09]*

4.2.6.2.1. Electromechanical system

The electromechanical conversion is ensured by 24 generator groups with a unit power of 10 MW, i.e. a total power of 240 MW, each span occupying a bay of the power plant. Directly located in the hydraulic conduit, they are made up of a horizontal Kaplan turbine with four swiveling blades, which is mechanically coupled by its axis to a synchronous alternator with a nominal output voltage of 3.5 kV. The alternator is located in a metal shell in the shape of a bulb (hence the name bulb-type unit) and rotates in air pressurized at 2 bars for refrigeration (Figure 4.73). The rotor has 64 salient poles. Their coils are supplied by a static excitation system.

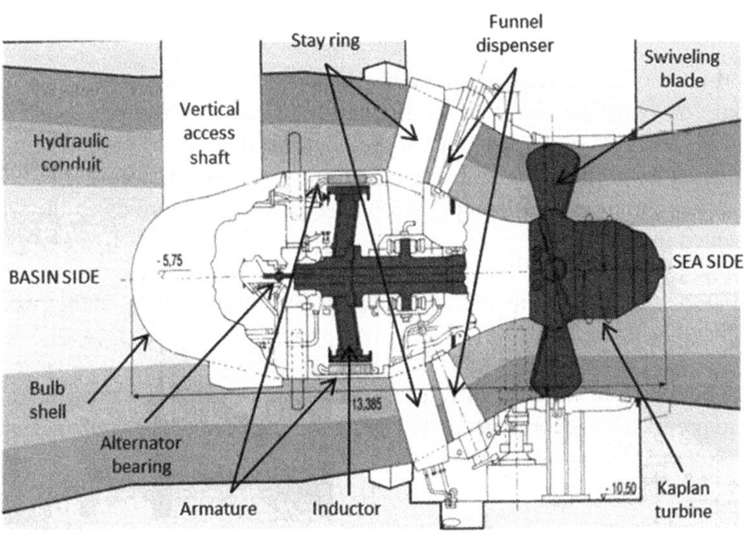

Figure 4.73. *Cross-section view of a bulb-type unit [DE 09]. For a color version of this figure, see www.iste.co.uk/robyns/electricity.zip*

Thanks to is swiveling blades of -5° to +35° and its symmetrical shape, the bulb-type unit can rotate equally in the two directions, with a constant rotational speed of 93.75 rpm, which is imposed by the network frequency, and a maximum flow rate of 275 m^3/s.

Several modes of operation are thus carried out for bulb-type units, according to the direction of the tide:

– Direct turbine operation: production of electricity in turbine mode with low tide, with coupling to the network of the synchronous actuator, via the drive of bulb-type groups by the water flow in the direction basin-sea.

– Reverse turbine operation: production of electricity in turbine mode with rising tide, by the water flow in the sea-basin direction.

– Direct pump operation: operation of bulb-type groups as a synchronous motor coupled with the network, in order to raise the level of the basin at the end of the high tide.

– Reverse pump operation: operation of the bulb-type groups as a synchronous motor coupled with the network to lower the level of the basin at the beginning of the high tide.

– Port operation: bulb-type units, uncoupled from the network do not rotate and let the water circulate freely in the periods when the gate of the mobile dam opens, in order to accelerate the filling of the basin at rising tide or the emptying at low tide.

In turbine mode, the rotation direction is determined by the water flow direction, depending on whether the tide is rising or lowering. Progressive starting is hydraulically ensured by the opening of the blades and the distributor. The coupling of the alternators to the network is carried out in synchronism according to the best practices. The power and efficiency of a bulb-type group according to the waterfall are presented in Figure 4.74.

In pump mode, the motor starting is also progressive, until we reach synchronism.

The combination of these different operation modes of the units enables the factory to produce any energy depending on the modes: simple or double effect.

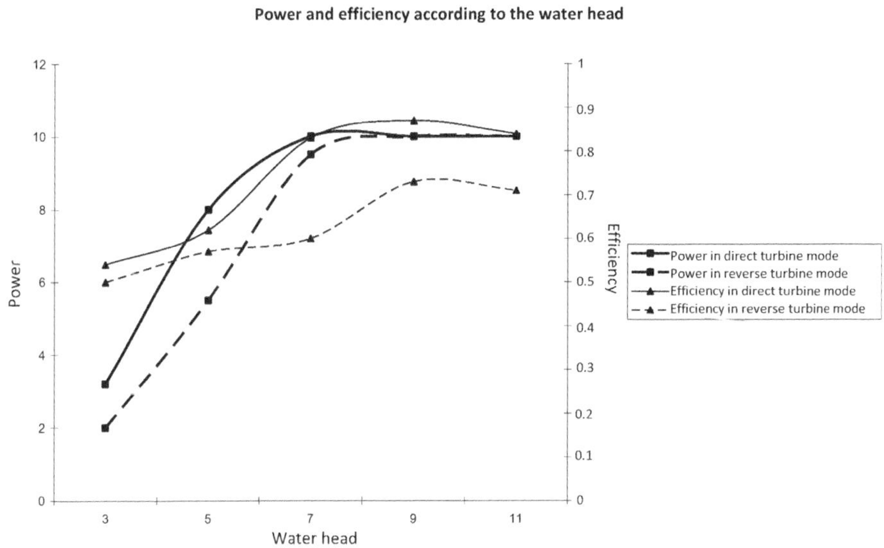

Figure 4.74. *Power (in MW) and efficiency of a bulb-type unit according to the operating modes and the water head (m) [EDF]*

4.2.6.2.2. Characteristics of electric production

The production directly depends on the tidal range (i.e. the difference of water height between a successive high sea and low water) and on the water height (i.e. the difference of level between the basin and the sea), which vary over time. The difference of water height must be at least 5 m to produce electricity.

Simple effect cycle – ebb generation

The turbines of the power plant operate in simple effect cycle during average tides or neap waters, when the coefficient is low and the tidal range is lower than 12 meters. The stages of this cycle are the most elementary ones. They are similar to those of the previously presented tide mill and are broken down as follows (Figure 4.75):

Stage 1: Filling – at rising tide, the opening of the six gates of the mobile dam allow the filling of the basin. The bulb-type units let the water circulate freely in the sea-basin direction and do not rotate (port operation).

Stage 2: Waiting or direct pump operation – when the high sea level is reached, the gates are closed and we wait for the tide to go out up to the low tide level, in order to obtain a sufficient water height. During the low tide, bulb-type units can be started at the end of filling to pump and artificially raise the level of the basin, in comparison to the maximum level reached by the tide.

Stage 3: Direct turbine operation – at low tide, thanks to the obtained water height, bulb-type units are driven by the water flow from the basin to the sea and produce electricity. Turbine operation stops when the water height is no longer sufficient.

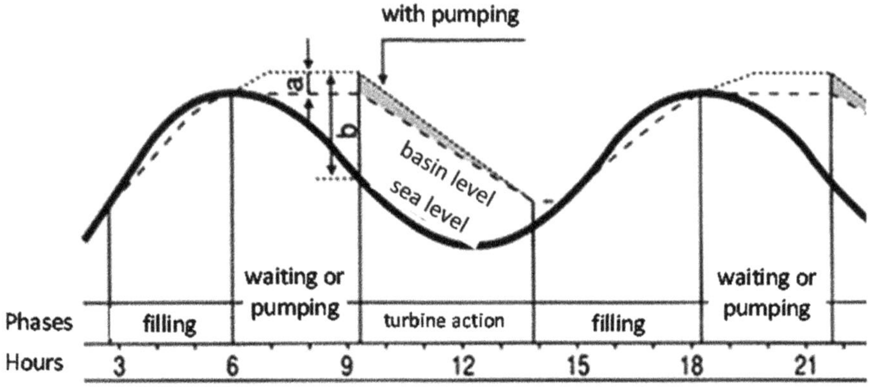

Figure 4.75. *Simple effect cycle at emptying – ebb generation [DE 09]*

The main operating difference in comparison to the tide mill lies in the possibility of replacing the waiting phase with a pump operation phase, which enables us to obtain a higher water height. This pump operation phase is only advantageous during filling, since it consumes electricity and is only profitable in off-peak hours. Indeed, synchronous machines are then operating as a coupled motor and thus consume energy. However, the operation remains interesting since it authorizes a more significant electricity production in comparison to a cycle with a waiting phase. The additional cubic meters thus pumped under a low head are subsequently usable, when the demand on the network will be more significant in peak hours, under a higher water head. We then estimate that the gain in the energy produced is twice as high as the energy absorbed during pump operation.

In the case of the simple effect cycle, the electric production of the power plant is intermittent, since it follows the rhythm of the tides and only produces at low tide, twice a day. To increase the exploitation time of the plant, the double effect operation enables us to produce electricity at the filling and emptying of the basin, four times a day.

Double effect cycle – ebb and flood generation

For high tides, with a coefficient higher than 105 and a tidal range higher than 12 meters, the double effect cycle enables us to produce at rising and low tide and is broken down as follows (Figure 4.76):

Stage 1: Waiting or reverse pump operation – at low tide, the gates of the mobile dam are closed and the basin, almost empty, is isolated from the sea. At rising tide, we wait for the level of the sea side to increase up to the water height required for turbine operation. To extend the turbine action time, we can use the bulb-type units in pump operation, not to raise the level of the basin as in the simple effect cycle, but to lower it even more.

Stage 2: Reverse turbine operation – with the water height sufficient for starting the bulb-type units being reached, the turbines are driven by the water flow from the sea to the basin and produce electricity.

Stage 3: Filling – when the difference of level between the sea and the basin is no longer sufficient, turbine action is stopped. The units let the water circulate freely in the sea-basin direction (port operation) and the gates of the mobile dam are open to accelerate basin filling.

Stage 4: Waiting and direct pump operation – idem, simple effect cycle.

Stage 5: Direct turbine operation – idem, simple effect cycle.

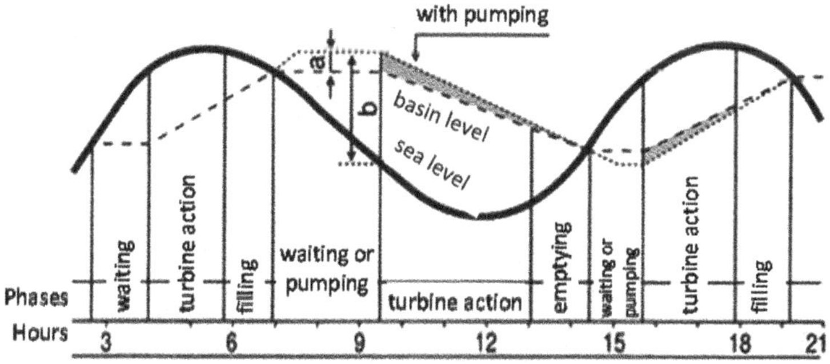

Figure 4.76. *Double effect cycle – ebb and flood generation [DE 09]*

In comparison to the simple effect cycle, the double effect cycle thus comprises two additional phases: a waiting or reverse pump operation phase at low tide and a reverse turbine operation phase when the basin is filling at rising tide.

A central computer carries out the required optimization calculations for the programming of operation cycles. The production optimization is managed by the calculation code AGRA (Algorithme de Gestion de la Rance, Management Algorithm from La Rance) which takes into account:

– the specific characteristics of each tide, which are predetermined by the "Service Hydrographique et Océanographique de la Marine" (SHOM, French Naval Hydrographic and Oceanographic Service);

– the cost in kWh over time;

– maintenance periods of the bulb-type units, which require us to take turbines out of the water, thanks to shut-off gates located upstream and downstream, as well as other electric devices (transformers, gates, automatisms, etc.);

– the possible unavailability of a machine, because of, for example, a breakdown.

The simple effect cycle is the most frequently used cycle and produces 450 MWh per tide and consumes 150 MWh for pump operation. The double effect cycle enables us to produce 400 MWh at rising tide and 900 MWh at low tide.

Between 1966 and 1996, the tidal power plant operated for 160,000 hours and produced 16 billion kWh, i.e. approximately 544 million KWh per year (deduced pump operation energy of 64.5 million kWh), a little bit less than the annual consumption of a city like Rennes.

The load factor is equal to 25% and is related to the periodicity and to the amplitude of the tides. Units are rotating 10 hours per day in turbine mode and possibly five hours per day in pump mode, i.e. between 9 and 14 hours of stopping/day. They produce 500 to 600 million kWh for 2,000 to 2,500 operation hours/day.

Figures 4.77 and 4.78 show examples of production profiles respectively for the simple effect cycle (with and without direct pump operation) and the double effect cycle (with and without direct and reverse pump operation).

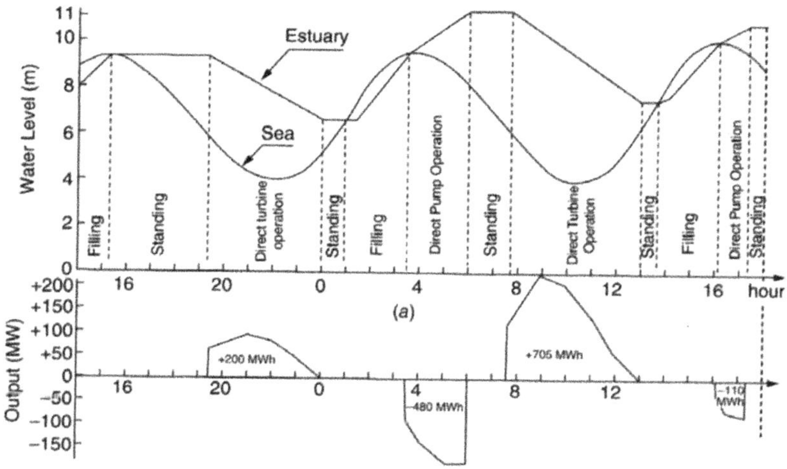

Figure 4.77. *Production profile of the simple effect cycle with and without direct pump operation [DE 09]*

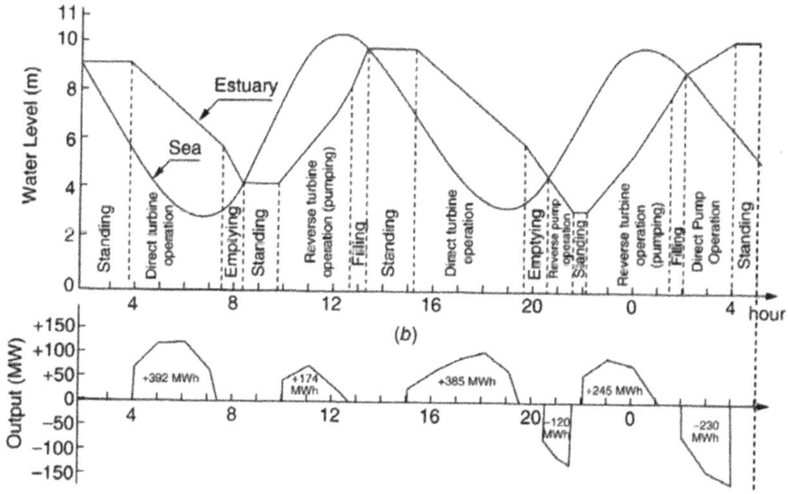

Figure 4.78. *Production profile of the double effect cycle with and without direct or reverse pump operation [DE 09]*

Operating principle of artificial lagoons

As a tidal power plant operating with a double effect cycle, an offshore artificial lagoon exploits the variation of the sea level to produce electricity with the help of low head turbines and a conventional electric generator.

Based on the Bélidor cycle, the production is obtained by exploiting the level difference between the inside lagoon basin, within a circular retaining structure built on the seabed, and the sea surrounding it according to a cycle that depends on the tide (Figure 4.79):

– the basin fills up at high tide;

– the basin empties at low tide;

– the bidirectional flow, resulting from the difference of water level between the basin and the sea, is exploited to activate the turbines.

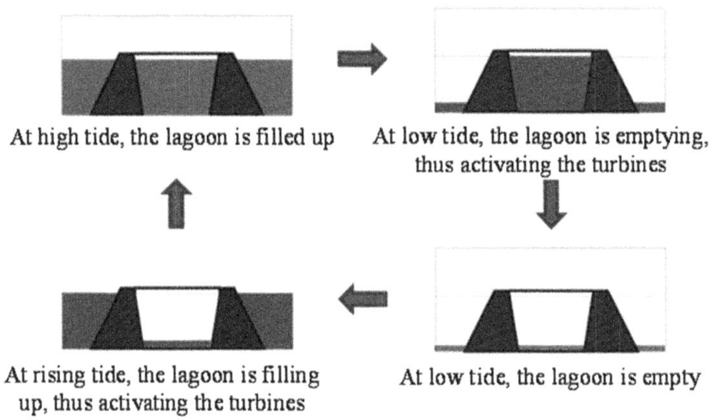

Figure 4.79. *Electric generation cycle of artificial lagoons [TID 11]*

We can increase the load factor and obtain a more regular electric production from the lagoon by adopting a multicellular retaining structure with several basins. In a configuration with two retaining basins, a high basin fills up at high tide and a low basin empties at low tide. The flow from one basin to another is obtained via the always positive level difference between the basins and can intervene at any moment of the tide.

4.2.6.2.3. Comparison of the potential tidal production with run-of-the-river hydropower

The operating principle of a tidal power plant is similar to that of a conventional hydraulic power plant. The main differences are caused by the fact that:

– the water flow is occurring during a tide cycle, hence intermittent production;

– operation is possible in the two directions;

– sea water density is slightly higher than that of soft water (about 2.5%).

4.2.7. *Exercise: estimation of the production of a simple effect tidal power*

A tidal power station is made up of the following elements:

– a cove or basin to store the water transferred at rising tide;

– a dike to retain the water in the basin and isolate it from the sea;

– a valve for opening or closing the basin;

– a turbine.

The single-effect operating principle of the plant is broken down into the following three stages:

– filling: at rising tide, the valve is open and the basin fills up. The turbine does not turn;

– waiting: at ebb tide, the valve is closed and there is a wait for the sea level to drop until the turbine is dry;

– emptying or turbining: after opening the valve, the driven turbine thus produces a motive force to convert the energy of the stored water at rising tide.

Let us consider an implementation site with an average tidal range (i.e. the difference in water height between a high tide and a successive low tide) over a year of 8 meters, and a basin of 500 m by 500 m.

1) Knowing that the average density ρ of seawater is approximately 1025 kg/m3, calculate the mass of the quantity of water stored in the basin.

2) Calculate the potential energy contained in the height of the water at high tide that can be recovered during the turbining, knowing that the level of the water in the basin is variable.

3) Knowing that the plant is located on the Atlantic coast, determine the potential energy recoverable daily.

4) Calculate the average daily power available.

5) Determine the daily power produced by the plant, knowing that the pressure drops are 6%, the efficiency of the turbine is 87%, and the efficiency of the alternator is 89%.

6) What type of turbine would work best for this plant? Justify your answer.

7) How could we increase the electrical energy produced by this plant (without new investment in civil engineering)?

1) $m = \rho\ V = \rho\ h\ S = 1025.8.500^2\ =\ 2{,}05.10^9$ kg

2) $E_{p\_average} = m\ g\ (h/2) = 2{,}05.10^9\ .9{,}81.(8/2) = 8{,}0442.10^{10}$ J (with an average height)

3) $E_{p\text{-}daily}\ =\ 2.E_{p\_average}\ =\ 2.8{,}0442.10^{10}$ J = $1{,}60884.10^{11}$ J (because of the two tides per day)

4) $P_j = E_{p\text{-}daily}/t = 1{,}60884.10^{11}\ /\ 86400$ s = 1862.083 kW

5) $P_u = P_j\ \eta_c\ \eta_t\ \eta_{alt}\ = 1862{,}08333.0{,}94.0{,}87.0.89 = 1355.30246$ kW = 1.355 MW

6) Kaplan: low head, suitable for tidal power plants because it is potentially reversible.

7) Recover energy from the rising tide.

4.3. References

[ABO 09] ABONNEL C., "EDF prépare le premier parc hydrolien pré-industriel de démonstration d'Europe au large de Paimpol-Bréhat", *Revue 3EI*, no. 59, pp. 33–38, 2009.

[ADE 04] ADEME, Energies des océans – Note de synthèse, Department of Renewable Energies, 2004.

[ADE 06] ADEME, "Guide pour le montage de projets de petite hydroélectricité", *Formation Hydro*, Grenoble, September 2006.

[ADE 10] ADEME, Feuille de route sur les énergies renouvelables marines, available at: https://www.ademe.fr/sites/default/files/assets/documents/6907-feuille-de-route-energies-marines-v220910.pdf, accessed November 2020, June 2010.

[AMO 09] AMORETTI R., "Production d'énergie électrique en Corse : les micro centrales", *Revue 3EI*, no. 58, pp. 73–78, 2009.

[AND 11] ANDRITZ HYDRO, Artist's impression of tidal turbine, available at: http://www.hammerfeststrom.com/, accessed December 2011.

[ANS 04] ANSEL A., ROBYNS B., "Production décentralisée d'électricité au moyen d'une micro centrale hydraulique au fil de l'eau", *Revue 3EI*, no. 39, pp. 46–57, 2004.

[ANS 06a] ANSEL A., ROBYNS B., "Small hydroelectricity: From fixed to variable speed electromechanical drives", *Electromotion*, vol. 13, no. 2, pp. 111–126, 2006.

[ANS 06b] ANSEL A., ROBYNS B., "Modelling and simulation of an autonomous variable speed micro hydropower station", *Mathematics and Computers in Simulation*, vol. 71, pp. 320–332, 2006.

[ANT 07] ANTHEAUME S., MAITRE T., ACHARD J.L., "Optimisation d'une ferme d'hydrolienne à axe de rotation transverse", *18ème Congrès Français de Mécanique*, Grenoble, August 2007.

[AQU 08] AQUARET, available at: http://www.aquaret.com/images/stories/aquaret/pdf/cswavewavedragon.pdf, accessed November 2020, 2008.

[AUB 09] AUBRY J., BABARIT A., AHMED H.B. *et al.*, "La récupération de l'énergie de la houle, Partie 2: Systèmes de récupération et aspects électriques", *La Revue 3EI*, no. 69, pp. 26–32, 2009.

[AUB 11] AUBRY J., Optimisation du dimensionnement d'une chaîne de conversion électrique directe incluant un système de lissage de production par supercondensateurs, Application au houlogénérateur SEAREV, PhD thesis, École Normale Supérieure de Cachan, 2011.

[BAB 11] BABARIT A., HALS J., KURNIAWAN A. *et al.*, The NumWEC project – Numerical estimation of energy delivery from a selection of wave energy converters, Finla report, December 2011.

[BAR 85] BARRUFET D.J., VECIANA Y., *Las Olas del Mar – Apuntes que se relacionan con el movimiento de las mismas y su aprovechamiento como fuerza motriz*, Imprenta de Redondo y Xumetra, Barcelona, 1885.

[BOY 96] BOYLE G., *Renewable Energy: Power for a Sustainable Future*, Oxford University Press, 1996.

[BRE 07] BREBAN S., NASSER M., ANSEL A. *et al.*, "Variable speed small hydro power plant connected to AC grid or isolated loads", *EPE Journal*, vol. 17, no. 4, pp. 29–36, 2007.

[BRE 09] BRETAGNE REGION, Des énergies marines en Bretagne : à nous de jouer !, Economic and Social Council report, Brittany, March 2009.

[BRY 05] BRYDEN I., "Current to tidal power", Energy and nanotechnology: Storage and grid workshop III, 15–16 November, 2005, *The Rice Energy Program of the Baker Institute and the Rice University Brown School of Engineering*, GEORGE R. (ed.), available at: http://www.rice.edu/energy/events/past/Bryden%20Nano%202005.pdf, accessed October 2011, 2005.

[CAL 07] CALCAGNO G., MOROSO A., "The Kobold marine turbine: From the testing model to the full scale prototype", *Tidal Energy Summit Conference*, London, November 2007.

[CAR 20] CANERGIE PROJECTS, available at: https://www.carnegiece.com/portfolio/, accessed December 2020.

[CHE 06] CHEVAILLER S., Comparative study and selection criteria of linears motors, PhD thesis, Ecole Polytechnique de Lausanne, 2006.

[CLE 08] CLÉMENT A., "Wave energy: Project and R&D updates at Ecole Centrale de Nantes and in the 'Pays de la Loire' French region", *OREG Spring Symposium*, Québec, April 2008.

[COR 15] CORDONNIER J., GORINTIN F., DE CAGNY A. *et al.*, "SEAREV: Case study of the development of a wave energy converter", *Renewable Energy* vol. 80, pp. 40–52, 2015.

[DAN 11] DANIELSSON O., *Division for Electricity*, Uppsala University, 2011.

[DCN 16] DCNS press release from 30 May 2016, available at: https://www.naval-group.com/fr/news/immersion-de-la-seconde-hydrolienne-du-projet-edf-de-paimpol-brehat/, accessed April 2016.

[DE 09] DE LALEU V., "La Rance Tidal Power Plant – 40 years operation feddback", *BHA Annual Conference*, Liverpool, October 2009.

[DRA 01] DRAGU C., SELS T., BELMANS R., "Small hydro power – state of the art and applications", *International Conference on Power Generation and Sustainable Development, (AIM)*, Liège, Belgium, pp. 265–270, October, 2001.

[ELE 06] ELECTRICITÉ AUTOMATIQUE FRANÇAISE, "La turbine de la très basse chute", *Innovation Technique*, no. 45, 2006.

[EDF] EDF, Presentation brochure, press releases and Memo guide for the La Rance tidal power plant issued by EDF.

[EDF 10] EDF, Press release by EDF, "Projet EDF Hydroliennes : Site de Paimpol-Bréhat. Point d'avancement du projet", September 2010.

[EDF 15] EDF, Video by EDF, "Le fonctionnement d'une centrale hydrolienne", available at: https://youtu.be/Nj3_BipfFVE, December 2015.

[EME 20] European Marine Energy Centre website, available at: http://www.emec.org.uk/, accessed November 2020.

[ENM 11] Tide navigation course, by Ecole Nationale de la Marine Marchande de Marseille, available at: http://dept.navigation.enmm.free.fr/les_marees.swf, accessed February 2011.

[ERE 09] EREMEEF L.R., "Turbines hydrauliques", *Techniques de l'Ingénieur*, BM 4-406, 2009.

[EUR 09] "Etat des énergies renouvelables en Europe", *Observ'ER*, ISSN2101-9622, December 2009.

[FAL 99] FALNES J., HALS J., *Wave Energy and Its Utilization: A Contribution to the EU Leonardo Pilot Project 1860 Alter ECO*, Trondheim, June 1999.

[GIB 55] GIBRAT R., "L'énergie des marées", *L'Astronomie*, vol. 69, pp. 449–469, 1955.

[HAC 20] Hydro Air Concept Energy (HACE) website, available at: http://hacewave.com/, accessed November 2020.

[HEM 99] HEMERY G., COULON J., "Centrales hydroélectriques et apport de la vitesse variable", *Revue de l'Electricité et de l'Electronique*, pp. 46–52, December 1999.

[HYD 20] HYDRO-GEN, available at: http://www.hydro-gen.fr, accessed April 2020.

[HYQ 19] HYDROQUEST, CMN press release from 15 July 2019, available at: https://www.hydroquest.net/wp-content/uploads/2019/07/HydroQuest-CMN_CP-Raccordement_VF 150719-1.pdf, accessed November 2020.

[HYQ 20] HYDROQUEST, available at https://www.hydroquest.fr/, accessed June 2021.

[KAN 04] KANT M., "Moteurs électriques à mouvement linéaire et composé", *Techniques de l'Ingénieur*, D3700, February 2004.

[KAT 03] KATO S., HOSHI N., OGUCHI K., "Small scale hydropower", *IEEE Industry Applications Magazine*, pp. 32–38, July–August 2003.

[KEL 00] KELBER C.R., SCHUMACHER W., "Adjustable speed constant frequency energy generation with doubly-fed induction machines", *VSSHy 2000 – European Conference Variable Speed in Small Hydro*, Grenoble, France, January 2000.

[KHA 17] KHAN N., KALAIR A., ABAS N. *et al.*, "Review of ocean tidal, wave and thermal energy technologies", *Renewable and Sustainable Energy Reviews*, vol. 72, pp. 590–604, 2017.

[KUN 94] KUNDUR P., *Power System Stability and Control, Electric Power Research Institute*, MacGraw-Hill, California, 1994.

[LED 07] LEDUC G., Variations temporelles des changements hydrologiques dans la zone du pacifique Est. Approches géochimiques, isotopiques et micropaleontologiques, Doctoral thesis, Université Aix-Marseille III, 2007.

[MAÎ 08] MAÎTRE T., "Projet HARVEST – Modélisation numérique de l'écoulement sub-cavitant et cavitant dans les hydroliennes à flux transverse", *OREG 2008 Spring Symposium*, Québec, April 2008.

[MAS 65] MASUDA Y., Ocean wave electric generator, United States Patent Office, no. 3. 200.255, August 1965.

[MCC 74] MCCORMICK M.E., "Analysis of a wave-energy conversion buoy", *AIAA Journal of Hydronautics*, vol. 8, pp. 77–82, 1974.

[MCC 81] MCCORMICK M.E., *Ocean Wave Energy Conversion*, Wiley, New York, 1981.

[MEY 20] MEYGEN, Tidal stream project, available at: https://simecatlantis.com/projects/meygen/, accessed November 2020.

[MOU 08] MOUSLIM H., BABARIT A., "SEAREV : système électrique autonome de récupération de l'énergie des vagues," *Techniques de l'Ingénieur*, 2008.

[MUL 09] MULTON B., "Marine energy resource conversion systems", in *Renewable Energy Technologies*, ISTE, London and John Wiley & Sons, New York, 2009.

[MUL 11] MULTON B., THIAUX Y., BEN AHMED H., "Consommation d'énergie, ressources énergétiques et place de l'électricité", *Techniques de l'Ingénieur*, D3900v2, February 2011.

[OWC 20] OWC, Wave Power Plant on the Island of Pico/Azores, available at: http://www.pico-owc.net/en/index.html, accessed November 2020

[PAC 95a] PACER, Petites centrales hydrauliques – Turbines hydrauliques, *Report of the Renewable Energies Action Program in Switzerland*, PACER, 1995.

[PAC 95b] PACER, Petites centrales hydrauliques – Générateurs et installations électriques, *Report of the Renewable Energies Action Program in Switzerland*, PACER, 1995.

[PAC 95c] PACER, Petites centrales hydrauliques – le choix, le dimensionnement et les essais de réception d'une mini-turbine, *Report of the Renewable Energies Action Program in Switzerland*, PACER, 1995.

[PÈR 23] PÈRE G., Description des machines et procédés spécifié dans les brevets d'invention, de perfectionnement et d'importation, Tome cinquième, invention patent, Madame Huzard, pp. 99–110, Paris, 1823.

[PER 03] PERRIN S., "Petites centrales hydrauliques", *Techniques de l'Ingénieur*, BM 4-166, November 2003.

[POL 02] POLINDER H., DAMEN M.E.C., GARDNER F., "Modelling and test results of the AWS linear PM generator system", *ICEM 2002 International Conference on Electrical Machines*, paper 121, CD-Rom, Belgium, 2002.

[POL 05] POLINDER H., MECROW B., JACK A. *et al.*, "Conventional and TFPM generators for direct-drive wave energy conversion", *IEEE Transactions on Energy Conversion*, vol. 20, no. 2, June 2005.

[RAH 10] RAHM M., "Ocean wave energy: Underwater substation system for wave energy converters", *Acta Universitatis Upsaliensis, Digital Comprehensive Summaries of Uppsala Dissertations from the Faculty of Science and Technology*, no. 711, Uppsala 2010.

[RÉS 08] RÉSÉLEC, French Advanced Extension Awards, available at: www.iufmrese.cict.fr, 2008.

[RSS 02a] RSS, "L'eau: de l'énergie en cascade", *Revue Systèmes Solaires*, no. 152, pp.12–22, 2002.

[RSS 02b] RSS, "Eau potable et…énergétique !", *Revue Systèmes Solaires*, no. 152, pp. 24–27, 2002.

[SAB 20] SABELLA, L'histoire, available at: https://www.sabella.bzh/fr/lentreprise/historique, accessed November 2020.

[SAI 76] SAINT JEAN B., *Electrotechnique et machines électriques*, Editions Eyrolles and Editions Lidec, 1976.

[SAL 75] SALTER S.H., Apparatus and method for extracting wave energy, United States Patent no. 3.928.967, December 1975.

[SEA 05] "SEAFLOW – World's first pilot project for the exploitation of marine currents at a commercial scale", Final Publishable Report, European Commission, 2005.

[SEA 10] "SeaGen environmental monitoring programme: biannual update", *Version 1 – SeaGen Biannual Environmental monitoring July 2009 – Jan 2010*, SeaGen, April 2010.

[SEM 20] SEM-REV, Centrale Nantes multi-function offshore test site, available at: https://sem-rev.ec-nantes.fr/, accessed December 2020.

[SSO 09] *Systèmes solaires*, no. 189, pp. 20–27, 2009.

[SWI 06] SWINGEDOUW D., Origine et impact climatique d'un changement de circulation thermohaline au cours des prochains siècles dans le modèle IPSL-CM4, PhD thesis, University Paris 6, 2006.

[TCT 05] THE CARBON TRUST, Marine energy challenge – oscillating water column wave energy converter – evaluation report, Arup Energy, 2005.

[TEB 03] THE ENGINEERING BUSINESS LTD, Stingray tidal stream energy device – phase 2, Report no. T/06/00218/00/REP URN 03/1433, 2003.

[TED 07] TEDD J., Testing, analysis and control of Wave Dragon, wave energy converter, PhD thesis, Department of Civil Engineering, Hydraulics and Coastal Engineering Laboratory, Aalborg University, 2007.

[TET 20a] TETHYS, Pico Power Plant project, available at: https://tethys.pnnl.gov/project-sites/pico-power-plant, accessed November 2020.

[TET 20b] TETHYS, Wave Dragon Project - Milla Fjord Site, available at: https://tethys.pnnl.gov/project-sites/wave-dragon-project-milla-fjord-site, accessed November 2020.

[TID 11] TIDAL ELECTRIC LTD, available at: http://tidalelectric.com/technology-illustrations-cycle.shtml, accessed October 2011.

[VIN 07] VINING J., Ocean wave energy converters: Overview, legal and economy aspects, and direct drive power take-off, Master of Science (Electrical Engineering) thesis, University of Wisconsin, Madison, January 2007.

[WAV 09] WAVE DRAGON, Prototype testing in Denmark, available at: http://www.wavedragon.net/prototype-testing-in-denmark/, accessed November 2020.

[WAN 19] WANG Z.J., WANG Z.W., A review on tidal power utilization and operation optimization, *IOP Conference Series: Earth and Environmental Science*, vol. 240, no. 5, 2019.

[WAV 20] WAVE HUB, Wave Hub test site, available at: https://www.wavehub.co.uk/our-projects/wave-hub, accessed December 2020.

5

Thermal Power Generation

5.1. Introduction

Most of the conventional electricity generation processes are based on thermodynamic conversions (steam and combustion cycles). These processes can be found, for example, in coal, gas or nuclear power plants. However, the heat required for the operation of the process can also be produced from renewable energy. This is particularly the case for geothermal power, thermodynamic solar power and cogeneration from biomass.

This chapter will describe electrical power plants, using these last three sources of heat production. For each of them, many similarities will be observed in the process, in particular for the thermodynamic and electromechanical conversion stages. However, readers will also have the chance during their reading to note the specific characteristics of the various energy sources [HOO 04].

5.2. Geothermal power

5.2.1. *Introduction*

The word "geothermal" comes from the Greek *geo*, which means Earth, and *thermo*, which means heat. This heat has two sources: the heat generated during the creation of the Earth and the heat produced by the radioactive decay of the isotopes found in the Earth's crust. The low thermal conductivity of the Earth has enabled the storage of this heat for billions of years.

Chapter written by Jonathan SPROOTEN.

The temperature gradient within the Earth is generally equal to 25°C per kilometer of depth. However, in some specific sites, this gradient can reach 1,000°C/km. These sites are exploited in order of the priorities of not having to go too far down and for an easier exploitation of the stored energy. Moreover, nowadays, down in order to obtain sufficient efficiency for electricity generation, only geothermal resources at a high temperature (T>90°C) are considered to be exploitable. Other more abundant resources are easily recoverable in heating applications.

5.2.2. *The resource*

We can distinguish four main types of geothermal resources according to their thermodynamic and hydrological characteristics:

– Hydrothermal resources with a liquid dominant: the reservoir pores are filled with water. In these reservoirs, the temperature can reach 360°C and is constant in the entire reservoir, due to convection flows. This energy is brought to the surface with the help of an extraction pump. After extraction of the energy contained in water, the latter is again injected into the reservoir with an injection pump.

– Hydrothermal resources with a steam dominant: in this type of resource, the fluid is in the form of steam at a temperature close to 230°C, which makes exploitation for electricity generation simple. Overheated steam is recovered by convection and directly crosses a turbo-expander, in which energy is converted into mechanical energy. However, these resources are rare and almost systematically completely exploited (or about to be exploited) for electricity generation.

– Hot dry rock resources: because of the heat capacity and the significant volume of deep rock, the quantity of stored energy in this rock is probably much higher than the energy of hydrothermal resources. Figure 5.1 represents a temperature mapping of European rocks at a depth of 5,000 meters. This type of resource represents a quite significant amount of energy, but is also quite difficult to recover due to the absence of a heat transfer fluid. One of the most widespread energy recovery processes consists of injecting water into deep zones of fractured rock, and then – with the help of an extraction pump – recovering this fluid that has been heated by the rock.

– Geopressurized resources: in these deep reservoirs, the water is confined by the pocket membranes. Over time, the movement of the walls, due to the weight of rocks located above the reservoir and the movement of the Earth's plates, leads to a decreasing of the pocket volume and thus to an increasing of the confined water temperature and pressure. This fact is even more effective because the pocket walls limit thermal conduction. This energy, in high pressure and high temperature is recovered by sinking a well through the pocket walls. In addition to thermal energy recovery, this type of resource also provides kinetic energy. The latter is recovered in turbines that are similar to those used in hydroelectric power plants [LAP 08].

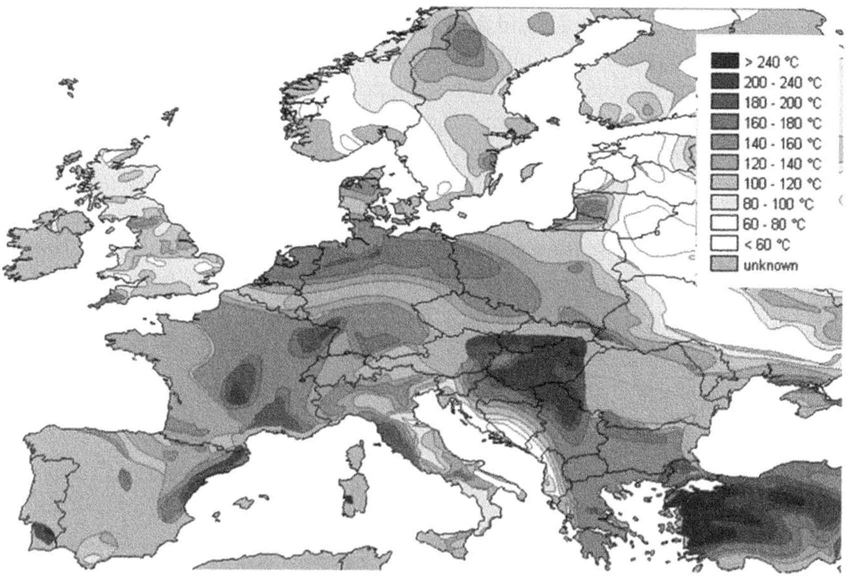

Figure 5.1. *Rock temperatures at deep locations in Europe [WIT 04]. For a color version of this figure, see www.iste.co.uk/robyns/electricity.zip*

5.2.3. *Fluid characteristics*

As mentioned in the previous section, the concepts of pressure, volume and temperature are essential to understanding the operating mode of thermal power plants. This section proposes establishing the relationship between these essential quantities and the fluid state (called solid, liquid, gaseous phase, etc.).

Let us consider a container full of gas placed in an environment at temperature T_0. If we compress this container, its volume decreases. If its temperature is maintained constant at T_0, the pressure of this gas increases according to the following law:

$$PV = ZnRT \quad [5.1]$$

where P, V and T are respectively the pressure, volume and absolute temperature of the gas, and where n is the number of moles of the gas, R is the universal constant of ideal gases and Z is the gas compressibility factor.

At constant temperature, the relationship is thus a hyperbola in the plane (P,V). Figure 5.2 represents, in 3D (P,V,T), the state of the considered fluid. This previously described trajectory corresponds to the hyperbola joining points A and B. By starting from point B, if we keep on compressing this gas, its pressure no longer varies, but its state changes. Liquid droplets then appear in the gas. From point C, if we keep on compressing this fluid, it becomes entirely liquid and the compression no longer modifies its volume (a liquid being not very compressible). Once again, if we keep on compressing the volume, the fluid changes its phase to become solid. Figure 5.2 in 3D can be experimentally obtained by repeating the same experiment at other temperatures. However, let us note that beyond a specific temperature, the previously described behavior no longer occurs and the state of the gas is then supercritical. In order to set orders of magnitudes, the critical point of water (beyond which the fluid is supercritical) is located at $T=374°C$ and $P=221$ *bars*.

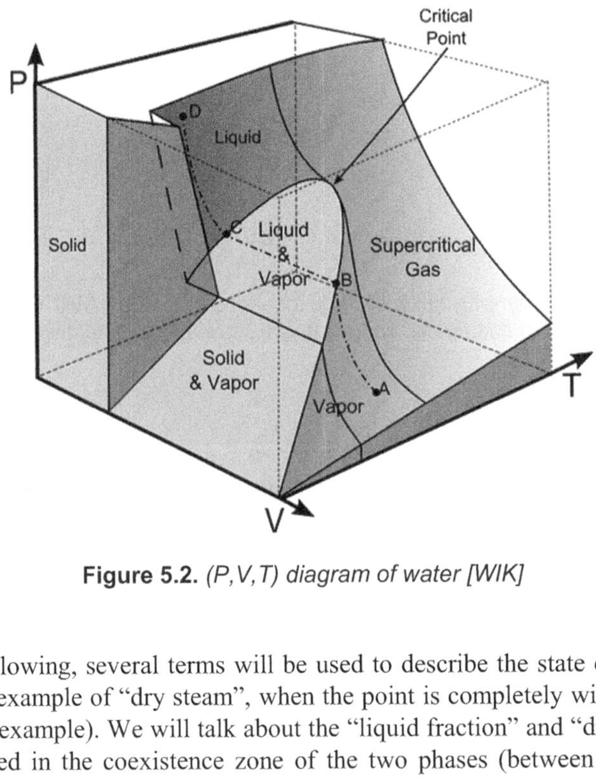

Figure 5.2. *(P,V,T) diagram of water [WIK]*

In the following, several terms will be used to describe the state of a fluid. We will talk for example of “dry steam”, when the point is completely within the steam zone (A, for example). We will talk about the “liquid fraction” and “dry fraction” of a fluid located in the coexistence zone of the two phases (between B and C, for example), in order to respectively identify the fluid parts that are at liquid and vapor state. Finally, the interface located between the mix liquid/steam and the vapor field will be called the “saturation curve” and the interface located between the liquid/steam mix and the liquid field will be called the “condensation curve”.

5.2.4. *The principle of geothermal power plants*

When the temperature of the thermal resource is higher than 175°C and when the resource is liquid, geothermal water under pressure is partially expanded in a reservoir, in order to produce steam. This principle is usually called "flash steam". The rest of the process is similar to that of geothermal resources that directly supply steam (hydrothermal power plants with a steam dominant). The obtained fluid is then separated into a dry fraction and liquid fraction. The dry steam fraction crosses a turbine supplying a driving force by expansion of this steam. The liquid fraction of the geothermal fluid is not used and is injected back into the well, together with the expanded steam condensate. This is the general principle of direct cycle power plants (Figure 5.3). In this figure, it is easy to understand that the expression "direct cycle" indicates that the geothermal fluid itself crosses the turbine [DI 05, LAP 08].

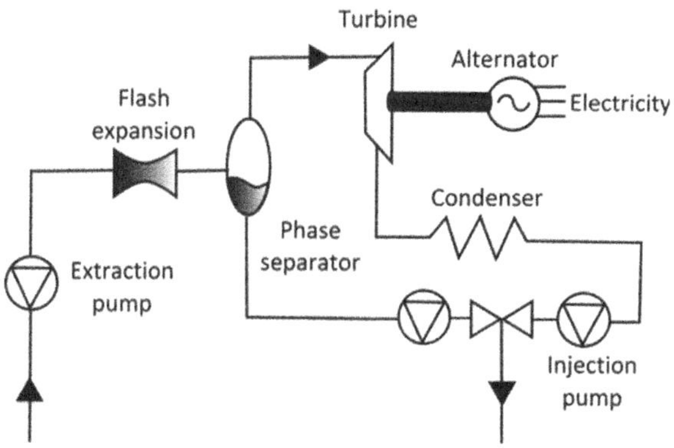

Figure 5.3. *Direct cycle power plant with "flash steam"*

In section 5.2.5, this thermodynamic cycle will be studied in more detail, in order to understand its important characteristics (including efficiency) and to identify the parameters influencing its value.

When the temperature of the thermal resource is lower than 175°C, the efficiency of this process decreases and a second heat transfer fluid is used. This fluid is often a hydrocarbon and is more volatile than water. It thus vaporizes at a lower temperature. This is the principle of binary cycle power plants, the operation diagram of which is represented in Figure 5.4. The geothermal fluid yields its energy to the second fluid, which then crosses a cycle similar to the previously presented thermodynamic cycle.

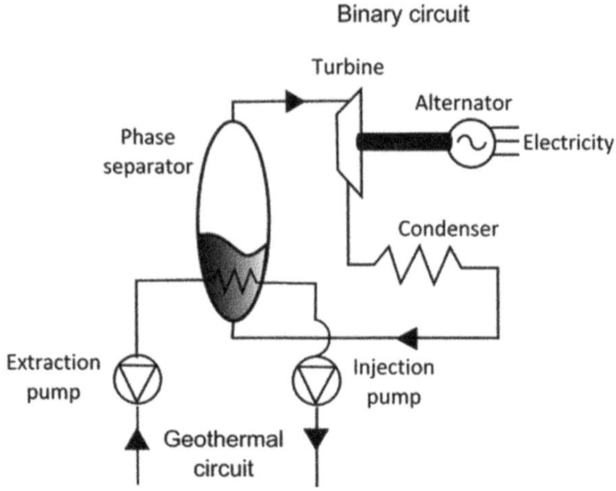

Figure 5.4. *Binary cycle power plant*

In practice, and in order to increase the energy conversion efficiency, the binary fluid is subjected to changes of phase during organic Rankine or Hirn thermodynamic cycles. These are described in more detail in section 5.2.5.

If finally, in the case of a direct cycle power plant, the quantity of energy at the turbine output is still significant, we can consider the combination of the two previous cycles. Rather than yielding this energy in the form of heat to the cold source in the condenser, this energy is used to evaporate a binary fluid, which will then go through a thermodynamic cycle. The operation diagram of this power plant is illustrated in Figure 5.5.

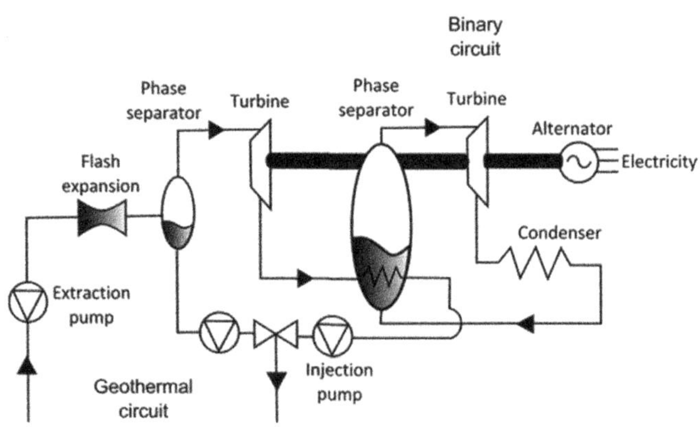

Figure 5.5. *Combined cycle power plant*

5.2.5. *Thermodynamic conversion*

5.2.5.1. *Introduction: Carnot cycle*

The thermodynamic conversion from heat to energy is carried out through a thermodynamic cycle, which can be broken down into four stages:

1) compression;

2) heating;

3) expansion;

4) cooling.

Let us consider that these stages are carried out in ideal operating conditions: i.e. an isentropic compression and expansion, a heat input at constant pressure and a heat extraction at constant pressure. Moreover, let us consider that the entire cycle occurs in the coexistence zone of the two phases (liquid and steam (wet steam)): there will thus be no phase change of the heat transfer fluid (here water) and isobars are merged with isotherms. This studied thermodynamic cycle is thus reversible and can be represented by Figure 5.6 in an entropic diagram of the temperature (T) according to the entropy (s) [BOU 98]. Let us recall that entropy (in J/K) is a state quantity characterizing the degree of disorder of a system on the microscopic level. During a state transformation, an increase in entropy characterizes the proportion of unused energy for obtaining work, which is transformed into heat at a lower temperature and is generally lost. An increase in entropy thus translates the irreversibility of a process. In a diagram T-s, the quantity of heat (Q) received from the outside environment by the system during a cycle is given by the area of this cycle.

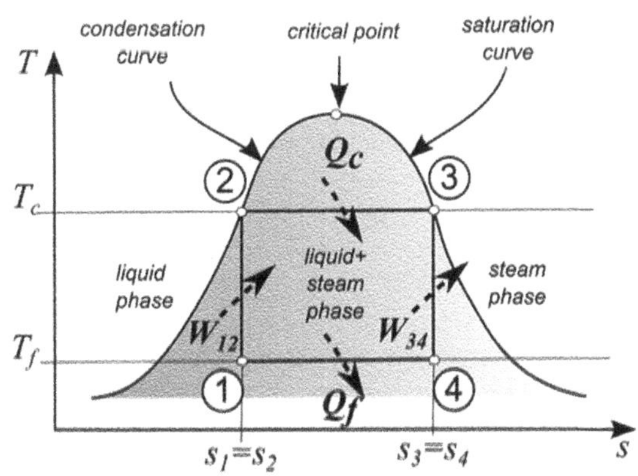

Figure 5.6. *T-s diagram of an ideal Carnot thermodynamic cycle*

In these ideal operating conditions, the cycle is called a "condensable steam Carnot cycle" and corresponds to the maximum efficiency of a thermodynamic conversion between two given extreme temperatures. Later on, this cycle will thus also be called the "ideal Carnot cycle".

The principle of energy conservation applied to a thermodynamic system can be written:

$$Q = W \tag{5.2}$$

where Q is the quantity of heat received by the system from the outside environment and is expressed in Joules, and where W is the mechanical energy supplied by the system to the outside environment.

The thermal efficiency of the cycle is expressed as the ratio of the difference (W) between the net energy flow (here the work producing expansion ($W_{34}>0$)) and the work consuming compression ($W_{12}<0$), on the expensive energy flow (here the quantity of heat supplied by the heat source ($Q_{23}>0$)).

We thus obtain:

$$\eta = \frac{W}{Q_{23}} \tag{5.3}$$

In the case of the ideal Carnot cycle presented in Figure 5.6, we can write:

$$\begin{cases} Q_{23} = T_c.(s_3 - s_2) \\ Q_{41} = T_f.(s_1 - s_4) = -T_f.(s_3 - s_2) \\ W = W_{34} + W_{12} = Q_{23} + Q_{41} \end{cases} \tag{5.4}$$

We thus obtain the thermal efficiency of an ideal Carnot cycle by:

$$\eta_{th\ Carnot\ ideal} = \frac{W}{Q_{23}} = \frac{W_{34} + W_{12}}{Q_{23}} = \frac{Q_{23} + Q_{41}}{Q_{23}} = \frac{T_c - T_f}{T_c} \tag{5.5}$$

If exchanges with the sources are not carried out at constant temperature and if compressions and expansions do not occur at constant entropy, the operation efficiency is reduced. In practice we will define the thermodynamic quality of the various cycles by the ratio of their thermal efficiency (η_{th}) with the thermal efficiency of the ideal Carnot cycle ($\eta_{th\ Carnot\ ideal}$), which have the same temperatures for hot and cold sources. This ratio of thermal efficiencies is called exergy efficiency (η_{ex}).

$$\eta_{ex} = \eta_{th} \frac{T_c}{T_c - T_f} \tag{5.6}$$

However, a thermodynamic Carnot cycle presents several major technological disadvantages:

– the compression of a mix of liquid and steam is quite dangerous (slugging, cavitation[1]) and requires a large compressor size [BOU 98];

– an expansion of a mix of liquid and steam has significant mechanical constraints (erosion and corrosion) in turbines.

To avoid this first drawback, we advise changing phase during the exchange with the cold source and carrying out compression while in the liquid phase. A cycle carrying out this operation is then called a Rankine cycle and is presented in section 5.2.5.2.

To avoid the second drawback, we advise carrying out a second phase change during the exchange with the hot source, and carrying out the expansion in the turbine in the steam phase. Cycles carrying out this operation are then called Hirn cycles or organic Rankine cycles. They are presented in section 5.2.5.2.

5.2.5.2. *Rankine and Hirn cycles*

The operating principle of most thermal power plants is based either on the Hirn cycle, or on the organic Rankine cycle. These two cycles are variations of a conventional Rankine cycle. The latter is made up of the four stages presented above, but during which a change of phase occurs so that compression takes place in the liquid phase. Figure 5.7 illustrates this cycle with a T-s diagram.

The compression stage corresponds to the transition ①-②. Heating starts in the liquid phase (②) and ends at the intersection with the saturation curve (③). The expansion stage is carried out in a liquid-vapor mix and cooling enables us to return to the liquid phase (from ④ to ①).

---

1 The presence of liquid in a compressor (steam compression) significantly increases the pressure at the end of compression (liquids not being very compressible). This leads to the destruction of flap gates (slugging). The presence of steam bubbles in a pump (liquid compression) creates mechanical constraints on the blades during the bubble implosion (cavitation).

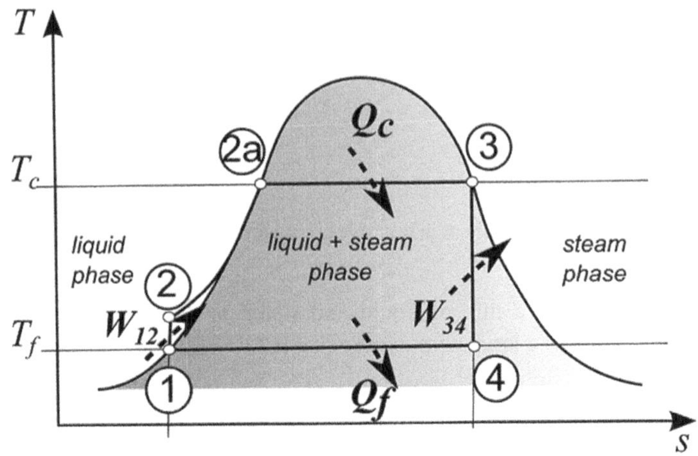

Figure 5.7. *T-s diagram of an ideal Rankine thermodynamic cycle*

The advantage of this cycle is that the energy required for the compression phase is reduced. The cycle exergy efficiency is thus increased despite the increase of the energy input by the hot source. In practice, the Rankine cycle has a thermal efficiency of about 80% of the Carnot efficiency.

The Hirn cycle enables the use of hot sources at high temperatures. It is based on a Rankine cycle, in which the stage of thermal exchange with the hot source carries on after having reached the saturation point. This cycle is represented in Figure 5.8. At the end of the heating stage (② to ③), we then obtain an overheated steam (③).

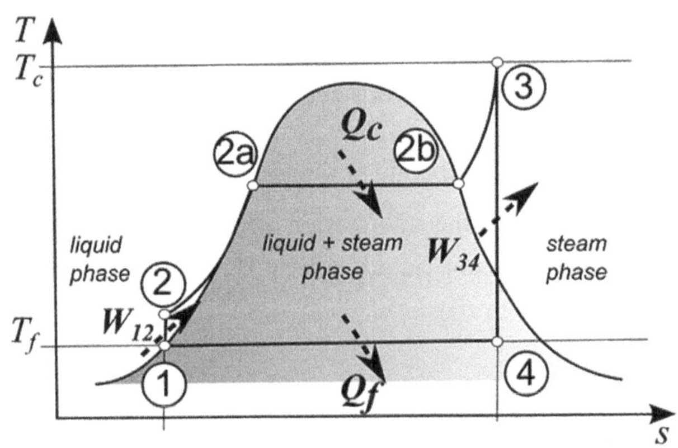

Figure 5.8. *T-s diagram of an ideal Hirn thermodynamic cycle*

In practice, because of the increase of irreversibilities, the exergy efficiency of this cycle is lower than the exergy efficiency of a Rankine cycle. However, as the temperature of the hot source is higher, the Carnot efficiency is higher and the cycle thermal efficiency is generally higher.

Indeed, the increase of recovered energy during the expansion phase is higher than the increase of the energy provided by the hot source. This thus increases the cycle thermal efficiency, despite the decrease of the exergy efficiency [LAL 05].

5.2.5.3. *Choosing the heat transfer fluid*

In the case of geothermal power plants, hot and cold extreme temperatures are respectively imposed by the geothermal source at temperature T_c and by the outside environment at temperature T_f.

Even if the value of the cold source temperature matters, the maximum thermal efficiency of the power plant highly depends on the hot source temperature. This justifies the fact that electricity generation with high temperature reservoirs is favored.

However, in the case of Rankine cycles, the necessity of carrying out compression in the liquid phase and starting expansion from dry steam requires the source temperatures to be lower than the critical temperature. Let us recall that for water, the critical temperature is 374°C and is reached at a pressure of 221 bar.

As shown in Figure 5.9, the diagram (T, s) of some organic fluids has a saturation liquid-vapor bell, whose slope is positive on the vapor side. This enables expansion in the turbine to occur in vapor phase without requiring a Hirn cycle with a lower exergy efficiency.

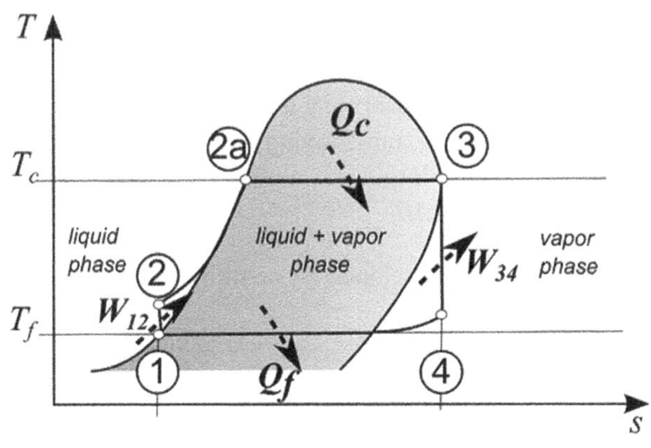

Figure 5.9. *T-s diagram of an ideal organic Rankine thermodynamic cycle*

For the two reasons that were previously mentioned, if the temperature of the hot source is too low to vaporize geothermal water, another heat transfer fluid is used. This is the principle of binary cycle power plants, which is illustrated in Figure 5.4. This fluid is an organic fluid, which has been chosen to have an evaporation point at a temperature lower than that of water. The expansion in the turbine can thus be carried out in the steam phase, without requiring thermodynamic fluid overheating.

Amongst the organic fluids, we find mineral oils, synthetic oils, halogenated fluids and ammonia. Most of these fluids also have the advantage of having a high molecular mass, which enables us to reach a high turbine efficiency at low rotational speeds for this turbine.

5.2.6. *Steam turbine*

The role of the turbine is to convert the energy of a thermodynamic fluid at high temperature and pressure into rotational energy, in order to drive an alternator. The turbine is made up of intake gates, which enable us to adjust the fluid flow rate, a casing with stationary blades (the nozzles) and a rotor with mobile blades. The steam is projected onto the blades and thus creates a force on them. This force comes from the variation of the kinetic (fluid speed) and potential (fluid pressure) energy during the turbine crossing.

Most turbines have several stages, i.e. after a first steam expansion through the blades, the lowest pressure steam is once again expanded by crossing through a second series of blades. Each series of blades is called a stage.

Modern turbines are made up of two types of stages:

– Impulse stages consist of transforming part of the potential energy (pressure) into kinetic energy, via stationary blades; and transforming kinetic energy into torque via mobile blades. An example of this type of stage is shown in Figure 5.10(a)).

– Reaction stages consist of transforming a little less potential energy into kinetic energy, via stationary blades; and transforming the remaining potential energy and the produced kinetic energy into torque via mobile blades. An example of this type of stage is shown in Figure 5.10(b)).

In steady state, the torque developed by a turbine can be determined by the equation:

$$C_m = kQ \quad [5.7]$$

where k is a constant specific to the turbine and Q is the mass flow of the thermodynamic fluid crossing the turbine, expressed in [kg/s] [KUN 94].

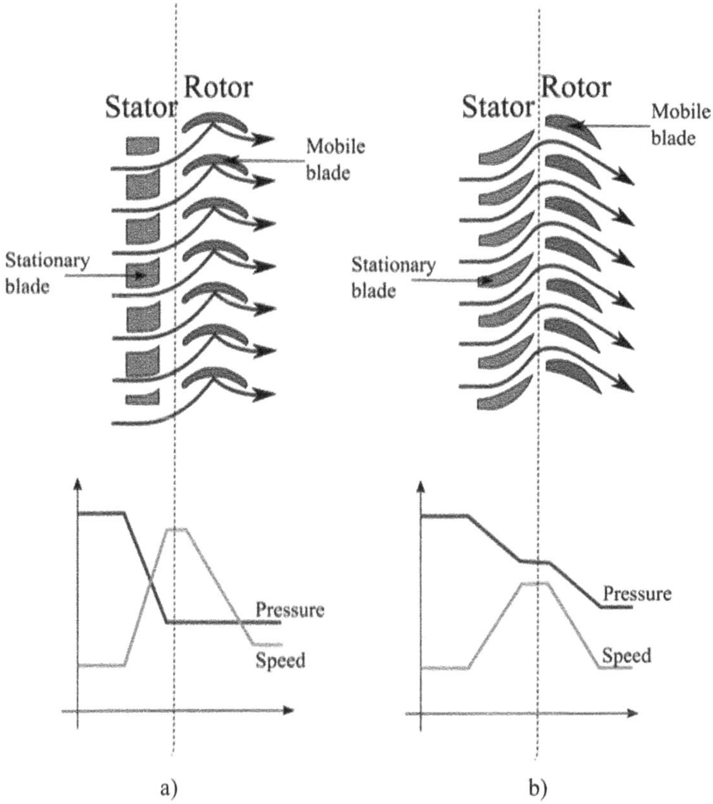

Figure 5.10. *Illustration of steam expansion in action (a) and reaction (b) turbines [WIK]. For a color version of this figure, see www.iste.co.uk/robyns/electricity.zip*

In the case of a thermodynamic cycle including several turbines placed on a single axis of rotation, the torque is obtained by the sum of the torques of each turbine; with the constant k_i and the flow rate Q_i of each turbine i that can be different.

A control valve is placed upstream of the first turbine (a high pressure turbine) to control the flow rate of the fluid entering the turbine.

This gate helps to adjust the turbine rotational speed (Ω) by adjusting the torque developed by the turbine as shown by the fundamental principle of dynamics:

$$J\frac{d\Omega}{dt} = C_m - C_{em} - C_f \qquad [5.8]$$

where J is the inertia of the masses in rotation (turbines, shaft, alternator), C_{em} is the electromagnetic torque created by the alternator during the generation of electrical power on the network and C_f is the torque representing the frictions applied to the set in rotation.

5.2.7. *The alternator*

For the considered power levels, the alternator of a thermal power plant is generally a synchronous machine directly connected to the electrical network.

The operating principle of this electrical machine was briefly explained for wind turbines in Chapter 3, section 3.5. We also saw, in Chapter 4, this machine as the main electromechanical conversion system of hydraulic power plants. The rest of this section will enable us to understand its use, in order to produce the desired electrical power.

5.2.7.1. *Equivalent diagram and modeling*

In the framework of electrical machines, it is often very complex to use the exact physics relations to determine behavior. Very often, equivalent electric circuits are used and help to reflect the main physical characteristics. Moreover, three-phase machines, which are very frequently used for the generation of power greater than 10 kW, have their output power equitably distributed (in normal operation) between the three phases. The study below will thus be carried out on only one phase and a factor 3 will be added to the equation of power to take into account the three phases. Therefore, the single-phase equivalent electric circuit in steady state of a synchronous machine is shown in Figure 5.11.

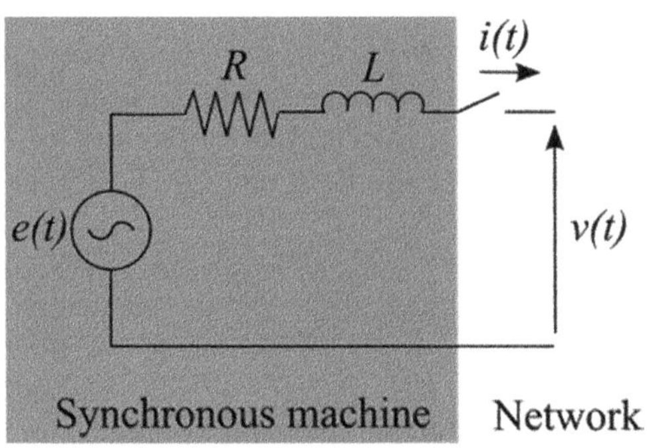

Figure 5.11. *Single-phase equivalent electric circuit of a synchronous machine connected to a phase of an electric network*

Again we can see a voltage source $e(t)$ connected to the induced e.m.f in series with an inductance L and a resistance R. The electrical quantities (current and voltage) of the two other phases are obtained by phase shifting the quantities of the first phase of $2\pi/3$ and $-2\pi/3$.

The equation of the electric circuit can be written as:

$$Ri(t)+L\frac{di}{dt}=e(t)-v(t) \qquad [5.9]$$

with $v(t)=\sqrt{2}V\sin(2\pi f t)$, the phase-to-ground voltage of the network; $e(t)=\sqrt{2}E\sin(2\pi f t+\delta)$, the induced voltage of the machine; f, the network frequency (50 Hz in Europe) and V, the RMS value of the network voltage.

Before connecting the synchronous machine to the electrical network, the amplitude, phase and frequency of the voltage $e(t)$ have to be equal to those of the network voltage $v(t)$. Therefore, the right side of the equation above disappears and the current is thus equal to zero. After the connection, if the amplitude or phase of the induced voltage increases, a current appears. The maximum value of the induced voltage will precede that of the network voltage. The phase shift between these two voltages is called δ, whereas the phase shift between the current and the network voltage is called ϕ.

The average value of the electrical power supplied by a phase of the machine is called the active power and is equal to:

$$\overline{p(t)}=\overline{v(t).i(t)}=VI\cos(\phi) \qquad [5.10]$$

For the three-phase system, the active power sent to the network is thus equal to:

$$P=3VI\cos(\phi) \qquad [5.11]$$

where V and I are respectively the RMS values of the phase-to-ground voltages of the network and the generated line currents. All the electrical quantities have the same frequency. Therefore, they can be represented by rotating vectors at the same speed and linked by equations of the equivalent diagram. We obtain the graph in Figure 5.12.

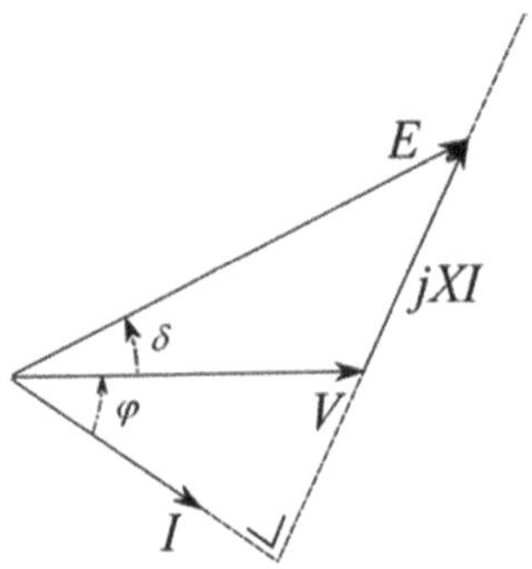

Figure 5.12. *Vector representation of the electrical quantities*

In Figure 5.12, the induced voltage of the machine (E) leads the voltage of the network (V). An equivalent circuit using these quantities can be carried out as is shown in Figure 5.13. On this equivalent circuit, the impedance of the winding at the frequency f, called the synchronous reactance, is equal to:

$$X = 2\pi fL = \omega L \tag{5.12}$$

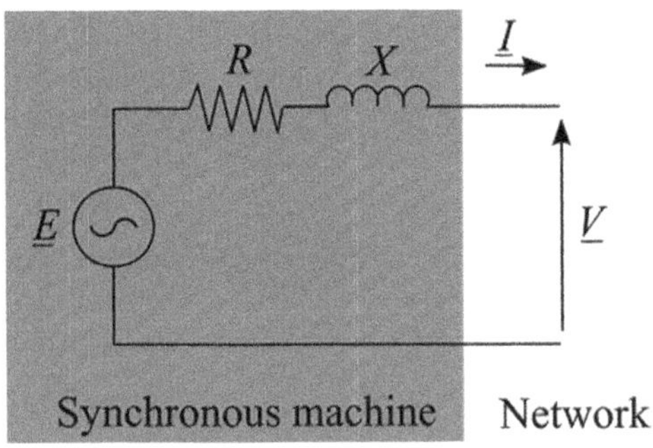

Figure 5.13. *Single-phase equivalent electric circuit of a synchronous machine in complex quantities*

Electrical machines are designed to minimize the losses and we can thus reasonably suppose that the resistance is very low. By assuming the resistance value to be negligible in front of the reactance, the following equation is obtained:

$$\underline{E} = \underline{V} + jX\underline{I} \tag{5.13}$$

with $j = \sqrt{-1}$

The vector representation implies the following relations:

$$E\sin(\delta)=XI\cos(\phi) \qquad [5.14]$$

The active power can then be expressed by:

$$P=3\frac{VE\sin(\delta)}{X} \qquad [5.15]$$

The reactive power sent to the network is equal to:

$$\mathrm{Q}=3VI\sin(\phi)=3\frac{EV\cos(\delta)-V^2}{X} \qquad [5.16]$$

If we assume the electrical machine to be without losses, power conservation implies that the mechanical power is equal to the converted electrical power. Moreover, under the assumption of steady state operation at constant rotational speed, the torque expression is equal to:

$$C_{em}=\frac{P}{\Omega}=\frac{P}{\frac{\omega}{p}}=\frac{3p}{\omega}VI\cos(\phi)=\frac{3p}{\omega}\frac{VE}{X}\sin(\delta)\,. \qquad [5.17]$$

5.2.7.2. *Control of the active and reactive powers*

Once the power plant is connected to the network, opening the steam valves is sufficient to put it into operation. The turbine driving torque tends to make the rotor revolve faster than the stator revolving field. The e.m.f $\underline{E}$, connected to the rotor, goes ahead of $\underline{V}$. The resulting increase of the internal angle δ leads to an increase in the braking torque C_{em} (given by equation [5.17]) and will stabilize the relative rotor position in comparison to the revolving field. The torque is thus maximum for $\delta=\pi/2$, as well as the generated power. The normal (stable) operating point of the machine is guaranteed for

$$0<\delta<\frac{\pi}{2}. \qquad [5.18]$$

How can we adjust the reactive power supplied by the alternator? To answer this question, we will disregard the machine stator resistance and trace (Figure 5.14), the locus of the complex number "stator current" $\underline{I}$ with the help of equation [5.13].

$$\underline{I}=\frac{\underline{E}}{jX}-\frac{\underline{V}}{jX} \qquad [5.19]$$

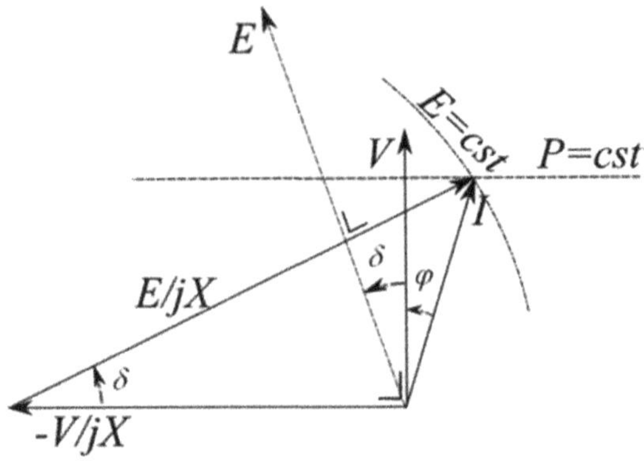

Figure 5.14. *Control of the reactive power*

If the opening of the turbine intake gates is not modified, the active power is constant and thus, at constant voltage, the real component (vertical component) of $\underline{I}$ is constant. When we modify the field (excitation) current, the module of $\underline{E}$ increases. These two last conditions impose that:

– the imaginary part (horizontal component) of $\underline{I}$ will increase and thus the reactive power generated by the machine will increase;

– the internal angle δ will decrease.

Conversely, when we want to decrease the reactive power supplied by the alternator, or even to consume reactive power, decreasing the machine field current is sufficient. During this maneuver, the internal angle δ of the machine will increase and it is important for it to be maintained lower than 90°, in order to avoid the machine stalling. To be sure of this, a link with equation [5.9] can be made. At constant active power and thus at constant torque, in Figure 5.15 we can observe that a decrease of the field current leads to an increase of the internal angle, so that the electromagnetic torque is equal to the shaft torque. If the field current is too low, the intersection of the two characteristics is no longer possible.

How can we adjust the active power supplied by the alternator? As previously shown, the increase of the steam flow rate of the turbine enables the increase of the active power. The real part of the current vector $\underline{I}$ is thus increased. However, as shown in Figure 5.14, if the field current is not modified, the imaginary part of the current related to the reactive power is also modified. It is thus necessary to slightly adjust the alternator field current, in order to maintain a constant reactive power.

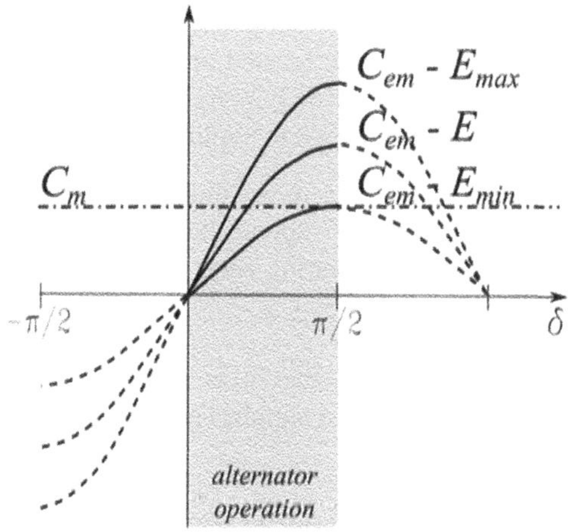

Figure 5.15. *Control of the reactive power – risk of stalling*

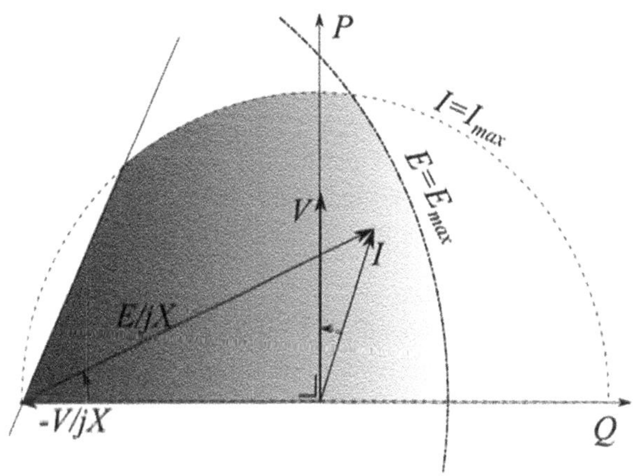

Figure 5.16. *Operating zone in the plane P-Q*

What are the alternator operating limits? Figure 5.16 summarizes the acceptable operating zone for the stator current of the alternator and thus for the active and reactive powers supplied by the machine:

– It seems obvious that the amplitude of this current must be lower than its nominal value in order to avoid an overheating of the stator windings. A first operation zone is thus a disk centered in (0,0) and of radius I_{max}.

– A second operation zone, corresponding to the constraint of having the internal angle lower than 90°, is the half-plane located on the right side of the vertical line going from the extremity of the vector of $\underline{V}/\underline{X}$. As illustrated in Figure 5.16, a safety margin is often used for the internal angle to avoid the machine stalling.

– As the field current corresponds to an induced voltage in the stator E, the thermal limit of the field circuit corresponds to a maximum value E_{max} for the induced voltage in the stator. The operating zone is thus a disk of radius E_{max} which is centered at the extremity of the vector $\underline{V}/\underline{X}$.

Figure 5.16 shows, for example, that it is not possible for an alternator to consume a lot of reactive power when it supplies active power.

5.3. Thermodynamic solar power generation

5.3.1. *Introduction*

In Chapter 2, we highlighted the significance of the solar energy resource. This can be used via photovoltaic conversion for a direct conversion of radiation into electricity. However, when a big power plant is considered, the solar energy can be used to heat a fluid, which will undergo a thermodynamic cycle similar to those previously presented in the context of geothermal power plants. In the literature, this type of power plant is called a "thermodynamic solar power plant", "solar thermal power plant" or "concentrated solar power (CSP) plant".

In order to reach an interesting efficiency for thermodynamic conversion, it is important to produce a high temperature fluid and thus to concentrate the energy supplied by the Sun's rays. Several types of power plants will thus be encountered in practice according to the methods used for the capture and concentration of this renewable primary energy. In the following sections we will thus present these methods and specify their field of use.

5.3.2. *The principle of concentration*

The incident solar flux at the Earth's surface is 1,000 W/m^2. This energy is not very dense and obtaining a significant temperature of thermodynamic fluid requires a large absorption surface. A first solution would consist of covering this surface with the help of the fluid, but the thermal losses of the fluid being proportional to the absorber surface, they would be significant. It is thus better to use an optical concentrator for the incident radiation in order to increase the energy density instead of the absorber surface. This concentrator will be made up of a significant collector surface and a system based on mirrors which reflect the radiation onto a smaller surface called an absorber. The concentrator is characterized by a geometric concentration factor C_g, i.e. the ratio between the incident power density and the power density at the concentrator output. This factor is also given by the ratio of the collector surface to the absorber surface. Its values range between 10 and several thousand, according to the technologies.

In order to determine the concentrator efficiency, let us quantify the power inputs P_{abs} and the power losses P_{losses} at the absorber, per unit of surface of the absorber.

$$P_{abs} = \alpha\tau R C_g P_{sol} \qquad [5.20]$$

where P_{sol} (in W/m²) is the incident sunlight per unit of surface area of the collector, α is the absorption coefficient of the absorber surface, τ is the transmission coefficient of the absorber, R is the reflection coefficient of the mirrors of the collector, which enables us to obtain the geometric concentration factor C_g.

The losses P_{losses} are the combination of losses by conduction-convection and losses by radiation.

$$W_{losses} = U\left(T_{abs} - T_{amb}\right) + \varepsilon\sigma\left(T_{abs}^4 - T_{amb}^4\right) \qquad [5.21]$$

where U is the overall heat transfer coefficient, T_{abs} and T_{amb} are respectively the absorber and the ambient temperatures (in K), ε is the absorber emissivity coefficient and σ is the Stefan-Boltzmann constant.

The quantity of heat received by the heat transfer fluid depends on the efficiency factor F of the absorber-fluid transfer. Therefore, the efficiency of the concentration system is given by:

$$\eta_r = \frac{F\left(P_{abs} - P_{losses}\right)}{C_g P_{sol}} = F\alpha\tau R - \frac{F\left(U\left(T_{abs} - T_{amb}\right) + \varepsilon\sigma\left(T_{abs}^4 - T_{amb}^4\right)\right)}{C_g P_{sol}} \qquad [5.22]$$

Let us consider the numerical application with the following characteristics: T_{amb} = 25°C; W_{sol} = 800 W/m; $F\alpha\tau R$ = 0.8; $F\varepsilon$ = 0.8; FU = 20 W/m² [PIT 07]. The concentrator efficiency according to the absorber temperature expressed in °C and for various values of the concentration factor is given in Figure 5.17.

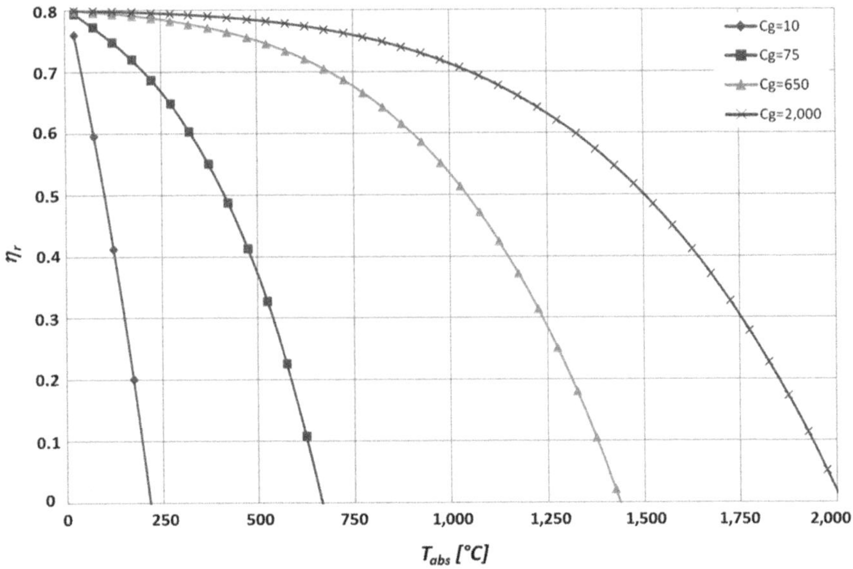

Figure 5.17. *Concentrator efficiency according to the absorber temperature. For a color version of this figure, see www.iste.co.uk/robyns/electricity.zip*

From this figure, we can note that the efficiency:

– decreases with the absorber temperature;

– increases with the concentration factor.

However, conclusions regarding the sizing of a thermodynamic system, which are only based on the concentrator efficiency, would be incorrect. Indeed, it is important to remember that the efficiency of the thermodynamic cycle increases with the temperature of the hot source (here the absorber), even if we have just seen that the efficiency of the concentrator decreases with the absorber temperature. An adaption of the concentrator and the thermodynamic cycle will thus have to be carried out, in order to maximize the efficiency of the set. Figure 5.18 illustrates this adaptation when the efficiency of the thermodynamic cycle is approached by the thermal efficiency of an ideal Carnot cycle.

If we carefully study this figure, we can note that depending on the hot source temperature, a concentration factor can be chosen in order to maximize the efficiency of the set. Moreover, the maximum value of this efficiency increases with the concentration factor [FER 08].

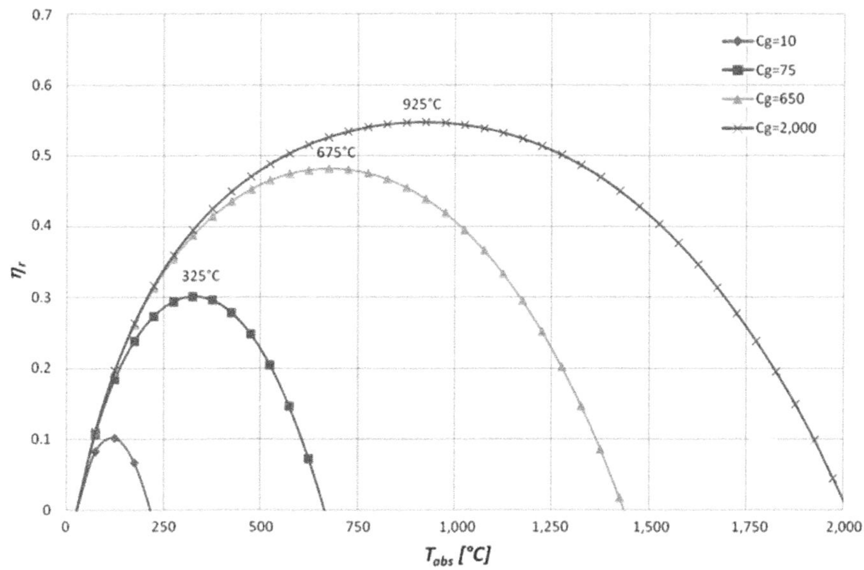

Figure 5.18. *Efficiency of the set (concentrator and thermodynamic cycle). For a color version of this figure, see www.iste.co.uk/robyns/electricity.zip*

There are currently three mature technologies for the industrial implementation of this process [FER 08]:

– The cylindro-parabolic (parabolic trough) collector, shown in Figure 5.19, enables us to carry out concentration factors from 70 to 80 and thus operates at hot source temperatures ranging from 270 to 450°C.

– The solar tower, shown in Figure 5.20, enables us to carry out concentration factors from 300 to 1,000 and thus operates at hot source temperatures ranging from 450 to 1,000°C.

– The parabolic dish collector, shown in Figure 5.21, enables us to carry out concentration factors from 1,000 to 3,000 and thus operates at hot source temperatures ranging from 600 to 1,200°C.

The last two technologies enable us to reach higher efficiencies, but are quite difficult to carry out (at a reasonable cost) for very high powers. The following sections will detail these various technologies.

An important characteristic of this category of electricity generation solution from the solar radiation relies on the possibility of storing heat upstream of the electricity generation process itself, i.e. with a lesser cost. It leads to the possibility of carrying out solar power plants. Therefore, the capacity factor[2] can be very high (close to that of conventional thermal fuel power plants). In the Spanish power plant Gemasolar (19.9 MW, 110 GWh per year), the capacity factor is about 75%.

Figure 5.19. *Solar power plant with a cylindro-parabolic concentrator of 33 MW-70 GWh per year (Kramer Junction Operating Company). For a color version of this figure, see www.iste.co.uk/robyns/electricity.zip*

Figure 5.20. *Solar power plant with tower concentrator of 11 MW-24 GWh (Abengoa Solar, SA). For a color version of this figure, see www.iste.co.uk/robyns/electricity.zip*

2 The capacity factor of a power plant is the ratio between the total energy produced over a given period and the total energy that the power would have produced over the same period while operating at nominal power.

Figure 5.21. *Solar power plant with parabolic concentrator of 8.4 kW [EUR 01]*

5.3.3. *Cylindro-parabolic design*

As revealed by the name of the technology, the collector has a parabolic invariant shape according to a direction z and is covered by mirrors. Figure 5.22 represents this configuration [FER 08].

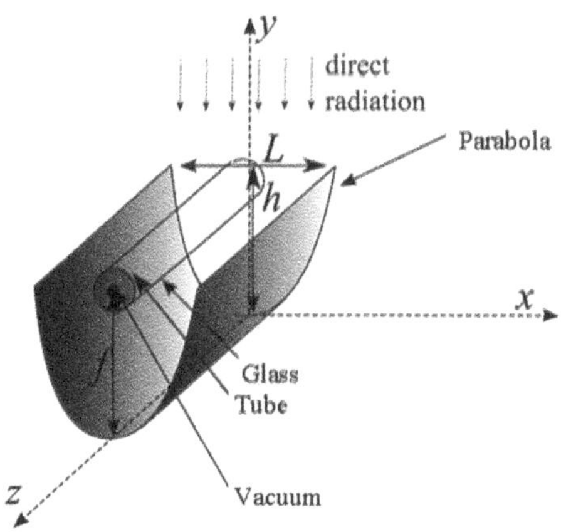

Figure 5.22. *Configuration of a cylindro-parabolic concentrator*

Any radiation parallel to the axis of the parabola will be directed towards its focal line. The principle of capture is thus mainly focused on the capture of direct radiation.

At the focal point, the absorber, a tube made from thermal conductive material, transmits this concentrated energy to a heat transfer fluid, typically synthetic oil. In order to avoid part of the collected heat being lost by conduction towards the outside, the tube is itself put in a vacuum glass tube.

A tracking mechanism is added to maintain the incident radiation parallel to the parabola axis. The axis z is thus placed according to the north-south direction, in order to reduce the tracking mechanism of the Sun to a rotation around this axis.

The geometric concentration factor can be calculated by:

$$C_g = \frac{A_{rec}}{A_{abs}} = \frac{L - d_R}{\pi . d_R} \qquad [5.23]$$

where L is the width of the parabola opening which defines the surface that is perpendicular to the Sun rays, i.e. the reception surface of the radiation, and d_R is the absorber diameter.

This choice leaves us a degree of freedom, which is the parabola form. This degree of freedom will allow us to find a compromise between several objectives:

– minimizing the surface of mirrors;

– increasing the tolerance on the orientation angle of the parabola axis in comparison to the axis of sun rays;

– decreasing the related sensitivity connected to the non-parallelism of the Sun's rays. The solid angle θ under which the Sun is seen is indeed 0.53°;

– decreasing the effect of the absorber shadow on the surface of the collector.

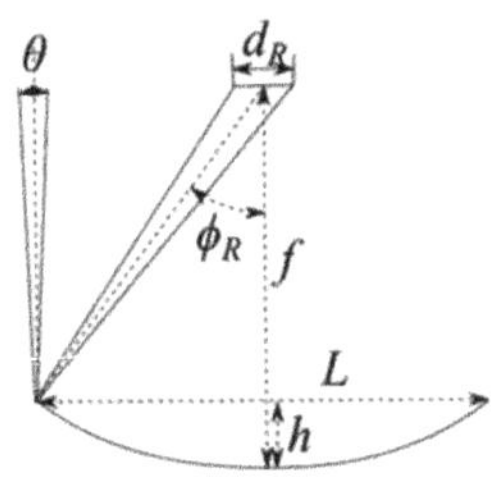

Figure 5.23. *Trajectory of the incident radiation on the absorber*

If we study the sensitivity to the non-parallelism of the Sun's rays, Figure 5.23 illustrates the trajectory of the rays on the collector. This leads to the following geometric relationship [SIN 03]:

$$d_R = \frac{f\theta}{\cos\varphi_R\left(1+\cos\varphi_R\right)} \tag{5.24}$$

where ϕ_R is the half opening angle of the parabola and f is the focal distance of the parabola.

By using relations [5.23] and [5.24], we can show that the concentration factor is given by:

$$C_g = \frac{1-\left(\sin\theta/\cos\varphi_R\right)}{\left(\sin\theta/\cos\varphi_R\right)\left(\varphi_R+\pi/2-\theta\right)} \tag{5.25}$$

Finally, the power collected per unit of length of the collector will depend on the solar power P_{sol} (in W/m²) and the parabola opening L. It is given by:

$$P_s = L.P_{sol} \tag{5.26}$$

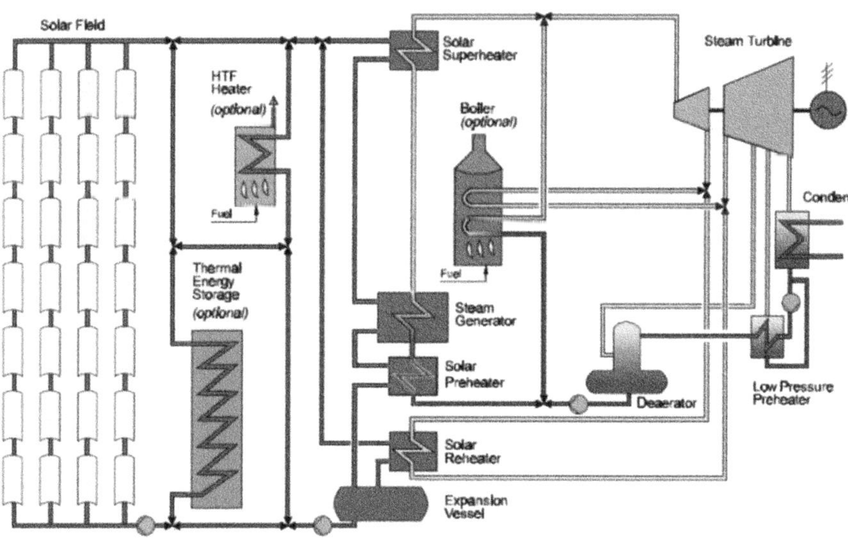

Figure 5.24. *Power plant with a cylindro-parabolic and Rankine thermodynamic cycle collector [PIL 96]. For a color version of this figure, see www.iste.co.uk/robyns/electricity.zip*

Figure 5.24 shows the entire cycle of a power plant with a cylindro-parabolic collector. The solar field made up of cylindro-parabolic collectors is represented on the left. The heat transfer fluid is used to heat water and produce steam. On the right side of the figure, this steam flows in several turbines (high pressure, medium pressure and low pressure) driving the alternator. Some elements of the process are optional, such as heat storage or a back-up heating system of the heat transfer fluid or steam, enabling the generation of electricity, when there is little or no sun.

This section would not be complete without mentioning a technological variant of the cylindro-parabolic collectors. Fresnel collectors illustrated in Figure 5.24 are made up of a set of plane mirrors, each equipped with a tracking mechanism following a north-south axis, and reflecting the radiation to a heat transfer tube. The efficiency is low, but the use of plane mirrors enables us to significantly reduce manufacturing costs.

Figure 5.25. *Prototype of a solar power plant with a Fresnel collector in Almeriá, Spain (Solar Power Group)*

5.3.4. *The solar tower*

The second technology used for the generation of electricity from solar heat is the solar tower. A large number of mirrors are placed north of a tower. The absorber is placed at the top of this tower. As the absorber is fixed, the transportation of the heat transfer fluid is quite simple.

Each mirror is equipped with a positioning system based on two axes of rotation enabling the solar radiation to be guided towards the absorber. Considering the significant distances between the mirrors and the absorber, the focal distance of a parabolic mirror must be significant. The surface of each mirror can thus be approached by a plane, which significantly reduces manufacturing costs.

In this configuration, illustrated in Figure 5.20, the concentration factor is higher than for the cylindro-parabolic capture and very high temperatures are reached by the heat transfer fluid.

5.3.5. *Parabolic dish design*

The last technology used for electricity generation from solar heat is parabolic dish collectors. The absorber is located at the focal point of a parabola. In order to guide solar radiation onto this absorber, a rotational device from east to west and another one from bottom to top is required. The obtained temperature, around 1,000°C, is higher than in the previous technologies. The efficiency of the device is thus significantly improved.

Considering the high temperatures obtained by this process and the constraints of the mobilities related to the Sun path, the thermal–mechanical–electrical conversion is carried out locally in the parabola cell. A Stirling engine is thus particularly adapted to this application.

A Stirling engine is an "external combustion" engine, that has at least one piston sliding in an hermetic cylinder and whose fluid, confined under pressure, is subjected to a thermodynamic cycle with:

– a heating phase at constant volume, due to the heat extracted from solar radiation. This phase is illustrated in Figure 5.26 by the transition ①-②. The temperature and pressure of the fluid increase;

– an isothermal expansion phase producing mechanical power. This is illustrated in Figure 5.26 by the transition ②-③. This mechanical power is transformed into electricity by an alternator;

– a cooling phase at constant volume. This is illustrated in Figure 5.26 by the transition ③-④. The cold source collects heat and the temperature and pressure of the fluid contained in the piston decrease;

– an isothermal compression phase. This is illustrated in Figure 5.26 by the transition④-①.

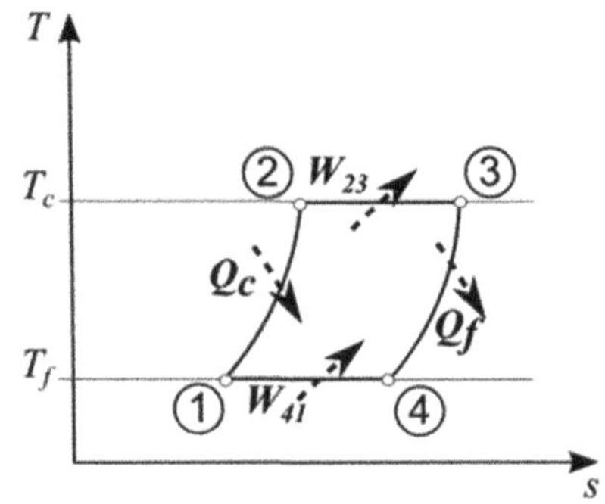

Figure 5.26. *Stirling thermodynamic cycle*

It is important to note that contrary to the Carnot, Rankine and Hirn cycles carried out in practice, the fluid contained in the piston is not subjected to any change of phase. We can show that the maximum theoretical efficiency of a Stirling engine is equal to the efficiency of an ideal Carnot cycle. It thus increases with the hot source temperature.

Several technologies can be used to develop such a machine. As an example, Figure 5.27 illustrates an "Alpha" Stirling engine. We can see an L-shape chamber containing the fluid, a heat exchanger with the hot source, a second exchanger with the cold source and a system enabling the conversion of the linear motion of the two pistons in a rotational movement. An alternator can be driven on the shaft in rotation.

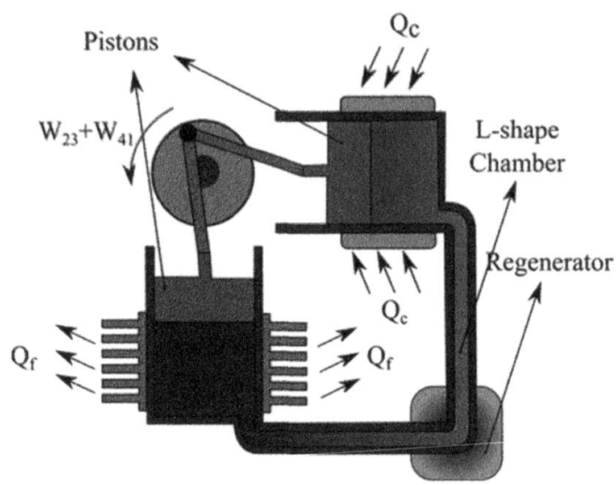

Figure 5.27. *"Alpha" Stirling engine [WIK]. For a color version of this figure, see www.iste.co.uk/robyns/electricity.zip*

In practice, in order to increase the efficiency of the thermodynamic conversion, a regenerator is placed between the hot and cold cylinders. This regenerator is made up of a heat exchanger between the chamber and a heat storage system. The regenerator allows the system to retain heat that would otherwise have been released into the atmosphere.

Moreover, to further increase the cycle efficiency, the heat transfer fluid is chosen for its low heat capacity and thus its significant pressure variation due to temperature changes, as well as its low viscosity. This reduces losses by friction. Helium and hydrogen are frequently used gases.

5.3.6. *Comparison of solar thermodynamic generations*

In the previous sections, three technologies were presented for electricity generation from the Sun, by using a thermodynamic cycle. Table 5.1 allows us to make comparisons.

	Cylindro-parabolic collector	**Solar tower**	**Parabolic dish collector**
Main application	Power plants connected to the electrical network Production of heat at medium temperature	Power plants connected to the electrical network Production of heat at high temperature	Isolated power plants or connected to the electrical network
Advantages	Medium efficiency Possibilities to add heat storage Possibility of high power plant	High efficiency Possibilities to add heat storage	High efficiency Modular operation: each parabola being independent
Drawbacks	Production of heat at medium temperature	Necessity of a high precision sun tracking mechanism which limits the distance mirrors-absorbers	Difficulty of association with heat storage High cost

Table 5.1. *Comparative table of technologies of electricity generation from thermal solar power*

5.4. Cogeneration by biomass

Traditionally, wood and agriculture waste can be burnt to generate heat. When this heat is in the form of steam at high temperature, it is possible to simultaneously generate heat (or cold) and electricity. This is the principle of cogeneration by biomass. Cogeneration enables the optimization of installation costs and energy efficiencies [LEV 96]. This will be the subject of the following sections.

5.4.1. *Origin of biomass – energy interests*

By definition, biomass designates all organic matter of plant, animal or fungus origin able to become sources of (renewable) energy by combustion, possibly after anaerobic digestion or chemical transformations [HOO 04]. The biomass concerned for electricity generation is mainly that of the first category. We can find lots of materials coming from different origins [SAB 09]:

– wood;

– wood by-products, waste resulting from the wood industry;

– traditional agriculture products;

– organic waste.

The CO_2 equivalent balance of a complete cycle of electricity generation by biomass is not always easy to determine. The combustion of biomass emits CO_2, whereas CO_2 is sequestered during the growth phase of the plantations. Moreover, depending on the resource used as biomass, CO_2 is also emitted by the growth process of the biomass (plantation, fertilizer, crop and transport) and must be attributed to the entire CO_2 balance. We thus distinguish two types of biomass resources:

– resources coming from waste or by-products;

– resources especially produced for the generation of electricity and heat.

Generally, we consider that electricity generation from biomass emits CO_2 in low quantity. In addition to the impacts related to CO_2 emissions, wood contains sulfur and nitrogen, which will produce SO_2 and NO_x during combustion. However, in comparison with coal, these emissions are lower by a factor of about 4.5 [HAQ 02] and are lower than all the other solutions based on fossil fuels.

5.4.2. *Cogeneration principle*

On the contrary to the conventional electricity generation of dispersing thermal energy into air or water in the form of waste, cogeneration is designed to valorize this energy. To do so, it is thus necessary to have heat outlets close to the generation site, because heat cannot be transported as easily as electricity over long distances. We distinguish three main categories of technological cogeneration solutions [LÉV 96]:

– *steam turbines* have been used for a long time by industries that have a significant need for heat and electricity (chemical industry, papermaking industry, sugar refinery, etc.);

– *combustion turbines, generally gas turbines*, which are widespread in industry and heat networks, as well as in sites with significant energy requirements and energy availability constraints (hospitals, etc.);

– *internal combustion thermal engines* consuming gas and/or fuel oil and sometimes biofuels. Their flexibility of use (variable load operation) and their correct efficiency for small units explains their predominance amongst small size generation units.

These various techniques simultaneously generate thermal energy and mechanical energy:

– the thermal energy is collected from the fumes and cooling systems of the engines or combustion turbines or from the expanded steam in steam turbine designs;

– the mechanical energy is generally transformed into electricity by driving an alternator, but can also directly drive compressors, fans, pumps, etc.

Cogeneration can also be carried out using Stirling engines or fuel cells [CRA 03]. The efficiencies of today's Stirling engines are however quite low, which seems to limit the potential of this technology, as it is the electricity part of the generated energy which enables us to make a profit out of these installations. Fuel cells could be an interesting technology for cogeneration, but they are still under development. Prices are still very high and lifespans still low in relation to the requirements of such applications.

Figures 5.28 and 5.29 compare the conversion of energy in a cogeneration facility with a separated generation of electrical energy in an electrical power plant from heat in a boiler. The considered cogeneration facility is based on a steam turbine design.

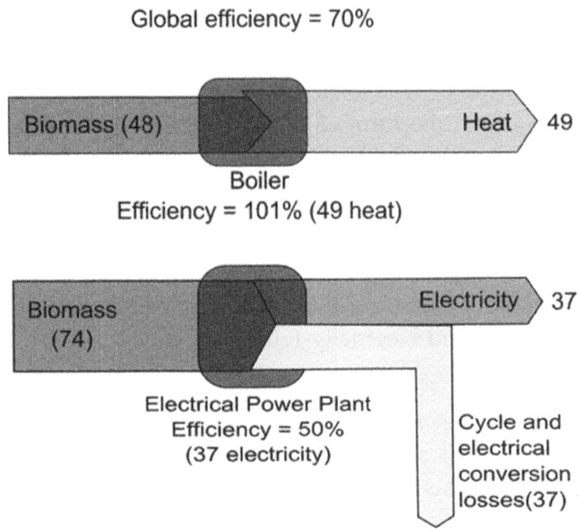

Figure 5.28. *Example of the energy balance for the separate generation of electricity and heat. For a color version of this figure, see www.iste.co.uk/robyns/electricity.zip*

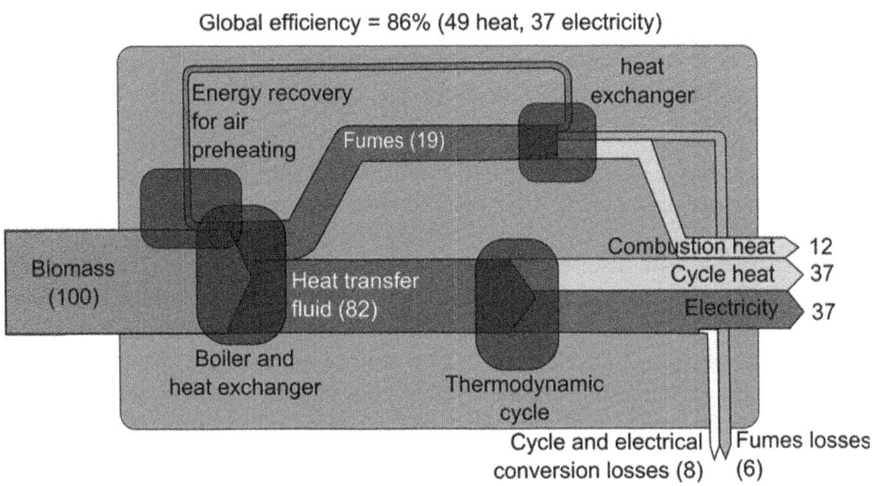

Figure 5.29. *Example of the energy balance for the combined generation of electricity and heat. For a color version of this figure, see www.iste.co.uk/robyns/electricity.zip*

The best conventional electricity power plants have an efficiency of about 50% (currently combined cycle power plants with combustion turbines and steam turbines reach efficiencies of about 60%, whereas conventional power plants with a steam cycle have an efficiency of about 40 to 45%). For heat generation, high thermal efficiency gas-fired boilers reach efficiencies higher than 100%. In order to obtain such efficiencies, it is necessary to condense the water vapor resulting from combustion and to then collect the energy of the water phase change, called the latent heat of water vaporization. The energies (kWh) mentioned in the following will thus be based on the Gross Calorific Value (GCV), which include the energy converted into heat by combustion (Net Calorific Value, NCV) and the latent heat of water vaporization [OBE 03].

Let us consider the example presented in Figure 5.28, where the electricity demand is 37 kWh and the heat demand is 49 kWh. The global efficiency of a system producing electrical energy and thermal energy separately is about 70%.

If heat and electricity generations are combined, one part of the losses related to the electricity generation is recovered in the form of heat. In the example in Figure 5.29, this efficiency reaches 86%. Let us note that to obtain energy savings from a cogeneration unit, heat requirements must be stable and continuous; electricity generation being considered subsidiary [CRA 03].

5.5. References

[BOU 98] BOUGARD J., *Thermodynamique Technique*, Presses Universitaires de Bruxelles, 1998.

[CRA 03] CRAPPE M., *Electric Power Systems*, ISTE Ltd, London and John Wiley & Sons, New York, 2008.

[DI 05] DI PIPPO R., *Geothermal Power Plants: Principles, Applications and Case Studies*, 1st ed., Elsevier Science, 2005.

[EUR 01] EURODISH – STIRLING, *A New Decentralized Solar Power Technology*, Schlaich Bergermann and Partner GbR, 2001.

[FER 08] A. FERRIERE, "Centrales solaires thermodynamiques", *Techniques de l'Ingénieur*, BE8903, 2008.

[HAQ 02] HAQ Z., Biomass for electricity generation, US Energy Information Administration Report, 2002.

[HOO 04] HOOGWIJK M.M., On the global and regional potential of renewable energy sources, PhD thesis, Utrecht University, 2004.

[KUN 94] KUNDUR P., *Power System Stability and Control*, McGraw-Hill, 1994.

[LAL 05] LALLEMAND A., "Production d'énergie électrique par centrales thermiques", *Techniques de l'Ingénieur*, D4002, 2005.

[LAP 08] LAPLAIGE P., LEMALE J., "Géothermie", *Techniques de l'Ingénieur*, BE8590v2, 2008.

[LÉV 96] LEVY C., "Les techniques de cogénération", *Techniques de l'Ingénieur*, B8910, 1996.

[OBE 03] OBERNBERGER, I., CARLSEN, H., BIEDERMANN, F. "State-of-the-art and future developments regarding small-scale biomass CHP Systems with a special focus on ORC and Stirling engine technologies", *Bioenergy 2003 Conference*, 2003.

[PIL 96] PILKINGTON SOLAR INTERNATIONAL, Status report on solar thermal power plants, 1996.

[PIT 07] PITZ-PAAL R., "High temperature solar concentrators in Solar energy conversion and photoenergy systems", in J. BLANCO GALVEZ, S. MALATO RODRIGUEZ (eds), *Encyclopedia of Life Support Systems*, Eolss Publishers, Oxford, UK, 2007.

[SAB 09] SABONNADIÈRE J.C., *Renewable Energy Technologies*, ISTE Ltd, London and John Wiley & Sons, 2009.

[SIN 03] SINGH MAHINDER B., SULAIMAN F., "Designing a solar thermal cylindrical parabolic trough concentrator by simulation", *International Rio Congress, World Climate and Energy Event*, Rio de Janeiro, 1–5 December 2003.

[WOR 78] WORLD ENERGY CONFERENCE, Conservation Commission, *World Energy Demand, 1985-2020: The Full Report to the Conservation Commission of the World Energy Conference*, IPC Science and Technology Press for the W.E.C, 1978.

[WIK] WIKIPEDIA COMMONS, library of freely usable media files, available at: https://commons.wikimedia.org/wiki/File:P-V-T_Diagram_(Water).fr.svg.

[WIT 04] WITTIG G., *Stand des europäischen hot-dry-rock Forschungsprojektes in Soultz-sous-forets*, BESTEC GmbH, 2004.

6

Integration of Decentralized Production into the Electrical Network

6.1. From a centralized network to a decentralized network

Historically, electrical networks developed in the 20th century were organized around a transmission network, receiving the electrical power produced by centralized units (nuclear, thermal or hydraulic power plants) and transporting this power to large consumers or to more modest domestic consumers via a distribution network [HAD 09, KUN 94].

Today, even if the structure of the electricity system – which includes producers, consumers, transmission and distribution networks – has not changed, the function of these actors has changed considerably with the presence of decentralized production units, active consumers (or "prosumers", individuals who both consume and produce) and an international electrical energy market.

6.1.1. *The transmission network*

The transmission network forms the backbone of a national and international (European) electrical network. It is through this network that electrical energy can be transmitted over medium or long distances, typically as exchange between countries and regions. The transmission network ensures the proper functioning of the electricity market by connecting production and consumption centers (including distribution networks). Its operation is also essential to ensuring the quality of the electricity supply. For instance, a problem on this network will likely have repercussions for the entire national area, and may also affect other European

---

Chapter written by Benoît ROBYNS and Jonathan SPROOTEN.

networks (there is the possibility of a cascade effect resulting from the sequential disconnection of key elements in the electrical system which, in worst case scenarios, can lead to system collapse or blackout).

Overhead lines or underground cables transport the electrical energy from exporting areas (including power plants) to consumption areas. In France, where the electric transmission network stretches for more than 100,000 km, it is necessary to increase the output voltages of the production units in order to reduce line losses. The voltage classification of this network is High Voltage (HV – more than 35 kV). Its voltage levels are standardized. In France, the following values are chosen: 400, 225, 90 and 63 kV (rarely, 150 and 45 kV). Figure 6.1 shows the French 400 kV and 225 kV transmission network. This network is highly meshed to ensure continuity of supply for the maximum number of consumers in the case of a line or cable outage.

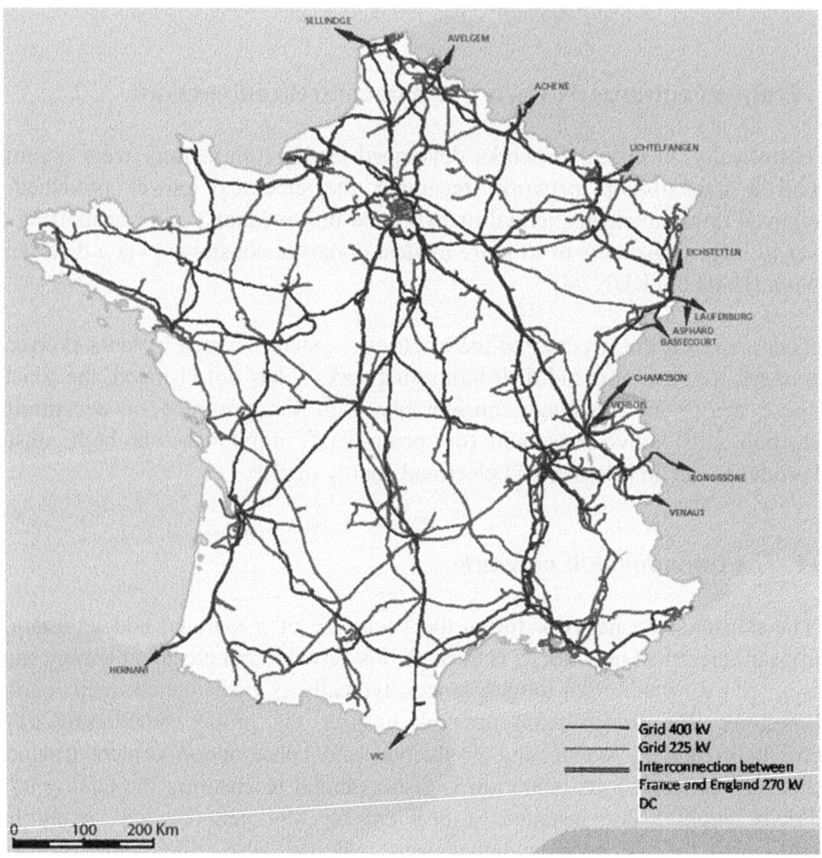

Figure 6.1. *French 400 kV and 225 kV transmission network (RTE)*

6.1.2. *The distribution network*

The distribution network supplies the majority of consumers and connects a large number of decentralized production sites. It is subdivided into a medium voltage network (MV or "HTA" in France with levels of 15 kV and 20 kV) and a low voltage network (LV with levels of 400 V and 230 V). This network is not meshed during normal operation. It is developed according to two types of configuration particular to overhead and underground networks respectively.

Overhead networks are mainly used in areas of low population density and therefore of low load density. The overhead line solution has a significantly lower investment cost than that based on underground cable and has more frequent service interruptions but most of these interruptions are shorter than those of underground networks. Figure 6.2 illustrates this type of antenna or radial network. In this configuration, in the event of an incident on a link, all consumers downstream of this link are no longer supplied.

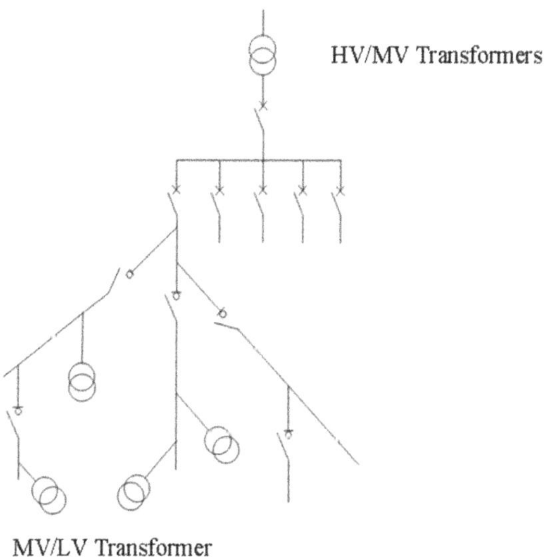

Figure 6.2. *Radial distribution network*

Underground electrical networks are mainly used in high density population or industry zones and therefore in high load density zones. Continuity of supply has been prioritized. Figure 6.3 shows an example of a looped or open loop diagram in which radial links are normally supplied by an HV/MV substation. However, when a fault switches off the connection to this substation, it enables supply via another substation by closing the appropriate circuit breakers.

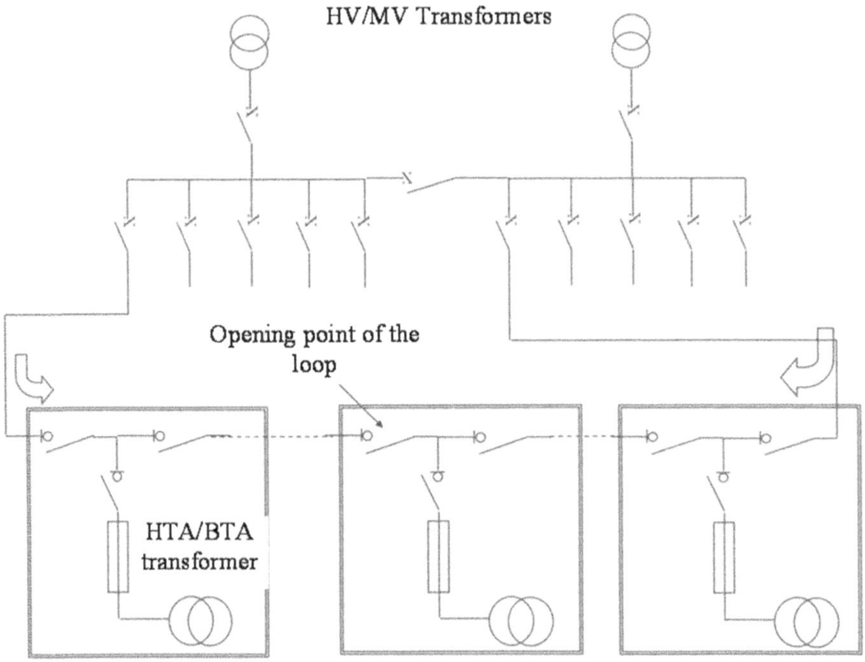

Figure 6.3. *Looped or open loop distribution network*

6.1.3. *Services for the electric system*

Ancillary services enable us to ensure the management of the network, i.e. the control of essential electrical quantities of the electrical network, that is, the frequency and voltage. Keeping the electric system in good working order is highly dependent on maintaining these quantities within a given range.

In a simplified manner, the frequency is determined by the rotation speed of the synchronous alternators (typically, conventional centralized nuclear, thermal or hydraulic production plants) while the voltage is determined by the output voltage of these same alternators and by the transformation ratio of transformers.

Figure 6.4 depicts the control principle of an alternator where control of the torque of the driving turbine, and thus of the speed, allows action on the frequency, and where the control of the excitation current (inductor field) allows for the adjustment of the voltage.

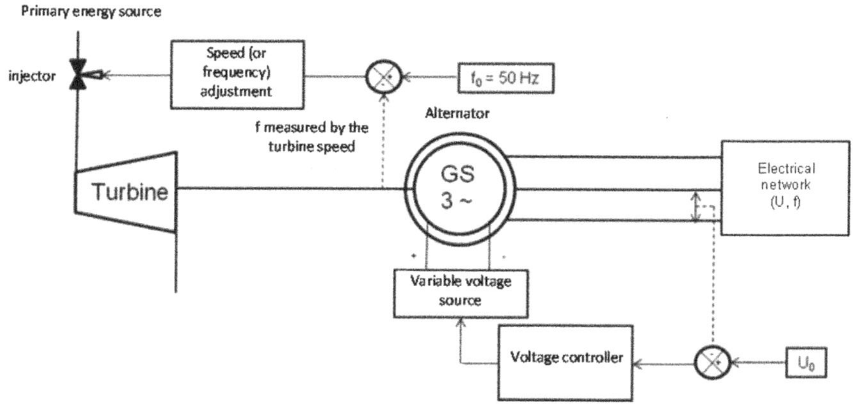

Figure 6.4. *Controllers of a conventional alternator*

In the context of an electrical network where several production and consumption units are connected, frequency variations are linked to variations in the active power balance, while the quality of the voltage plane is affected by voltage dips in network elements, mainly due (at least in transmission networks) to the transit of reactive power.

Conventionally, centralized production was a major contributor to these services both in the steady state, due to their ability to adjust active and reactive power, and in transient regimes, due to their mechanical inertia (which counteract rapid variations in speed), and due to their contribution to short-circuit current in the event of a large voltage sag.

6.1.3.1. *Frequency control*

Frequency variation is caused by an imbalance between the production and consumption of active power (including losses):

– if production is equal to consumption, the frequency remains stable (50 Hz for the European network);

– if production is higher than consumption, the frequency increases because, during the transient control regime, the surplus of energy is stored in the form of kinetic energy in the rotating units, the speed of which increases slightly;

– if production is lower than consumption, the frequency decreases because the energy deficit is taken from the kinetic energy of the rotating units, thereby decreasing their speed.

One can then observe that the behavior of the electrical system is largely conditioned by the fact that part of the electrical energy is produced by synchronous machines directly coupled to the network and driven by turbines (thermal or hydraulic). Note that production systems connected to the grid via an electronic power converter do not naturally provide inertia to the electrical system, which is therefore more sensitive to power imbalances. With less inertia, a greater variation in speed, and therefore frequency, will be necessary in order for the excess energy to be stored as kinetic energy. On the other hand, systems using power electronics in their interface with the network are far less sensitive to frequency quality.

To stabilize and restore the frequency, and therefore achieve high operating reliability of the electrical system, three levels of power and therefore frequency control have been hierarchically defined:

– Primary (automatic) control ensures the balance between production and consumption by adapting the power set-point of the production units. However, this control within the generator maintains a static error on the frequency.

– Secondary (automatic) control, returns the frequency to its set-point.

– Tertiary (manual) control, adjusts the production schedule, exchanges and power transits again and aims at maintaining the control margin.

Primary frequency control is carried out automatically at the level of the production units. It provides rapid, decentralized correction within a few seconds of production–consumption mismatches. This adjustment follows a linear relationship between the frequency (imposed by the speed of the rotating units) and the power which is expressed as follows:

$$\frac{P-P_0}{P_n} = -\frac{1}{\delta}\frac{f-f_0}{f_0} \qquad [6.1]$$

P: power produced by the group;

P_n: nominal power of the group;

P_0: programmed power;

δ: droop of the controller ($\approx 4\%$);

f: network frequency;

f_0: nominal network frequency.

If we aggregate all the power variation behaviors related to the frequency variation at the level of an electrical system, expression [6.1] can be rewritten in a more generic form:

$$\Delta P = -K\Delta f \qquad [6.2]$$

where K is called the primary regulating energy of the system in MW/Hz, and $\Delta P = P - P_0$ the power variation induced by the frequency variation $\Delta f = f - f_0$. For the European grid in continental Europe K = 29,000 MW/Hz, which means that if we deviate by 2,900 MW from the programmed power, the frequency deviates by 0.1 Hz. The contribution of France towards this energy regulating is K = 5,000 MW/Hz Figure 6.5 illustrates the principle of primary regulation.

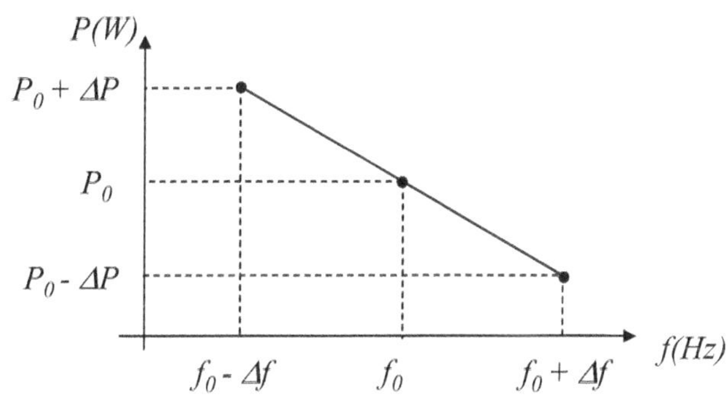

Figure 6.5. *Primary frequency control*

The fast adaptation of production to consumption carried out by primary control leaves a frequency deviation in comparison to the nominal frequency at the end of the process. Moreover, as the frequency is common to the interconnected networks (for example the synchronous network of Continental Europe), this control causes transit variations between countries. They are the expression of "solidarity" between networks to contain hazards (losses of production units, unanticipated consumptions, etc.). The secondary control role is then, in a few minutes (15 minutes in Continental Europe), to return the frequency to its nominal value and to return exchanges between partners to their contractual values. This is the expression of the "responsibility" principle: the hazard correction is the responsibility of the network, where it occurred.

6.1.3.2. *Voltage control*

The transits of active and reactive energy in networks are the source of voltage sag. For the very simple case of a load supplied through a single-phase line by a voltage source (Figure 6.6), we can approximatively write, on the basis of the Fresnel diagram representing the voltages as a function of the current, that the voltage sag in the line ($\Delta V = V_1 - V_2$) is equal to:

$$\Delta V \cong \frac{rP + xQ}{V_2} \qquad [6.3]$$

with r the conductor resistance in Ω, x the line reactance in Ω, and P and Q, respectively, the active and reactive powers transiting in the line.

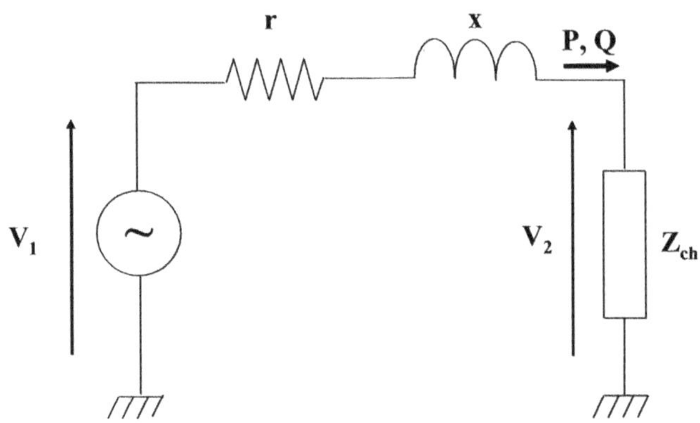

Figure 6.6. *Simplified equivalent diagram of a single-phase line*

For extra high voltage (EHV) lines, $x \geq 10r$; expression [6.3] can thus be simplified as follows:

$$\Delta V \cong \frac{xQ}{V_2} \qquad [6.4]$$

Voltage control can thus be carried out locally to avoid the transit of reactive energy. There are three levels of voltage control:

– Primary (automatic) control ensures maintained voltage by production units connected to 400 kV and to 225 kV.

– Secondary (automatic) control returns the voltages of various network zones to their set-point.

– Tertiary (manual) control defines the set-point value of the zone controllers.

6.1.4. *Actors of a liberalized system*

The liberalization of the electricity market, implemented within the European Union at the beginning of the 21st century, has led to the management of electricity production, subject to competition, separate from the management of electrical networks, transmission and distribution, because the infrastructure of the latter cannot be multiplied. In France, the transmission network is managed by RTE (*Réseau de Transmission de l'Électricité*) while the distribution networks are managed in the form of concessions awarded to an operator, Enedis being the principal though not the only one. Strong competition is emerging among these operators. The CRE (*Commission de Régulation de l'Électricité*) is responsible for ensuring that the competition mechanisms that are put in place are respected, while regulating competition so that it is not unfavorable to consumers, and does not jeopardize an infrastructure that is vital to the economy and the security of the country. Liberalization does not simplify the management of the electricity system, and requires the establishment of a contractual structure for the transfer of responsibility, and new market mechanisms which must integrate the characteristics of new decentralized sources.

The conditions for connecting the production facilities to the electrical network, and in particular the technical constraints, are defined by regulatory legislation such as decrees and orders [FRE 08a, FRE 08b, FRE 08c, FRE 10a, FRE 10b, RFG 16]. These technical constraints are mainly dependant on the the power being connected.

At the same time, in France, decrees govern the target voltages for the connection of production units, as well as the responsibilities for developing and operating these voltage levels. Table 6.1 gives the voltage levels required to connect facilities of given power outputs for the French electrical network

Network type	Voltage range (standard)	French voltage levels	Power
LV	LV	230 V single phase	$S \leq 18$ kVA
LV	LV	400V triphased	$S \leq 250$ kVA
MV (HTA)	$1 \text{ kV} < U \leq 50 \text{ kV}$	15 kV, 20 kV	$P \leq 12$ MW[1]
HV (HTB1)	$50 \text{ kV} < U \leq 130 \text{ kV}$	63 kV, 90 kV	$P \leq 50$ MW
HV (HTB2)	$130 \text{ kV} < U \leq 350 \text{ kV}$	150 kV, 225 kV	$P \leq 250$ MW
EHV (HTB3)	$350 \text{ kV} < U \leq 500 \text{ kV}$	400 kV	$P > 250$ MW

Table 6.1. *Connection voltage levels of facilities according to their power*

1 As an exception, production facilities of up to 17 MW can be connected in MV (HTA) if the network permits and if it is in the interest of society.

The development, operation and, therefore, verification of the compliance of production facilities with these voltage levels involves:

– the transmission network for voltage levels greater than or equal to 63 kV (HTB domain);

– the distribution network for voltage levels less than or equal to 20 kV (HTA and LV domain).

6.1.5. *Roles of decentralized production in network management*

The development of decentralized production, as defined in Chapter 1, has changed the situation considerably, because its generally reduced power means that it is often connected to the distribution network. The remainder of this chapter develops the impact of the integration of this production, in particular at the level of the distribution network, but also at the level of the transmission network, based on the constraints of connecting such production units to these networks. The difficult-to-forecast nature of certain sources, such as wind power and photovoltaics, obviously greatly complicates the management of these networks [ROB 04, ROB 06].

6.2. Connection constraints and usage checks

6.2.1. *Voltage management*

6.2.1.1. *Connection to the distribution network*

Given the weak mesh structure of the distribution networks, the low X/R ratio of the network elements, and the absence of an automatic voltage regulator at the lowest voltage levels, connection of a production plant may modify the voltage plane by modifying the power flows. This should not prevent the network operator from ensuring the quality of the voltage it delivers to all of its users. The voltage ranges to be ensured are defined by normative texts (of the order of ±5% of the contractual voltage). Therefore, any power plant must be able to operate within these ranges and, depending on their power, contribute to the management of the voltage of the network.

The European network codes have proposed the following categorization of power plants.

Smaller type A plants often need to be capable of maintaining reactive power or the power factor at a given value or within a range defined at the time of connection of the plant. Operational communication with the grid operator is often not foreseen. In Europe, depending on the country, this concerns power plants of up to 100 kW or even 1 MW.

Larger type B plants, typically up to a few MW or a few tens of MW, must be able to adjust their reactive power at the request of the network operator.

To illustrate the impact of a decentralized production unit on the voltage plane of a distribution network, the connection of a 55 kW photovoltaic unit in an LV network is considered in Figures 6.7 and 6.8 [PAN 03]. When the PV generator is placed at the end of the line at node 4, the voltage along the line is increased compared to a situation without a generator. A similar result is obtained when the photovoltaic generator is replaced by a conventional generator, such as a diesel generator. This situation can be interesting in certain cases, but can also induce excessively high voltages requiring the disconnection of the production units in periods of low load and high production.

In order to increase the hosting capacity of the network for these production units, these can contribute to voltage management by supplying or absorbing reactive power. However, these must be controllable, which is not easy to achieve, especially for small power units, even if it is increasingly technically possible thanks to the presence of static converters that are equipped with pulse width modulation.

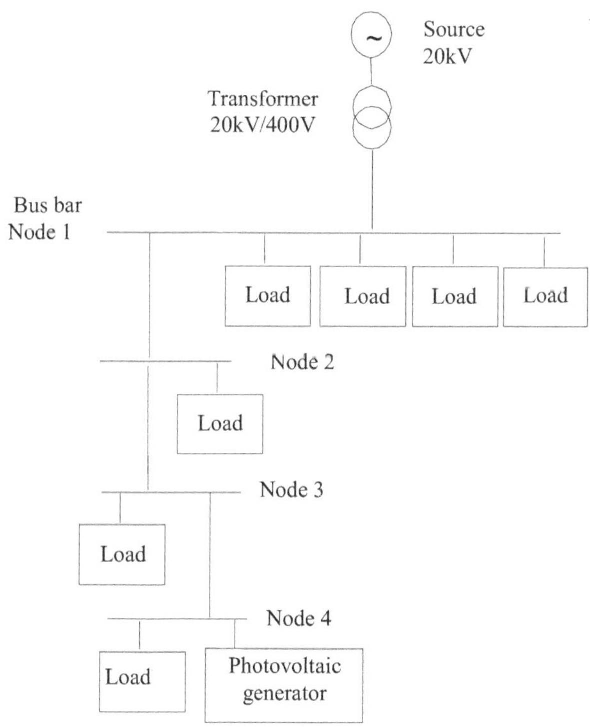

Figure 6.7. *Radial structure of a low voltage network*

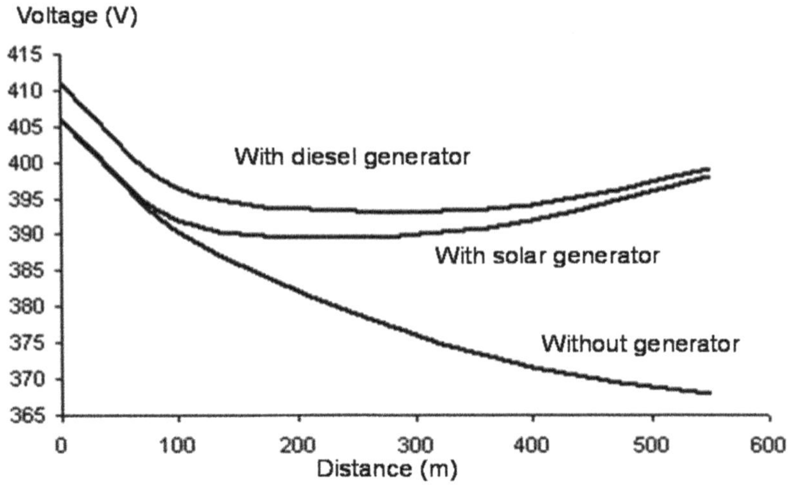

Figure 6.8. *Voltage plan from node 1 to node 4*

Voltage variations created by changes in active power flows can be compensated for by changes in reactive power. For example, the European CENELEC standards propose that small production units, using power electronics in their interface with the network, operate at an inductive power factor of 0.95 when the output power is greater than 50% of nominal power.

Larger power plants must be able to adjust their active power within the following range (these values may vary from country to country):

– supply $Q = 0.5\ S_n$ and absorb $Q = 0.1\ S_n$ for power substations that have an installed power of $1\ MW < P_n \leq 10\ MW$;

– supply $Q = 0.6\ S_n$ and absorb $Q = 0.2\ S_n$ for power substations that have an installed power of $P_n > 10\ MW$.

6.2.1.2. *Connection to the transmission network*

All larger power production units (type C and D) typically connected to the transmission network, including wind farms, must be able to operate within an operating range determined by a graph with the voltage (U) on the ordinate (vertical y-axis), and on the abscissa (horizontal x-axis), the ratio between the reactive power and the maximum active power (Q/P_{max}). An example of an operating domain, the trapezoid [U,Q], is shown in Figure 6.9.

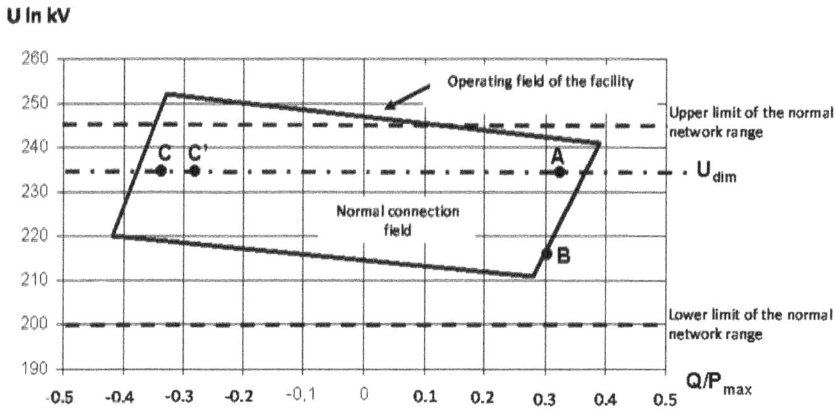

Figure 6.9. *Example of the normal operating range of a production facility*

Within this operating range, these production units must ensure voltage and/or reactive power control at the connection point. Two types of primary adjustment are possible:

– constant reactive power regulation;

– voltage regulation to a value that varies linearly, as a function of the reactive power (or voltage for the current sources) with an adjustable droop.

In addition, within the operating range defined above, the generating unit must be able to move the primary voltage control characteristic so as to shift the initial operating point to a setpoint that has been slaved to the commands coming from the secondary voltage control.

6.2.1.3. *Dynamic contribution to the control of the voltage magnitude*

Nowadays, the vast majority of production units using power electronics in their interface with the network are based on a control of the current injected into the network and provide only a limited short-circuit current in the event of a fault, barely greater than the nominal current of the facility.

This can disrupt the operation of protection devices of the electrical system and can degrade the dynamic behavior of the network voltage. More robust network protection systems, or a "grid-forming" type of control for these sources, are being considered by many network operators and manufacturers in order to meet these challenges. For this type of source, the grid connection inverter controls the complex voltage at the connection point and contributes to the control of the network frequency [PAO 20], which is not the natural approach with wind and photovoltaic sources which, up until recently, essentially behaved as current injectors into the electrical network (control of P and Q injected into the network).

6.2.2. *Frequency control*

6.2.2.1 *Network connection*

In order to correctly manage the frequency of the network, two actions are necessary: the first is to maintain the balance between production and consumption, as explained earlier in this chapter; and the second is to ensure the robustness of the system to limited variations of its frequency.

A disconnection during grid incidents could worsen a production–consumption imbalance, potentially triggering a "domino effect" and accelerate the occurrence of a major grid incident. For example, during the Italian blackout on 28 September, 2003, 3,400 MW of decentralized production was disconnected when the network frequency reached 49 Hz [UCT 04], which in effect aggravated the incident.

In order to avoid a simultaneous triggering of all or part of the wind production on a fault which is normally eliminated, all production plants (of any size) are typically required to be able to remain connected to the network in the event of voltage variation (see previous section), frequency variation and in the event of a rate of change of frequency (RoCoF).

For example, production facilities are typically required to remain connected for limited periods of time in the exceptional frequency ranges, located between 47 and 52 Hz or for a RoCoF of 2 Hz/s calculated over a rolling window of 500 ms.

In addition, production plants are requested to remain connected to the network as long as the voltage sag remains above a profile similar to that shown in Figure 6.10 (valid for regional 225 kV transmission networks, 90 kV and 63 kV). For example, the order [FRE 08a] specifies the voltage template to be observed during a voltage sag on the MV network. The objective is also to prevent a local short circuit, eventually eliminated by the safeguards a little too late, from leading to a cascade and having an impact on the stability of the entire interconnected system.

The fast variations of the power generated by the wind turbines (that can reach a few hundred kW in several dozen seconds), as for the load variations, can induce network frequency fluctuations and activate the primary control of the production units. However, as long as the penetration level of the wind power remains low, this influence can be considered to be negligible. It has significantly the same effect as sudden consumption variations. In the opposite case, in order to ensure the network stability, participation in the primary control of the wind turbines will have to be considered. Nowadays, when the production is higher than consumption and thus when the frequency is higher than 50 Hz, wind turbines might be requested to reduce their production [ACK 05].

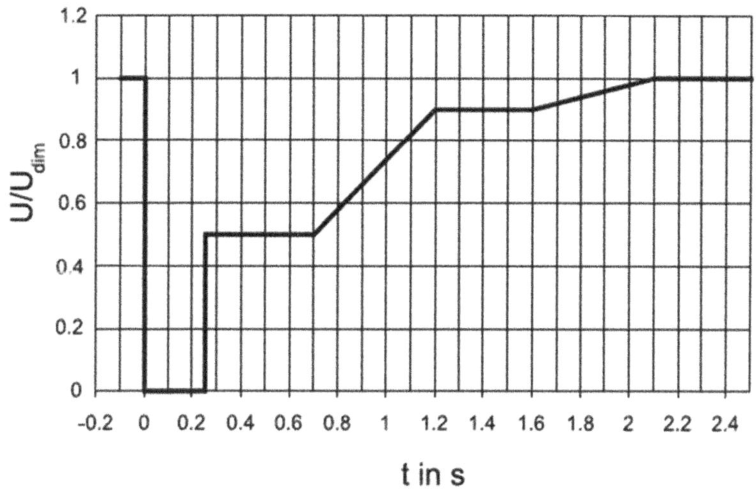

Figure 6.10. *Example of a voltage template defining the conditions needed to maintain production of wind turbines connected to the distribution network*

The participation in frequency control is not required for production facilities based on variable and difficult to forecast energy sources such as wind farms. It is, however, interesting to note the constraints imposed on conventional facilities:

– facilities of a power higher or equal to 40 MW must participate in the primary frequency control;

– facilities of a power higher or equal to 120 MW must participate in the secondary frequency control.

Failure to participate in frequency control could limit the penetration level of wind power, as this control is then shifted to traditional groups. In the event of high wind penetration, the requirements in terms of participation in the frequency control of wind power will therefore change. Obviously, the problem being the existence of a reserve of power associated with wind energy thereby ensures this control. In the long term, a similar evolution could also be seen for the photovoltaic power in the significant networks, knowing that this is already the case in some insular areas.

6.2.2.2. *Dynamic contribution to frequency control*

In addition to the primary and secondary frequency control needs, the decrease in the inertia of the electrical system creates additional dynamic challenges.

In the case of an electrical system with low inertia, the primary control being relatively slow, transient frequency excursion phenomena (frequency nadir) can be observed. Figure 6.11 [SHA 11] illustrates, for the Texas 60 Hz system, the system response to the loss of about 890 MW of power for two operational situations, characterized by different inertias (lower inertia for the dotted curve and higher inertia for the continuous line curve).

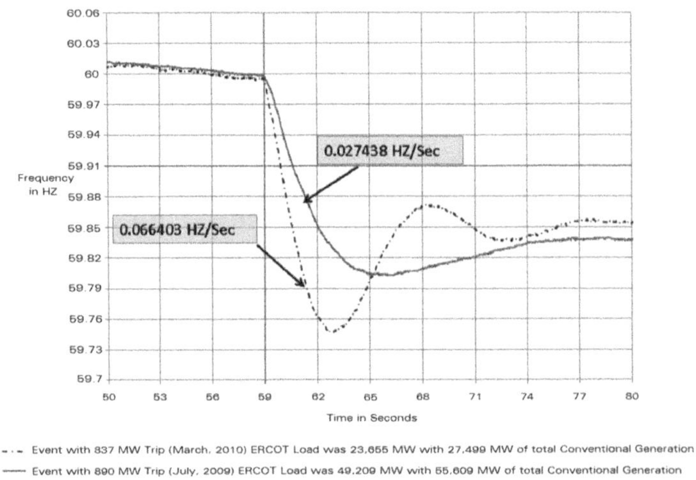

Figure 6.11. *Inertial response of the Texan system [SHA 11]. For a color version of this figure, see www.iste.co.uk/robyns/electricity.zip*

In the event that the frequency nadir goes outside the exceptional frequency ranges for which the power plants must remain connected to the grid (see section 6.2.2.1), a cascade of inadvertent disconnection of production units could be triggered.

In order to overcome this problem a fast frequency support, typically based on the measurement of the rate of change of frequency (RoCoF), can be provided by production units using power electronics in their interface with the network. As with the principle of primary frequency control, the output power of the unit is adjusted proportionally to the derivative of the frequency.

Finally, a change of frequency rate that is too high can also disrupt the control of sources using power electronics in their interface with the network and cause them to be inadvertently disconnected from the network. More robust control systems of the "grid-forming" type [PAO 20] are being considered by many network operators and manufacturers in order to overcome these challenges.

6.2.3. *Quality of the electric wave*

6.2.3.1. *Tolerance to voltage wave quality*

Since a slight harmonic disturbance of the voltage wave supplied by the electrical network is possible, it is important that the operation of decentralized production using power electronics in their interface with the network is not disturbed by this voltage quality. A tolerance threshold is therefore typically defined in connection contracts or network codes.

6.2.3.2. *Current harmonics*

Decentralized production using power electronics in their interface with the network generate harmonic currents that are liable to induce harmonic voltages in the network. Harmonics are frequencies that are added to the sine waves of the voltage and current with 50 or 60 Hz fundamental frequency. These frequencies are generated by the operating principle of electronic power converters generating rectangular waves, full waves, or pulse width modulation (Chapter 2, sections 2.5.3 and 2.5.4).

Fully controlled converters equipped with IGBT transistors, currently the most widely used, generate high frequency harmonics (several kHz). However, these harmonics can be quite easily eliminated, unlike current converters with ignition delay control, equipped with thyristors, which generate low frequency harmonics that require substantial filters to eliminate them.

These harmonic currents must thus be limited to a threshold expressed in % (value of the sum of the harmonic currents of rank h of the facility brought back to the value of the facility nominal current) [FRA 02]. Limit values for facilities of more than 100 kVA are thus given in Table 6.2.

Uneven ranks	**Boundary rate (%)**	**Even ranks**	**Boundary rate (%)**
3	4%	2	2%
5 and 7	5%	4	1%
9	2%	> 4	0.5%
11 and 13	3%		
> 13	2%		

Table 6.2. *Recommended boundaries (brought back to the facility nominal current) for current harmonics*

6.2.3.3. *Voltage fluctuations*

The flicker is induced by the power fluctuations causing low frequency voltage variations (lower than 25 Hz). This flickering effect can be disruptive with electric lights. These power fluctuations appear during the starting and stopping of the production and during changes of the production regime. In the case of wind power, these power fluctuations also come from wind speed variations or from the tower shadow effect.

Figure 6.12 shows a recording of the power generated by a wind turbine at a fixed speed of 300 kW (old wind power site in Dunkirk) subjected to an average wind of 10 m/s. This recording shows that this power can be subjected to variations of more than 100 kW in 3 seconds. We have to check that these power variations do not lead to fast voltage fluctuation levels (flickering), which are unacceptable for other network users. The acceptable levels and the frequency are standardized [FRA 02].

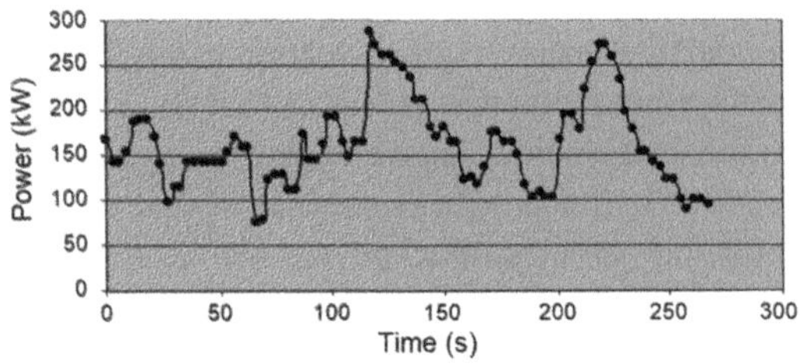

Figure 6.12. *Example of power generated by a wind turbine at a fixed speed of 300 kW*

6.2.3.4. *Coupling of production facilities to the network*

Other constraints related to the connection of the production facility to the distribution network must be taken into account. Let us outline some of these constraints and some orders of magnitude:

– the production facilities are only coupled to the distribution network if the latter is in service (except in special cases such as network reconstitution, etc.);

– the coupling of synchronous machines is only allowed with maximum voltage deviations of 10%, maximum frequency deviations of 0.1 Hz and maximum phase deviations of 10°;

– the voltage variation at the connection point must not exceed a few percent over 0.5 seconds;

– the increase and decrease in power must not exceed a few MW/minute. Note that the reasons for this connection constraint is both with respect to the quality of the voltage wave and the impact of a simultaneous reconnection of a large number of production units on the frequency of the electrical system. Simultaneous reconnection can, for example, be the logical consequence for the simultaneous disconnection of these same units as a consequence to large variations in voltage or frequency.

6.2.4. *Protection and short-circuiting of the electrical system*

In terms of short-circuit current it is of paramount importance that the protection and breaking systems (circuit breakers) of the network and of the production units ensure the rapid elimination of the short-circuits so as to protect people and equipment. These systems are based on safeguards designed to detect the fault, determine its location, and send an order to open circuit breakers and breaking devices.

The production units help to increase the short-circuit current in the vicinity of their connection point. This increase of the short-circuit current may not in any case exceed the limits of the MV devices of the substations and the network [ROB 04].

The distribution network protection plan is designed on the basis that, in the event of a fault, the short-circuit current is supplied exclusively by the upstream source and that no source connected on the distribution network can affect this short-circuit current. The connection of production facilities is contrary to this founding principle, which can lead to malfunctions as soon as the short-circuit current contributions of the production units reach the same order of magnitude as the network fault currents [FRA 02].

6.2.5. *Decoupling protection*

As is the case for all the elements of the electric system, a dispersed production facility must be protected by a set of protection relays. For example, a disconnection protection must be installed on the level of the production facility, in order to:

– detect the network losses, i.e. a plant islanding (deliberate: works, or unintentional: LV, MV, HV faults, risk of false coupling);

– disconnect the production facility, if the previous cases have been detected.

6.2.6. *Other limitations*

At the request of the network operator, and within the framework of agreements, the production facility can participate in the reconstitution of the network, that is to say, in the resupply of the network step by step.

If the power of the production facility is not marginal, the planned generation schedule of the latter must be communicated to the network operator upon its request. It may then be necessary to set up a communication link between the operator and the production facility in order to better exchange operating information.

If decentralised power generation must be connected to a network that is not linked to an interconnected network, additional constraints must be observed.

6.3. The challenges of integrating decentralized power generation

In the early 2000s, the biggest challenges to the integration of decentralized power generation sources into power grids were related to the fact that:

– The connection requirements for production units were not suitable for electronic-based power technologies, and in particular, not suitable for the size of these units. A large number of small, first generation, decentralised power generators were therefore connected into the system, and they did not comply with the requirements presented in this chapter, due to the fact that these were only established later.

– The time required to complete these projects is much less than the time required to strengthen the network. For example, the connection of a centralized production unit of several hundred MW can take several years, ranging from the request to the network operator as to the feasibility of a connection at a given location up until the commissioning of the production unit. The necessary network investments are often localized and can often be made on time. By way of illustration, between 2009 and 2010, 9 GW of photovoltaic production was connected to the Italian distribution network.

These days, certain challenges still persist that can slow down the capacity of the network to accommodate these productions.

6.3.1. *Defense and infrastructure reconstruction plan for the electricity system*

One of the important elements of the defense plan for the electricity system, in the event of a significant drop in frequency, consists of the ability to disconnect consumption and therefore to restore the production–consumption balance. This disconnection is typically done at the HV/MV interface. The presence of decentralized production in MV and LV makes this disconnection less and less effective, since each unwanted disconnection of decentralized production by these load shedding relays means more consumers that must be disconnected.

The ability of these relays to act only on consumption is a challenge which, with a high penetration level of decentralised generation, can only be solved by integrating frequency load shedding functionality directly into consumption devices [WAN 17].

Rebuilding the system also poses challenges. Flexible power plants (black start power plants) supply pockets of the network whose consumption gradually increases. It is important that these pockets do not become energy producers, which could lead to the loss of stability of black start power plants, which are unable to absorb energy. This could happen in the event of a high penetration level of decentralized generation and in the absence of any management of these productions [SPR 14].

Finally, the possibility of operating as an island, following on from a fault, allows for decentralized generation to continue supplying their consumers by isolating them from the network, thereby reducing the number of customers affected by the fault. Islanded operation is, however, currently prohibited in distribution networks due to concerns surrounding the safety of personnel and equipment. Allowing islanded operation will require reviewing the operation and protection strategies of the distribution networks, as well as the control strategies of the production units, or even the addition of electrical energy storage systems, in instances when the primary source is fluctuating greatly.

6.3.2. *Production forecasting for extreme weather conditions*

One of the better known challenges facing wind and photovoltaic production is related to uncertainty in predicting wind speed or levels of sunlight. Although the quality of weather forecasts is continuously improving, production reserves need to be kept on controllable units in order to overcome these uncertainties, especially to meet demand during peak periods. The interconnection of networks makes it possible to mitigate the impact of forecast uncertainties significantly.

However, managing extreme weather situations, such as storms, remains problematic. As illustrated in Chapter 3, Figures 3.27 and 3.37, wind power production under high wind is very nearly constant up until the cut-off speed (or stopping speed) is reached. At this point, the entire wind power output in an area drops to zero. This problem is particularly palpable in the context of offshore wind farms, densely equipped with high power turbines. Preventive wind power reduction or soft cut-off features are the best alternatives available today. Figure 6.13

illustrates two methods for shutting down a wind turbine in the event of storm winds. In Figure 6.13(a), the conventional method stops the wind turbine suddenly when the wind speed reaches V3 (typically 25 m/s, with a hysteresis loop for restarting the wind turbine once the wind speed decreases), while in Figure 6.13(b), the speed of the turbine and therefore the power generated by the wind turbine are progressively reduced as the wind speed varies between V3 and V4 (typically between 25 and 34 m/s).

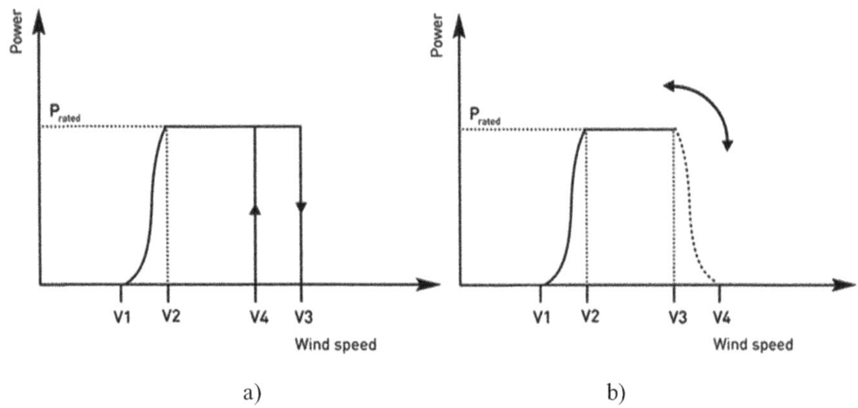

Figure 6.13. *Power curve of a wind turbine as a function of wind speed: (a) imposes a sudden cut-off of the wind turbine in the event of a storm wind, and (b) makes use of a soft cut-off of the wind turbine for high winds (source: https://www.enercon.de/en/technology/wec-features/)*

Similar problems can also be present for photovoltaic production in the context of a snowstorm, or solar eclipse, as was the case for Europe in March 2015 and the USA in August 2017.

6.3.3. *Network hosting capacity and protection*

The capacity of electricity transmission lines and substations is limited. This limitation can be an important problem in the case of wind power, because production sites (windy sites) are sometimes far from the consumption sites or from the substation of the connection point. This is also the case for photovoltaic and hydraulic power and for cogeneration; the latter depending on heat consumers at a reasonable distance from the substation (maximum several km, because heat is not easily transported over long distances). The adaptation and strengthening of substations can concern MV-HV substations following the development of decentralized production in the distribution network (modification of safeguards, increase of the short-circuit power, etc.), but also the internal infrastructure of this network. To prevent the congestion of transmission lines and to ensure the network safety, new lines should be built, especially at the interconnections between the networks that are managed by various operators. However, we should note that the strengthening delay of a substation can reach 5 years and that the construction delay of a new line can reach 10 years and be the subject of significant opposition from local populations.

Regarding network protection, beyond the operation of safeguards in the event of a fault, it is also important that the installation of a new production unit does not lead to unwanted tripping of the protection systems since the connection of a new production unit inevitably modifies the power flow. Typically, on a radial distribution network, energy flows from upstream (connection point to the transmission network) to downstream (loads). Connecting a production unit to a feeder can reverse the direction of the power flow if it produces more power than that being consumed by the loads connected to the same feeder. This can lead to a malfunction of directional safeguards, for example. This reversal of power flows on the network is illustrated by means of arrows in Figure 6.14. This may require modification of network safeguards.

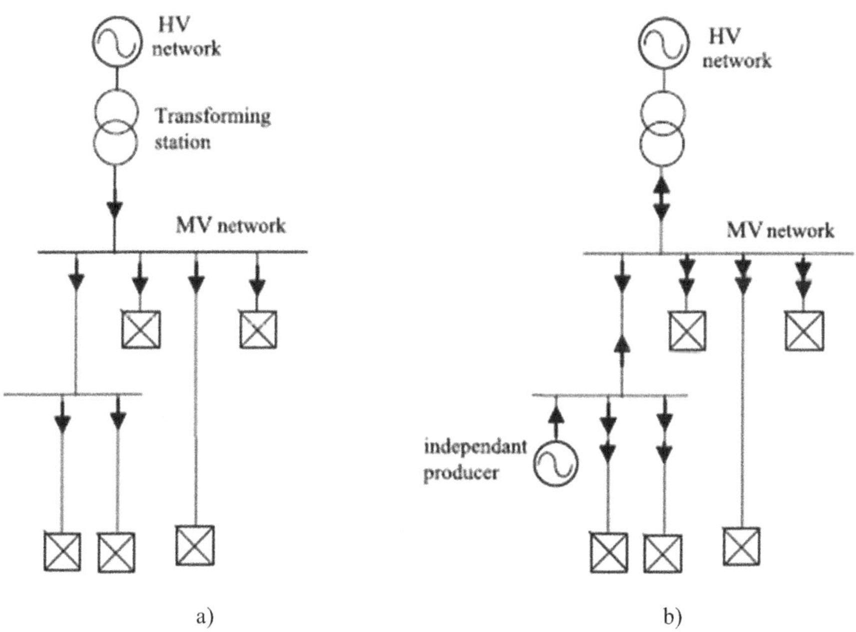

Figure 6.14. *The introduction of generators on (a) a feeder of a distribution network can (b) reverse the flow of power*

Compliance with connection constraints can therefore, in certain cases, lead to network reinforcement requiring the change of conductors, transformers and safeguards, the creation of a dedicated feeder, or even a new substation.

6.4. Perspectives for better integration into networks

To significantly increase the penetration level of the decentralized production units based on renewable energies in the future, three types of evolution will be required. They involve several actions at:

– the source level;

– the network level;

– the consumer level.

The actions between these various system components will require a certain degree of coordination, raising the question of the degree of decentralization that is indeed desirable or acceptable, and the necessity of a communication system between these components. These issues are not purely technological, but include economical and sociological aspects.

6.4.1. *Actions at the source level*

Increasing the penetration level of decentralized production units is therefore possible if this type of source:

– takes part in network management (ancillary services, remote controllability);

– can operate during islanding;

– has an increased and reliable availability, despite the unpredictability of the primary energy source, when it concerns wind and photovoltaic power renewable energies.

These objectives can be achieved by:

– using the possibilities offered by power electronics;

– developing new strategies of control and supervision;

– imagining adapted structures for decentralized production;

– developing and/or exploiting energy storing possibilities in the short and long term (there are already hydraulic pumping energy storage stations, but their use is not always optimal to answer these issues);

– developing multi-source systems with an integrated and optimized energy management.

Participation of decentralized production units to ancillary services is increasingly encouraged, even imposed.

Participation in voltage regulation by absorbing or injecting reactive power is becoming common in both MV and HV networks.

Participation in frequency control by adapting the generated active power is possible in the case of production units made up of a synchronous generator, which is directly coupled to the network, as in conventional power plants (including large hydraulics). This situation is frequently encountered in cogeneration systems, but participation in frequency control is, however, not that common because the electric power involved is rarely significant (on the European network scale) and because this production is mainly correlated to heat demand, in order to ensure a high production efficiency.

In the case of production units connected to the network via power electronics converters (modern high power wind turbines, photovoltaic facilities, gas turbines), participation in frequency control, as in conventional power plants, is not immediate. Indeed, the primary frequency control in conventional power plants is based on the natural link existing between the active power variations generated by an alternator and its rotational speed, which determines the frequency of generated voltages and currents. Obviously, there is no such link when the power is controlled via a power electronics converter, since the operating frequency of this converter is determined by the control and not by the rotational speed of a rotating unit. Moreover, this latter speed is directly associated with its kinetic energy. New "grid-forming" strategies for controlling electronic power converters now allow for these sources to operate in a stable manner in the event of significant disturbance in weak networks and therefore to participate in its management [PAO 20].

To enable difficult-to-forecast production units based on fluctuating sources (wind and photovoltaic power) to fully participate in ancillary services, we will have to associate electric energy storage systems with them. Electricity storage represents additional investment costs and requires the transformation of electricity into another form of energy (chemical, mechanical, thermal, etc.), which leads to energy losses, which, even if they remain quite low, represent some operating costs. Electric energy storage constitutes one of the main components of a future sustainable development. There are several concepts enabling short and long-term storage [BAR 04, MAR 98, DEL 09, ROB 15, ROB 17, ROB 19]. They have to be implemented on a large scale in order to reduce costs. Concrete implementations show that significant storage is possible. In parallel with the electricity production from a nuclear origin, 4,200 MW of hydraulic storage systems have been developed in France. These systems pump the water of a lower basin towards an upper basin to store the energy. In turbine operation, the water flows from the upper basin towards the lower basin in order to produce electricity.

In 2003, a battery energy storage able to supply 40 MW in 7 minutes (or 27 MW in 15 minutes) was installed in Alaska, in order to support the network [REE 03]. A kinetic storage system enabling the supply of 1 MW in 15 minutes was proposed by an American firm. This last form of storage seems particularly well adapted to an association with wind turbines, because of its large dynamics, efficiency and lifespan, which are similar to those of wind turbines [HEB 02]. This type of short-term storage enables us to dynamically smooth the power generated by the wind turbine [LEC 03a, LEC 03b, ROB 05, CIM 06, CIM 10, ROB 15], or could even enable wind turbines to participate in the primary frequency control of the network to which the wind turbine is connected [LEC 04, DAV 04, DAV 06, PIE 09, ROB 15]. The association of storage systems to renewable energy sources is evidently an additional cost, which could, in the future, be compensated for by the financial valuation of the supplied ancillary services [STE 04, DEL 09, FRE 10c, ROB 15].

The development of multisource systems, associating wind or photovoltaic power with other difficult to forecast (for example, hydraulics) or conventional sources and the energy storage with an integrated and optimized management of the energy will also constitute a solution that will enable a high penetration level of renewable energies in the networks [SPR 09, ROB 15].

6.4.2. *Actions at the network level*

6.4.2.1. *Congestion treatment*

The management of congestions, i.e. of electrical transmission lines transiting their maximum power, is essential to guarantee the safety of the electric system and people. A congested line leads to a significant temperature rise of the conductors, which, if it is prolonged and repeated, can lead to arcing risks (when the distance between the line and the ground is reduced by the line thermal elongation, which is prevented by the detection of exceeding a maximal acceptable current for each line, varying according to weather conditions) and jeopardize people or equipment safety. Congestion can be caused by a load transfer during the outage due to grid equipment (lines, transformers, etc.), production units or an evolution of the consumption and production. To manage congestion, network operators have implemented several measures that can be classified according to various characteristics.

Preventive measures carried out D-1 consist of modifying production schedules in order to prevent the congestion that was identified by planning studies.

Corrective measures are carried out on the same day. They consist of preventing any unexpected congestion. Measures can mainly be of two types: modification of the network topology or modification of the production schedules of the groups located in the congestion zone.

Therefore, the network operator can impose operation constraints on the various production units in order to guarantee the electric systems safety, preventively or in real-time.

In France, congestion occurring following connections of new production means are managed as follows. In the zone where the congestion is predicted, operation constraints are applied to production units in the reverse order of their connection to the network. Therefore, the first limited group will be the last to be installed. This principle is equally applied to all types of technology. Its economic logic enables new actors to only pay for the works necessary for their electric connection; in exchange for this reduced connection cost, we can sometimes impose operation constraints on producers, in the case of congestion on the network. There are other methods based, for example, on a market logic.

Amongst operation constraints, we can find, for example, partial or total production limitations. These constraints can be applied to periods of several hours and are determined with the help of planning studies carried out at D-1. The exact duration time of the congestion cannot always be exactly predicted. In the case of a network only integrating predictable productions, load predictions of the electric networks resulting from the various production are reliable and hazards are mainly caused by consumption prediction errors. However, in the case of a massive integration of variable and difficult to forecast (at least locally) production (wind and photovoltaic power, etc.), the network loading prediction is made difficult, because it depends on errors in the prediction of the production of each site. These errors can be about 20 to 50% for the prediction of production in D-1 of a wind farm [PIW 05].

The study presented in [VER 09, VER 11] shows that the method for congestions management in D-1 planning, based on the rule stipulating that the last producer connected to the network will be the first to be disconnected, can, in the case of wind power, induce a loss of 50% of the production. This loss can be considerably reduced by taking into account the real influence of each wind farm on power transits in the network and by managing the wind turbine disconnections in real-time, thereby enabling the reduction of the dependence on uncertainties concerning wind predictions. However, automatic supervision requires the instrumentation and remote control of wind farms, in order to be able to follow the requested reductions in real-time.

6.4.2.2. *Smart grids*

For over a century, network management has been based on a centralized approach using limited means of communication, especially in distribution networks. The implementation and use of new communication technologies associated with advanced management means will increase the intelligence level of the networks and will contribute to a safe increase in the penetration level of the variable and difficult-to-forecast productions, all the while increasing the energy efficiency of these networks. The intelligence level of the system and networks has two aspects [ADE 09]. The first corresponds to the intensification of the deployment on transmission and distribution networks of a telecommunication network, of devices and equipment enabling an increase of the remote controllability and automation of network management. The second includes an advanced management of the (centralized and decentralized) production and of the load, notably leading to the development of new products and services by producers and suppliers for:

– the network operator to increase degrees of freedom in network piloting (e.g. creation of power flexibility reserves, increase of the network capacities thanks to a more thorough operation);

– the end customer, who could benefit from service offers and tariff systems enabling them to adopt ambitious behaviors in relation to the control of the instant requests for electricity and to the integration of renewable energies.

Two approaches can be considered in terms of communication: “Internet” or “enlightened regulation” [ADE 09].

The “Internet” model assumes efficient, global use of transmission and distribution networks, but distributes control of the network to every point in the system.

The “enlightened regulation” or “Internet of energy” model postulates a system wherein the functions and interventions of the various actors in the system are framed by more stringent protocols than in the “Internet” model [FAS 19, ELI 20].

Figure 6.15 shows an example of all Internet communication.

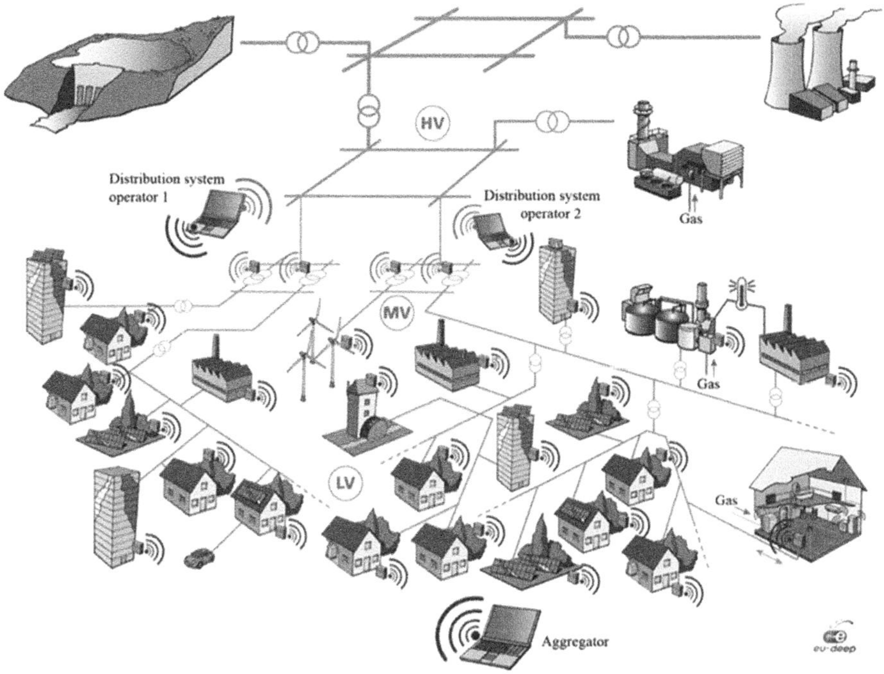

Figure 6.15. *Smart grid with Internet communication (source: European project eu-deep)*

6.4.2.3. *Distribution network architecture*

New network architectures, such as cluster architecture, would enable an increase of the efficiency, safety and availability of electric networks

This type of architecture consists of gathering together the various producers and consumers around a medium voltage network that is able to operate decoupled (islanded) operation from the rest of the network. This system is thus made up of at least one decentralized production unit of renewable energy (DPRE), one conventional decentralized production unit (CDP) and possibly one storage unit, the set being connected to an external distribution network which enables energy consumption or production (Figure 6.16). Such a set must enable customers to have access to quality and reliability, to satisfy power requirements and consist of an insular grid. The customers connected to this grid have the choice of supply amongst various decentralized producers, hence the expression "cluster network". In this cluster, the existence and operation of a structured market of sale and purchase is made easier from a technological point of view. Production surplus can be either partially stored or sold back to the centralized network or to another cluster via the centralized network (Figure 6.17).

From the point of view of the centralized management of clusters, the global architecture has the advantage of being reconfigurable. Any cluster causing instability of the centralized network can be isolated and the network is therefore less vulnerable.

This type of architecture could be built from a reorganization of the current distribution networks. A cluster could be made up of one or several buildings (commercial buildings, offices, factories, etc.) or of a residential area.

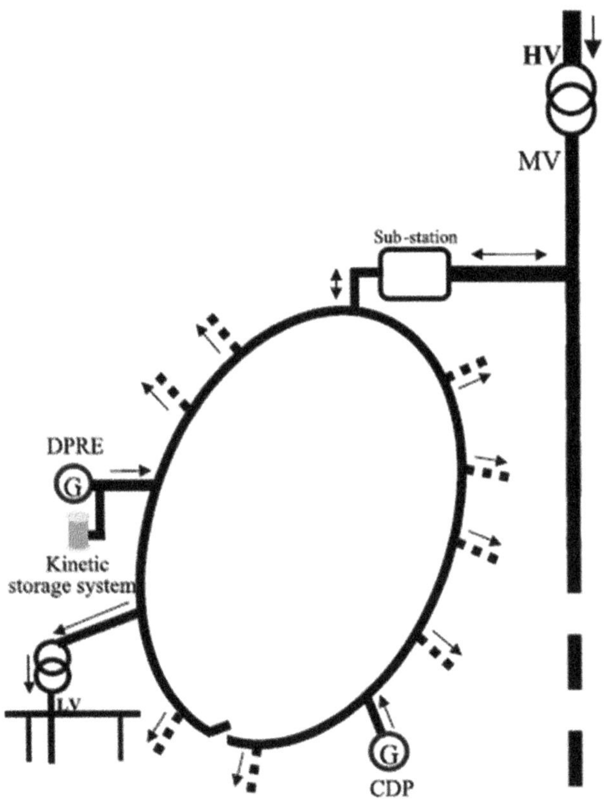

Figure 6.16. *General composition of a cluster (DPRE: decentralized production of renewable energy; CDP: controllable decentralized production)*

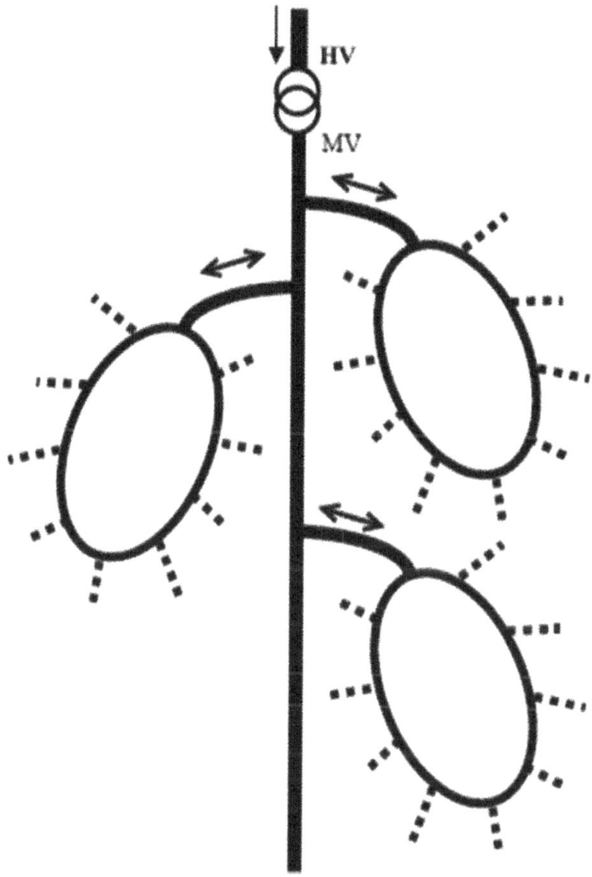

Figure 6.17. *Cluster network trees around centralized production*

6.4.2.4. *Shared or multiple storage services*

Energy storage enables us to compensate for the difficult-to-forecast variations of the production of renewable origin, in order to ensure an available power level. This storage can provide various services, which will depend on its positioning in electric networks.

In electrical networks, several storage development paths can be distinguished [ROB 15]:

– centralized storage, providing services to the balance responsible parties (BRP) (e.g. hydraulic storage connected to the transmission network and associated with the wind power generation portfolio) and also to the network operator for congestion management;

– decentralized storage, whether or not it is associated with energy consuming or producing processes (e.g. V2G or Vehicle-to-Grid type electric vehicle [ROB 19], etc.).

To make storage profitable, one of the approaches consists of pooling the services that can be provided by a storage system for various actors (operators, producers, consumers, etc.) [DEL 09, ROB 15]. These services are as follows:

– local precise and dynamic voltage control;

– support of the network in degraded operation;

– energizing network parts;

– reactive compensation for network operators (and customers);

– reduction of transmission losses;

– power quality;

– energy displacement and support to the existing production fleet;

– primary frequency control and frequency stability of the insular grids;

– congestion management [SPA 20];

– supporting participation in ancillary services;

– valuation of load or generation curtailment;

– guarantee of a production profile;

– peak shaving;

– consumption displacement;

– supply quality/continuity.

6.4.2.5. *European integration of electrical systems*

Creating a large interconnected power system helps limit the impact of local uncertainty on forecasts of wind speed or sunlight. The interconnection of production sites, subject to different conditions at any given time, makes it possible to limit the variability of total wind production or photovoltaic origins. This surface effect increases as the geographic area covered by the interconnected electricity grid and market system expands.

6.4.3. *Actions at the consumer level*

In the past, the presence of little flexibility and communication with consumption, and the opposite with centralized production, defined the concept of "load following": consumption being considered fatal meant that production was adapted hour by hour to keep up with demand.

Figures 6.18 and 6.19 show typical profiles for domestic and commercial consumers. They illustrate the variable nature of consumption depending on the time of day, the seasons and the type of load.

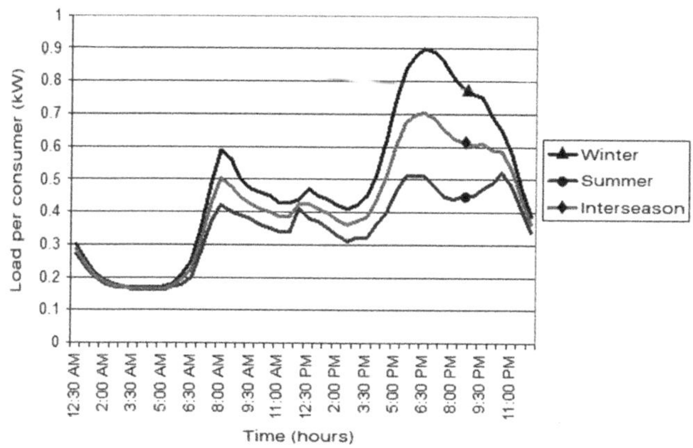

Figure 6.18. *Typical profiles of domestic consumers without electric heating (RTE)*

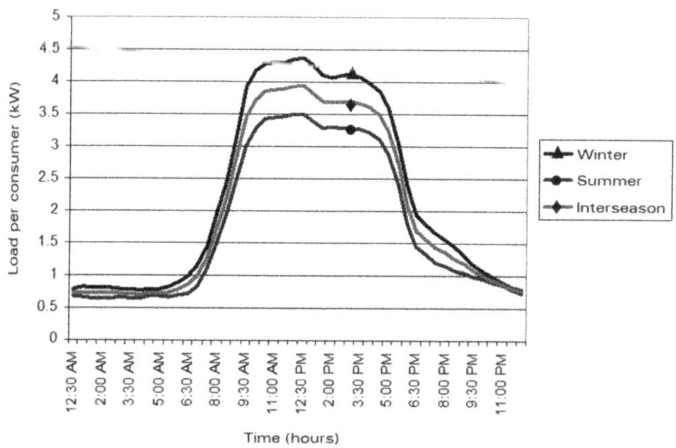

Figure 6.19. *Typical profiles of tertiary consumers and artisans (RTE)*

Today, maximizing the use of renewable production is seen as key to achieving our CO_2 emission reduction targets. At the same time, the arrival of new means of communication of the load and its flexibility (dynamic control of electric vehicle charging, local storage capacities, etc.) makes it possible to evolve towards a concept of "renewable resource tracking" or the "management of power demand". The objective is to shift the consumption of certain loads over time when production based on renewable energies is available, preferably locally, but also to organize hourly consumption in order to make better use of the electrical network, while avoiding, for example, the congestion of certain network lines.

Controlling power demand will therefore allow for more efficient use of electrical networks but also, in certain cases, a better match between consumption and the production characteristics of decentralized sources.

6.4.3.1. *Positive energy buildings (commercial buildings, residential areas, habitats, micro grids)*

In France, buildings represent 45% of the total energy consumed. This is thus a significant issue, aiming to reduce the consumption of buildings, in order to obtain zero consumption buildings or production buildings (negative consumption or positive energy buildings).

The expression "positive energy building" must be understood as a building that, throughout the year, has a lower energy consumption than its production. Nevertheless, positive energy buildings can sometimes have a consumption higher than the production and must therefore rely on the network to maintain a supply–demand balance on a building scale.

This objective of positive energy buildings will require, amongst other things, a massive integration of renewable energies into these buildings, probably the addition of energy storage means and a "smart" management of the energy.

The development of self-production and self-consumption of renewable energy (section 1.5) encourages the development of zero or positive energy buildings. This development will be further accentuated by the development of renewable energy communities, making it possible to pool local renewable energy production between several local consumers with different and potentially complementary consumption profiles, as well as possible storage systems [ROB 19].

6.4.3.2. *Electric vehicles*

The massive shift of the fleet of private vehicles to plug-in hybrid and/or entirely electric (battery) vehicles will have a significant impact on electrical networks. From an electrical point of view, these vehicles have specific loads whose recharging could be optimized by taking into account the loading state of the network, the necessary speed for recharging and the availability of decentralized production sources. These vehicles will be modular loads and, if possible, intelligent loads [ROB 16], but also storage systems whose management, coupled with that of energy-producing buildings, appears relevant for the future (concepts like Vehicle-to-Grid, or V2G, and Vehicle-to-Home, or V2H, [ROB 19]). By 2030, the goal is to reach 30 million electric vehicles in Europe.

6.5. References

[ACK 05] ACKERMANN T., *Wind Power in Power Systems*, John Wiley & Sons, New York, 2005.

[ADE 09] ADEME, Feuille de route sur les réseaux et systèmes électriques intelligents intégrant les énergies renouvelables, ADEME, June 2009.

[BAR 04] BARTON J.P., INFIELD D.G., "Energy storage and its use with intermittent renewable energy", *IEEE Transactions on Energy Conversion*, vol. 19, no. 2, pp. 441–448, June 2004.

[CIM 06] CIMUCA G., SAUDEMONT C., ROBYNS B. *et al.*, "Control and performance evaluation of a flywheel energy storage system associated to a variable speed wind generator", *IEEE Transactions on Industrial Electronics*, vol. 53, no. 4, pp. 1074–1085, August 2006.

[CIM 10] CIMUCA G., BREBAN S., RADULESCU M. *et al.*, "Design and control strategies of an induction machine-based flywheel energy-storage system associated to a variable-speed wind generator", *IEEE Transactions on Energy Conversion*, vol. 25, no. 2, pp. 526–534, June 2010.

[CRA 08] CRAPPE M., *Electric Power Systems*, ISTE Ltd, London and John Wiley & Sons, New York, 2008.

[DAV 04] DAVIGNY A., LECLERCQ L., ANSEL A. *et al.*, "Wind and storage system based dispersed generation contribution to power grid ancillary services and network reliability", *Proceedings of the 2th International Conference on Securing Critical Infrastructures, CRIS 2004*, Grenoble, France, 25–27 October, 2004.

[DAV 06] DAVIGNY A., ROBYNS B., "Fuzzy logic based supervisor of a wind farm including storage system and able to work in islanding mode", *Proceedings of the 32nd Annual Conference of the IEEE Industrial Electronics Society (IECON-2006)*, Paris, France, 7–10 November, 2006.

[DEL 09] DELILLE G., FRANÇOIS B., MALARANGE G., "Construction d'une offre de services du stockage pour les réseaux de distribution dans un contexte réglementaire dérégulé", *European Journal of Electrical Engineering*, vol. 12, nos 5–6, pp. 733–762, 2009.

[ELI 20] ELIA GROUP INNOVATION, "IO.Energy ecosystem enters its 2nd phase and plans to test new energy services", Elia Group Innovation, Press release, October 2020.

[EUR 01] EUROPEAN COMMISSION, Energy storage – A key technology for decentralized power, power quality and clean transport, Report, EUR 19978, Luxembourg, 2001.

[FAS 19] FASSIAUX J.,' "IO.Energy: Sixty organisations to launch eight projects to help shape the energy landscape of the future", Elia Group Innovation, Press release, June 2019.

[FRA 02] FRAISSE J.-L., "Le raccordement de la production décentralisée en HTA et BT", *Revue REE*, no. 7, pp. 34–46, July 2002.

[FRE 08a] FRENCH MINISTRY OF ECOLOGY, SUSTAINABLE DEVELOPMENT, TRANSPORTATION AND HOUSING, Prescriptions techniques de conception et de fonctionnement pour le raccordement à un réseau public de distribution d'électricité en basse tension ou en moyenne tension d'une installation de production d'énergie électrique, Decree, 23 April 2008.

[FRE 08b] FRENCH MINISTRY OF ECOLOGY, SUSTAINABLE DEVELOPMENT, TRANSPORTATION AND HOUSING, Prescriptions techniques de conception et de fonctionnement pour le raccordement à un réseau public de transport d'électricité d'une installation de production d'énergie électrique, Decree, 23 April 2008.

[FRE 08c] FRENCH MINISTRY OF ECOLOGY, SUSTAINABLE DEVELOPMENT, TRANSPORTATION AND HOUSING, Prescriptions techniques générales de conception et de fonctionnement pour le raccordement d'installations de production aux réseaux publics d'électricité, Decree, 23 April 2008.

[FRE 10a] FRENCH MINISTRY OF ECOLOGY, SUSTAINABLE DEVELOPMENT, TRANSPORTATION AND HOUSING, Modalités du contrôle des performances des installations de production raccordées en basse tension aux réseaux publics de distribution, Decree, 29 March 2010.

[FRE 10b] FRENCH MINISTRY OF ECOLOGY, SUSTAINABLE DEVELOPMENT, TRANSPORTATION AND HOUSING, Modalités du contrôle des performances des installations de production raccordées aux réseaux publics d'électricité en moyenne tension (HTA) et en haute tension (HTB), Decree, 6 July 2010.

[FRE 10c] FRENCH MINISTRY OF ECOLOGY, SUSTAINABLE DEVELOPMENT, TRANSPORTATION AND HOUSING, Prescriptions techniques de conception et de fonctionnement pour le raccordement à un réseau public de distribution d'électricité en basse ou en moyenne tension d'une installation de production d'énergie électrique, Decree amending and supplementing Article 22 of the Decree of 23 April 2008, 24 November 2010.

[HAD 09] HADJSAÏD N., SABONNADIÈRE J.C., *Power Systems and Restructuring*, ISTE Ltd, London and John Wiley & Sons, New York, 2009.

[HEB 02] HEBNER R., BENO J., WALLS A., "Flywheel batteries come around again", *IEEE Spectrum*, pp. 46–51, April 2002.

[JEN 00] JENKINS N., ALLAN R., CROSSLEY P. *et al.*, *Embedded Generation*, The Institution of Electrical Engineers, London, 2000.

[KUN 94] KUNDUR P., *Power System Stability and Control*, MacGraw-Hill, Electric Power Research Institute, Palo Alto, 1994.

[LEC 03a] LECLERCQ L., ROBYNS B., GRAVE J.M., "Control based on fuzzy logic of a flywheel energy storage system associated with wind and diesel generators", *Mathematics and Computers in Simulation*, vol. 63, pp. 271–280, 2003.

[LEC 03b] LECLERCQ L., SAUDEMONT C., ROBYNS B. *et al.*, "Flywheel energy storage system to improve the integration of wind generators into a grid", *Electromotion*, vol. 10, no. 4, pp. 641–646, 2003.

[LEC 04] LECLERCQ L., DAVIGNY A., ANSEL A. *et al.*, "Grid connected or islanded operation of variable speed wind generators associated with flywheel energy storage systems", *Proceedings of the 11th International Power Electronics and Motion Control Conference, EPE-PEMC 2004*, Riga, Latvia, 2–4 September, 2004.

[MAR 98] MARQUET A., LEVILLAIN C., DAVRIU A. *et al.*, "Stockage d'électricité dans les systèmes électriques", *Techniques de l'Ingénieur*, D4030, May 1998.

[PAN 03] PANKOW Y., FRANÇOIS B., ROBYNS B. *et al.*, "Analysis of a photovoltaic generator integrated in a low voltage network", *Proceedings of the 17th International Conference on Electricity Distribution, CIRED 2003*, Barcelona, Spain, 12–15 May, 2003.

[PAO 20] PAOLONE M., MONTI A., GAUNT T. *et al.*, "Fundamentals of power systems modelling in the presence of converter-interfaced generation", *21st Power Systems Computation Conference (IEEE-PSCC)*, Porto, Portugal, 29 June – 3 July, 2020.

[PET 14] PETROVIC V., BOTTASO C.L., "Wind turbine optimal control during storms", *Journal of Physics: Conference Series*, 524, 2014.

[PIE 11] PIERQUIN J., DAVIGNY A., ROBYNS B., "Current and voltage control strategies using resonant correctors: Examples of fixed frequency applications", in MONMASSON E. (ed.), *Power Electronic Converters – PWM Strategies and Current Control Techniques*, ISTE Ltd, London and John Wiley & Sons, New York, pp. 449–486.

[PIW 05] PIWHO R., OSBORN D., GRAMLICH R. *et al.*, "Wind energy delivery issues", *IEEE Power and Energy Magazine*, vol. 3, no. 6, pp. 45–56, 2005.

[REE 03] REE, "Accumulateur: 40 MW pendant 7 minutes", *Revue de l'Electricité et de l'Electronique, REE*, no. 10, p. 8, November 2003.

[RFG 16] RFG, Network Code on Requirements for Generators (RfG NC), Commission Regulation, (EU) 2016/631, 14 April 2016.

[ROB 04] ROBYNS B., BASTARD P., "Production décentralisée d'électricité : contexte et enjeux techniques", *Revue 3EI*, no. 39, pp. 5–13, December 2004.

[ROB 05] ROBYNS B., ANSEL A., DAVIGNY A. *et al.*, "Apport du stockage de l'énergie à l'intégration des éoliennes dans les réseaux électriques, contribution aux services systèmes", *Revue de l'Electricité et de l'Electronique, REE*, no. 5, pp. 75–85, May 2005.

[ROB 06] ROBYNS B., DAVIGNY A., SAUDEMONT C. *et al.*, "Impact de l'éolien sur le réseau de transport et la qualité de l'énergie", *Journal sur l'enseignement des sciences et technologies de l'information et des systèmes, J3eA*, vol. 5, no. 1, 2006.

[ROB 15] ROBYNS B., FRANÇOIS B., DELILLE G. *et al.*, *Energy Storage in Electric Power Grids*, ISTE Ltd, London and John Wiley & Sons, New York, 2015.

[ROB 16] ROBYNS B., SAUDEMONT C., HISSEL D. *et al.*, *Electrical Energy Storage in Transportation Systems*, ISTE Ltd, London and John Wiley & Sons, New York, 2016.

[ROB 19] ROBYNS B., DAVIGNY A., BARRY H. *et al.*, *Electrical Energy Storage for Buildings in Smart Grids*, ISTE Ltd, London and John Wiley & Sons, New York, 2019.

[RTE 04] RTE, *Mémento de la sûreté du système électrique*, no. 2, pp. 32–41, available at: www.rte-france.com, 2004.

[SER 04] SERPU S., BÉSANGER Y., HADJSAÏD N., "Performance control for better power systems security in a re-regulated environment: A survey", *Proceedings of the 2th International Conference on Securing Critical Infrastructures, CRIS 2004*, Grenoble, France, 25–27 October, 2004.

[SHA 11] SHARMA S., HUANG S., SARMA N.D.R., "System inertial frequency response estimation and impact of renewable resources in ERCOT interconnection", *IEEE Power and Energy Society General Meeting*, Detroit, USA, 2011, pp. 1–6, 2011.

[SPA 20] SPAES J., BELLINI E. "France's 'virtual power line' takes shape", *PV Magazine*, October 2020.

[SPR 09] SPROOTEN J., COURTECUISSE V., ROBYNS B. *et al.*, "Méthodologie de développement de superviseurs à logique floue de centrales multi sources à base d'énergie renouvelable", *European Journal of Electrical Engineering*, vol. 12, nos. 5–6, pp. 553–583, 2009.

[SPR 14] SPROOTEN J., GUNST T., MESTDAG C. *et al.*, "Power System Restoration with high penetration level of renewable generation – New challenges and strategies", *Smart Grid Saudi Arabia*, December 2014.

[UCT 04] UCTE, Final report of the Investigation Committee on the 28 September 2003 blackout in Italy, available at: www.ucte.org, April 2004.

[VER 09] VERGNOL A., SPROOTEN J., ROBYNS B. *et al.*, "Gestion des congestions dans un réseau intégrant de l'énergie éolienne", *Revue 3EI*, no. 59, pp. 63–72, 2009.

[VER 11] VERGNOL A., SPROOTEN J., ROBYNS B. *et al.*, "Line overload alleviation through corrective control in presence of wind energy", *Electric Power Systems Research*, vol. 81, pp. 1583–1591, July 2011.

[WAN 17] WANG J., ZHANG H., ZHOU Y., "Intelligent under frequency and under voltage load shedding method based on the active participation of smart appliances", *IEEE Transactions on Smart Grid*, vol. 8, no. 1, pp. 353–361, January 2017.

List of Authors

Arnaud DAVIGNY
L2EP
Junia Graduate School of Engineering
Lille
France

Bruno FRANÇOIS
L2EP
Centrale Lille Institute
Villeneuve d'Ascq
France

Antoine HENNETON
Junia Graduate School of Engineering
Lille
France

Benoît ROBYNS
L2EP
Junia Graduate School of Engineering
Lille
France

Jonathan SPROOTEN
Elia Transmission Belgium
Brussels
Belgium

Index

A, B, C

D, E

F, G, H

I, L, M

N, O, P

R, S, T

V, W

Other titles from

iSTE

in

Energy

2020

BOISGIBAULT Louis, AL KABBANI Fahad
Energy Transition in Metropolises, Rural Areas and Deserts

SOUALHI Abdenour, RAZIK Hubert
Electrical Systems 1: From Diagnosis to Prognosis
Electrical Systems 2: From Diagnosis to Prognosis

2019

BENALLOU Abdelhanine
Energy Transfers by Convection
(Energy Engineering Set – Volume 3)
Energy Transfers by Radiation
(Energy Engineering Set – Volume 4)
Mass Transfers and Physical Data Estimation
(Energy Engineering Set – Volume 5)

LACHAL Bernard
Energy Transition

ROBYNS Benoît, DAVIGNY Arnaud, BARRY Hervé, KAZMIERCZAK Sabine, SAUDEMONT Christophe, ABBES Dhaker, FRANÇOIS Bruno
Electrical Energy Storage for Buildings in Smart Grids

2018

BENALLOU Abdelhanine
Energy and Mass Transfers: Balance Sheet Approach and Basic Concepts
(Energy Engineering Set – Volume 1)
Energy Transfers by Conduction
(Energy Engineering Set – Volume 2)

JEMEÏ Samir
Hybridization, Diagnostic and Prognostic of Proton Exchange Membrane Fuel Cells: Durability and Reliability

RUFFINE Livio, BROSETA Daniel, DESMEDT Arnaud
Gas Hydrates 2: Geoscience Issues and Potential Industrial Applications

VIALLET Virginie, FLEUTOT Benoit
Inorganic Massive Batteries
(Energy Storage – Batteries, Supercapacitors Set – Volume 4)

2017

BROSETA Daniel, RUFFINE Livio, DESMEDT Arnaud
Gas Hydrates 1: Fundamentals, Characterization and Modeling

LA SCALA Massimo
From Smart Grids to Smart Cities: New Challenges in Optimizing Energy Grids

MOLINA Géraldine, MUSY Marjorie, LEFRANC Margot
Building Professionals Facing the Energy Efficiency Challenge

SIMON Patrice, BROUSSE Thierry, FAVIER Frédéric
Supercapacitors Based on Carbon or Pseudocapacitive Materials
(Energy Storage – Batteries, Supercapacitors Set – Volume 3)

2016

ALLARD Bruno
Power Systems-on-Chip: Practical Aspects of Design

ANDRE Michel, SAMARAS Zissis
Energy and Environment

DUFOUR Anthony
Thermochemical Conversion of Biomass for the Production of Energy and Chemicals

2015

CROGUENNEC Laurence, MONCONDUIT Laure, DEDRYVÈRE Rémi
Electrodes for Li-ion Batteries
(Energy Storage – Batteries, Supercapacitors Set – Volume 2)

LEPRINCE-WANG Yamin
Piezoelectric ZnO Nanostructure for Energy Harvesting
(Nanotechnologies for Energy Recovery Set – Volume 1)

ROBYNS Benoît, FRANÇOIS Bruno, DELILLE Gauthier, SAUDEMONT Christophe
Energy Storage in Electric Power Grids

ROSSI Carole
Al-based Energetic Nanomaterials
(Nanotechnologies for Energy Recovery Set – Volume 2)

TARASCON Jean-Marie, SIMON Patrice
Electrochemical Energy Storage
(Energy Storage – Batteries, Supercapacitors Set – Volume 1)

2013

LALOUI Lyesse, DI DONNA Alice
Energy Geostructures: Innovation in Underground Engineering

2012

BECKERS Benoit
Solar Energy at Urban Scale

ROBYNS Benoît, DAVIGNY Arnaud, FRANÇOIS Bruno, HENNETON Antoine, SPROOTEN Jonathan
Electricity Production from Renewable Energies

2011

GAO Fei, BLUNIER Benjamin, MIRAOUI Abdellatif
Proton Exchange Membrane Fuel Cell Modeling

MULTON Bernard
Marine Renewable Energy Handbook

2010

BRUNET Yves
Energy Storage

2009

SABONNADIÈRE Jean-Claude
Low Emission Power Generation Technologies and Energy Management

SABONNADIÈRE Jean-Claude
Renewable Energy Technologies